MOLECULAR BIOLOGY IN BLOOD TRANSFUSION

DEVELOPMENTS IN HEMATOLOGY AND IMMUNOLOGY

Volume 35

The titles published in this series are listed at the end of this volume.

Molecular Biology in Blood Transfusion

Proceedings of the Twenty-Fourth International Symposium on Blood Transfusion, Groningen 1999, organized by the Blood Bank Noord Nederland

edited by

C. TH. SMIT SIBINGA
Blood Bank Noord Nederland,
Groningen, The Netherlands

and

H. G. KLEIN
National Institutes of Health,
Bethesda, MD, U.S.A.

KLUWER ACADEMIC PUBLISHERS
DORDRECHT / BOSTON / LONDON

A C.I.P. Catalogue record for this book is available from the Library of Congress

ISBN 0-7923-6534-8

Published by Kluwer Academic Publishers,
P.O. Box 17, 3300 AA Dordrecht, The Netherlands.

Sold and distributed in North, Central and South America
by Kluwer Academic Publishers,
101 Philip Drive, Norwell, MA 02061, U.S.A.

In all other countries, sold and distributed
by Kluwer Academic Publishers,
P.O. Box 322, 3300 AH Dordrecht, The Netherlands.

Printed on acid-free paper

Printed in the Netherlands.

Dr. Cees Th. Smit Sibinga
Initiator and Organisor

Dr. Harvey G. Klein
Chairman

Prof. Dr. Siebrand Poppema
Opening Address

Baxter

Acknowledgement

This publication has been made possible through the support of Baxter, which is gratefully acknowledged.

CONTENTS

III. Developments in Processing of Cells and Therapeutic Proteins

IV. New Horizons in Clinical Medicine

MODERATORS AND SPEAKERS

Moderators

H.G. Klein (chairman)	–	National Institutes of Health, Bethesda, MD, USA
G.D.M. Beun	–	Blood Bank Noord Nederland, Groningen, NL
M.P. Busch	–	Irwin Memorial Blood Centers, San Francisco, CA, USA
C.L. van der Poel	–	Sanguin Blood Supply Foundation, Amsterdam, NL
C.Th. Smit Sibinga	–	Blood Bank Noord Nederland, Groningen, NL
J.E. Wagner	–	University of Minnesota, Minneapolis, MN, USA

Speakers

R.W. Allen	–	H.A. Chapman Institute of Medical Genetics, Tulsa, OK, USA
D.J. Anstee	–	International Blood Group Reference Laboratory, Bristol, UK
J. Coste	–	Molecular Biology Laboratory ETS-LR, Montpellier, F
A.L. Drayer	–	Blood Bank Noord Nederland, Groningen, NL
W.H. Dzik	–	Massachusetts General Hospital, Boston, MA, USA
H.A. Fields	–	CDC, Hepatitis Branch, Atlanta, GA, USA
G.A.P. Hospers	–	University Hospital Groningen, Groningen, NL
A.D. Michelson	–	University of Massachusetts, Worcester, MA, USA
M.S. Mincheff	–	Naval Medical Research Center, Bethesda, MD, USA

M.E. Reid	–	New York Blood Center, New York, NY, USA
E. Robinet	–	Blood Transfusion Service Franche-Comité, Besançon, F
C.E. van der Schoot	–	CLB, Amsterdam, NL
J.W. Semple	–	Blood Bank St. Michael's, Toronto, Ontario, C
S.V. Seremetis	–	Haemophilia Treatment Center, New York, NY, USA
D.L. Siegel	–	University Pennsylvania Medical Center, Philadelphia, PA, USA
C.E.M. Voorter	–	Academic Hospital Maastricht, Maastricht, NL

FOREWORD

The international symposia on blood transfusion in Groningen have taken place without interruption since 1976. Each year Dr. Smit Sibinga and his team have not only organized a meeting on timely topics in blood transfusion but also succeeded in attracting prominent chairmen and speakers. It has already been determined that next year the 25th meeting will take place. The subject will be Transfusion Medicine 'Quo Vadis; what has been achieved and what is to be expected?', and will be chaired by professor John D. Cash from Edinburgh.

The subject of the 24th symposium that begins today is:
Molecular Biology in Blood Transfusion
and will be chaired by Dr. Harvey G. Klein (NIH).

It is a special pleasure for me as Dean of the Faculty of Medical Sciences to open this meeting. We have many things to be proud of in this city. The Academic Hospital is one of the most modern in the world. The medical curriculum, termed C2000, is highly acclaimed in this country by medical students as well as our medical colleagues. We strive very hard to ensure that a few years from now the research programs of the Faculty will also rank in the top. Important steps have already been taken by Faculty and Hospital to focus our best research efforts in Research Institutes and Graduate Schools, such as the Groningen University Institute for Drug Exploration (GUIDE) and the School for Behavioural and Cognitive Neuroresearch (BCN).

The Blood Bank Noord Nederland (BBNN) is another of the institutes to be proud of. The number of volunteer donors is currently approximately 85,000. The organisation carries out its mission of motivating the population to donate blood, to procure, work up and preserve blood and plasma from well tested donors, to perform appropriate quality control and distribute the products to hospitals throughout the Northern part of the Netherlands. In addition, the Blood Bank provides clinical consultation and has active research and educational programs.

The BBNN is unique for Europe for having been accredited by the American Association of Blood Banks since 1981 and also has been recognized by the WHO and the WFH since 1993 as a training ground for fellows in Blood Transfusion.

Dr. Smit Sibinga and his team collaborate withaccredited several research groups within the Faculty of Medicine of the University of Groningen as well as the Technical University Twente. It is a fortunate development that in the next few years the Blood Bank will return to the grounds of the Hospital and Faculty who already are taking firm steps towards an integration of their activities.

Someone also to be proud of is Dr. Smit Sibinga himself. In my travels around the world I encounter many people that know him but also know that he is from Groningen. As such he has definitely contributed his share in putting Groningen on the map.

In this year's meeting an extraordinary range of subjects will be covered, including diagnostic aspects, virology, QC, cell and protein processing and also a session on new horizons in Clinical Medicine. Some 130 participants from 31 different countries and all continents have come to Groningen once again to share with each other the progress in this field.

I wish you a successful meeting and a pleasant stay in the city of Groningen.

Prof. Dr. S. Poppema
Dean Faculty of Medical Sciences
University of Groningen
Groningen, NL

I. DIAGNOSTIC ASPECTS

MOLECULAR BIOLOGY IN BLOOD TRANSFUSION – INTRODUCTION

H.G. Klein[1]

There is an oft-repeated joke that ends with the punch line, " I'd rather jump from a speeding car than hear another lecture on the importance of molecular biology." Indeed it is almost impossible to attend the most clinical lecture today without hearing that the future belongs to those who understand and apply molecular techniques. Yet how many practicing physicians appreciate the degree to which molecular sciences have already entered our lives to what extent the concepts and technology have begun to revolutionize the practice of medicine. When I trained in transfusion medicine, a discipline then known as "blood banking and immunohematology", we traced population migrations by assessing with agglutination reactions the presence or absence of blood group antigens in hundreds of samples collected in the field. Today we pinpoint the evolutionary divergence of hominids from apes by measuring the base pair differences between Homo sapiens and the chimpanzee. Two decades ago we used serum antibodies from multiparous women in crude cytotoxicity assays to calculate the likelihood that the random man might be implicated in a case of disputed paternity. Today with no more than a spot of dried cells, we ascertain that only O.J. Simpson, or his long lost identical twin, could have shed the blood detected at 875 South Bundy in Santa Monica. Such is the power of molecular biology.

Molecular biology begins with the gene. The basic concepts of the genetic code and gene expression and regulation are almost 40 years old [1]. Today every biology student knows that genes are the molecules of deoxyribonucleic acid (DNA) that code biologic information in sequences of nucleotides and through the processes of cell division, transmit this information to their cell progeny and to future generations. Genetic information directs the production of amino acids and their incorporation into proteins. The techniques of molecular biology derive from attempts to understand the rules that govern this system of biologic information management by investigating the structure, function, and control of genes. In the field of medicine, harnessing these techniques provides powerful tools for diagnosis and therapy.

It should come as no surprise that Transfusion Medicine, with its traditional foundation in human genetics, should become increasingly interested in and dependent upon the techniques of molecular biology. Perhaps it is surprising only that its practitioners have not pioneered the discovery and application of

---

1. Dept Transfusion Medicine, Warren G. Magnuson Clinical Center, National Institutes of Health, Bethesda, MD, USA.

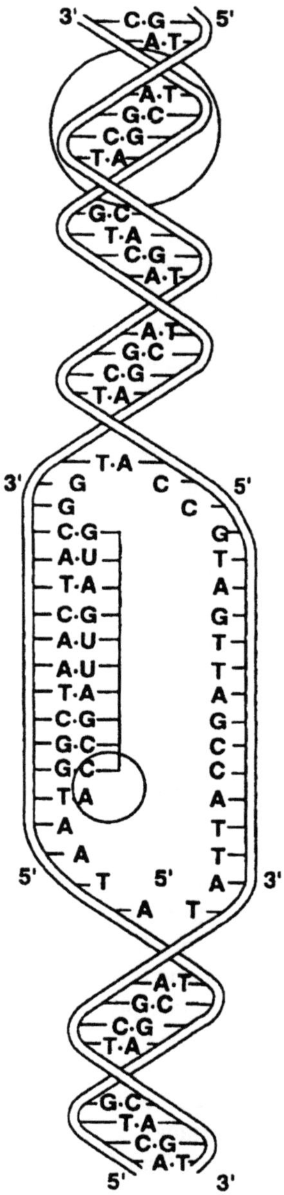

Figure 1. The double helical structure of DNA. Complementary purine and pyrimidine nucleotides bond to one another internal to the sugar phosphase backbone. Adapted from Berliner N and Benz EJ Jr. Anatomy and physiology of the gene. In: Hoffman R, Benz EJ Jr, Shatill S, Furie B, Cohen H (eds). Hematology: Basic Principles and Practice. New York: Churchill Livingstone, 1999:3.

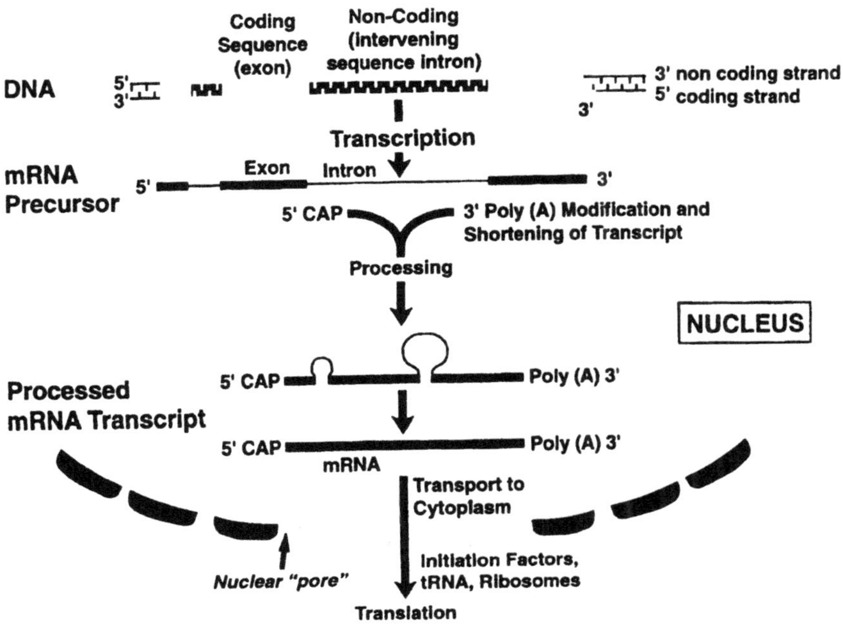

Figure 2. Gene expression. The DNA molecule is transcribed into mRNA, which is transported to the cytoplasm to direct protein synthesis. Adapted from Berliner N and Benz EJ Jr. Anatomy and physiology of the gene. In: Hoffman R, Benz EJ Jr, Shatill S, Furie B, Cohen H (eds). Hematology:Basic Principles and Practice. New York: Churchill Livingstone, 1999:5.

these tools. After all, the genetic basis of blood group inheritance was demonstrated in 1910, and the basic concepts that defined transfusion and transplantation compatibility for almost a century depended upon the inheritance of polymorphic cell surface proteins [2]. We now appear to be making up for some of this lost time. This symposium will touch upon a variety of areas in which molecular biology is revolutionizing transfusion sciences and point to future challenges and opportunities. Time and space limitations preclude any hope of being encyclopedic, so our goal is to provide a stimulating overview for the eclectic reader.

Gene Structure and Regulation

Research in the last quarter of a century has clarified much of the molecular structure of genes and begun to unravel issues of inheritance, function, and regulation. To understand the basic tools of molecular biology, it is helpful to refer to the elementary "Watson-Crick rules" proposed in the 1950's. The human genome consists of 23 pairs of double-stranded DNA molecules complexed with a variety of other nuclear proteins to form 46 chromosomes. Chromosomes are composed of subunits called nucleotides. Each nucleotide consists of a backbone deoxyribose sugar and a purine or pyrimidine base bound to it by a phosphate group. (Figure 1) There are about three billion base pairs in the human genome. Genes are sequences of bases arranged longitudinally along the DNA strand, and a DNA molecule is usually described by its number of bases or nucleotides. The ends of the DNA strands are distinguished by their phosphodiester bond linkage, either 3' or 5', and DNA strands coil about one another in anticomplimentary fashion, one strand running 3' to 5' while the other runs 5' to 3' Cellular enzymes read the genetic information in one direction only. Genetic information flows from DNA to RNA to protein assembly according to what have become known as Watson-Crick rules. According to these rules, specific bases are paired in the complementary strands, adenine to thymine, and guanine to cytosine in DNA; in double stranded molecules containing RNA, uracil replaces thymine. Enzymes that copy DNA and RNA, so-called polymerases, follow these rules of complementarity as they construct daughter sequences or strands along the nucleic acid template. This process results in accurate replication of the genome as cells divide, so-called conservative replication.

Information stored in these nucleic acid sequences control the quantity and quality of proteins synthesized by cells, as well as the timing and sequence of production. The aggregate of this process is what is often referred to as gene expression. Information moves from the chromosomes in the nucleus via complementary messenger RNA (mRNA) to the ribosomes in the cytoplasm, where the message is translated into single amino acids, polypeptides and eventually, proteins. (Figure 2) The process is elegant in its simplicity, and devilish in the details. The original mRNA transcript, pre-mRNA, is modified on nuclear organelles called spliceosomes to remove non-coding RNA sequences or introns, that lie between the gene sequences or exons. Specific base sequences initiate and terminate splicing, leaving variable numbers of exons to be translated. Control of the splicing (or alternative splicing) process is complex and the target of a great deal of basic research.

Since each cell contains a complete set of information contained in the chromosomes, one of the most puzzling aspects of gene expression is how specialized cell recognize which genes to express and in what sequences in order to coordinate the synthesis and interaction of those proteins that define the cell. There are DNA sequences located close to the genes, so-called cis-acting factors such as promoters, enhancers, and silencers. In addition, there are numerous nuclear binding proteins known as transcription factors or trans-acting factors that bind to these sequences and either directly influence structural genes, or influence other sequences or DNA conformation to expose control elements. Some of these factors control single genes, while others influence gene clusters. The amount of mRNA that accumulates in the cell is further modified by post-transcription and post-translational factors that further regulate the amount of protein that a cell produces.

Selected Tools of Molecular Biology

Restriction Enzymes and Blotting

Components of the gene-to-protein process, when applied in accordance with Watson-Crick rules provide a set of powerful tools for exploring biology at the basic level. However these tools are no longer restricted to the province of the laboratory researcher.

Molecular techniques have evolved into practical methods for diagnosis and therapy. Among the most widely used techniques is the application of restriction enzymes or restriction endonucleases. These bacterial enzymes recognize singular short sequences of double-stranded DNA and cleave the molecule with exquisite specificity. Hundreds of these enzymes have been identified and isolated, each identifying a unique sequence of bases. Using combinations of endonucleases, investigators can digest large segments of DNA into small, reproducible fragments. Restriction fragments can be used to identify silent point mutations, or polymorphisms in human genomic DNA. Small, characterized DNA sequences excised with restriction enzymes can also be radiolabeled and used as probes. Restriction fragmentlength polymorphisms (RFLP's) produced by restriction enzyme digestion are identified by incubation with radiolabeled probes followed by electrophoresis using an autoradiographic technique described by E.M. Southern, so-called Southern blotting (Figure 3). Southern blots can detect extremely small fragment differences when mutations alter restriction sites on DNA. The analogous procedure applied to RNA, without apologies for the scientific community's attempt at humor, is termed Northern blotting.

The human genome is characterized by specific repetitive sequences of DNA, sequences that distinguish the human genome from that of other mammalian species. Digestion with appropriate endonucleases and analysis with standard methods can thus distinguish human DNA. Similarly, other tandemly repeated DNA sequences, so-called variable number of tandem repeats, or VNTR's, can be used to identify individuals with remarkable precision. Recog-

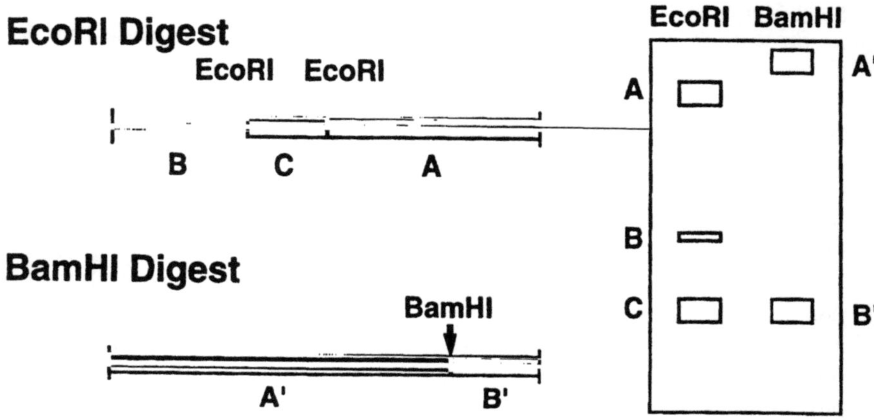

Figure 3. Southern blotting. DNA digested by restriction enzymes. The fragments are separated by gel electrophoresis, denatured into single strands, blotted onto cellulose, and identified by annealing with radiolabeled probes.

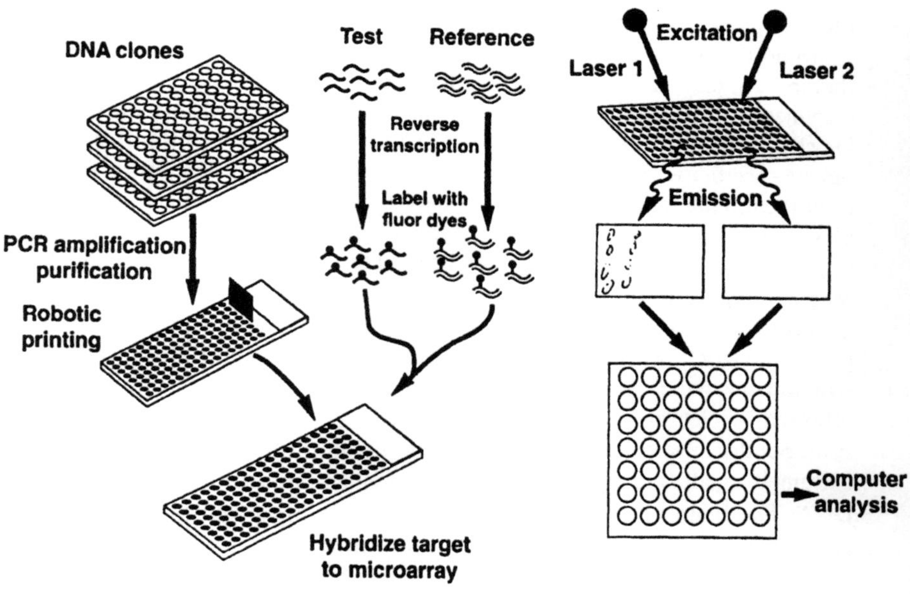

Figure 4. Microarray technology. Amplified cDNA clones are coated by high-speed robotics onto plates, incubated with fluorescent labeled test and reference samples, and activated by lasers. The pattern of activation is computer analyzed to determine relative gene expression.

nizing the variable number of these normal repeated sequences, laboratories can fingerprint DNA by enzymatic digestion and pattern recognition of VNTR's.

Hybridization and amplification

A number of powerful molecular techniques have been based on the Watson-Crick complementarity rules and the use of enzymes to amplify sequences of DNA and RNA. The best known and most widely employed of these methods is the polymerase chain reaction or PCR, which allows the simple and reproducible amplification of a fragment of DNA that lies between two defined points [3]. If the nucleic acid sequence for a particular region of a genome is known, a complementary oligonucleotide primer can be synthesized, incubated with the target DNA or RNA, and using a DNA polymerase (5' to 3') or a reverse transcriptase (3' to 5'), complementary strands can be produced. Automated systems are now available that use thermostable polymerases and repeated cycles of heating and cooling to denature and dissociated double-stranded DNA so that the sequences can be repeatedly amplified. After thirty or so cycles, up to a million copies of the target can be transcribed and analyzed or sequenced. Numerous amplification strategies are now in use, but all use the same basic principles. A variety of different polymerases are available so that mRNA can be transcribed into populations of complementary single-stranded DNA (cDNA) and libraries of double-stranded DNA can be produced from cDNA templates. Amplification of RNA with these techniques employs a reverse transcriptase and is referred to as rtPCR.

Gene Chips and Microarrays

With several million cDNA sequences available in GenBank, much of the interest of the molecular biologist is turning from the study of gene structure to what has been termed functional genomics [4]. Technologies are evolving to study the simultaneous expression of large numbers of genes. These methods are sophisticated extensions of the Southern blotting technique. The general principle involves immobilization of large numbers of cDNA clones or synthetic oligonucleotides on glass (microarrays) or on silica wafers (chips). The immobilized targets can then be probed with cDNA from the test cells and from reference cells that have been labeled with different fluorescent markers. Laser excitation of the hybridized product provides fluorescent signals that can be computer analyzed to compare the relationship between the test cell and the reference cell. (Figure 4) This technique should advance fingerprinting of gene structure to fingerprinting of gene expression [5].

Gene Isolation and Insertion

Microbial geneticists have developed several methods for inserting nucleic acid into cells. These methods have become the mainstay of gene cloning and early gene therapy techniques. Small, circular DNA molecules called plasmids can be used to infect cells. These plasmids can be cut with restriction enzymes to render them linear and foreign genes, for example drug resistance genes or human gene sequences can be inserted with the use of enzymes called ligases. Plasmids provide an important technique for inserting these genes into bacterial hosts, which can then produce large amounts of the foreign genes. These kinds of techniques are used for marking clones and for producing recombinant proteins such as insulin, erythropoietin, and hemoglobin. DNA can also be inserted into bacterial cells by using infective viruses known as bacteriophages. Foreign genes are inserted into phage DNA with restriction enzymes and ligases, and the virus inserts its modified nucleic acid into the appropriate bacteria. Retroviruses, adenoviruses, and adeno-associated viruses have also been used as vectors to deliver genes into mammalian cells.

Human genes are cloned with the techniques that have been described. Genomiç DNA is digested with restriction enzymes, or mRNA from expressed genes is isolated and converted to double-stranded DNA using transcription enzymes. Plasmids or phage vectors are used to introduce the DNA fragments into bacterial hosts that lack some selective phenotypic factor. For example, the recombinant DNA molecules may have an antibiotic resistance gene inserted so that bacterial colonies that do not contain this gene can be eliminated by exposure to neomycin. Each colony then contains a single human DNA fragment (along with the selection gene) and the colony of interest can be located among the numerous other colonies by exposure to a DNA probe or to an antibody, should the gene of interest be expressing a known protein. Colonies containing the gene of interest can then be grown in great quantity.

Applying Molecular Techniques to Transfusion Medicine

Most Transfusion Medicine practitioners were introduced to practical molecular biology during the AIDS epidemic of the 1980's. Amplification techniques were employed to identify viral sequences from plasma and mononuclear cells, to quantify viral load, and to sequence strains of the human immunodeficiency virus (HIV). Amplification techniques for HIV have now been adapted to the screening of blood for HIV and hepatitis C virus (HCV), and methods are available to detect hepatitis B virus (HBV), HTLV I/II, and cytomegalovirus (CMV) as well [6]. While none of these assays is licensed for screening of donated blood in the United States at this time, Phase III testing is underway and licensure is expected imminently.

PCR-based technology for genotyping developed rapidly during the late 1980's, in part as a result of the rapid growth of the National Marrow Donor Program in the United States. HLA typing for transplantation was performed by serologic testing exclusively when the program was initiated in 1986, but a decade later, virtually all class II testing was performed by molecular methods and HLA class I typing has advanced rapidly as probes have become commercially available [7]. As is so often the case, technology revealed the enormous complexity and diversity of the HLA loci, and presented the biologic challenge of determining the clinical importance of the genetic heterogeneity disclosed by amplification and sequencing [8]. Genomic sequences responsible for antigens on other hematopoietic cells, red cells, granulocytes, and platelets, are being investigated actively, and are revealing numerous surprises involving the relationship between nucleotide sequences and antigen expression [9-11]. It seems almost comical now that within the last two decades serologic identification of blood cell antigens formed the basis for forensic identification. Molecular techniques have spawned an industry that has all but removed paternity testing and forensics from the blood bank laboratory.

Recombinant technology has had a major impact on blood transfusion. Recombinant human erythropoietin has effectively substituted for red cell transfusions for most patients with chronic renal disease and for other patient groups ranging from those with cancer and AIDS to those undergoing orthopedic surgery. Other recombinant cytokines have proved useful for harvesting granulocytes for transfusion and early progenitor cells for stem cell transplantation. Recombinant interleukins and interferons have been used in experimental cellular immune therapies. Recombinant Factor VIII is replacing the plasma-derived fractions, and Factors IX and VIIa are commercially available and growing in use. Human hemoglobin has been produced on a scale appropriate for clinical use as an oxygen therapeutic, and recombinant technology has enabled production of hemoglobin variants with diverse oxygen affinity characteristics. Research into recombinant albumin production and immunoglobulin engineering is being actively pursued [12].

Molecular biology has already begun to influence therapy for disorders traditionally treated with transfusion components. The first cellular gene therapy trial met with at least limited success for a patient with severe combined immunodeficiency disease related to adenosine deaminase deficiency (ADA) [13]. Cellular gene therapy is now in clinical trials for a wide variety of inherited, neoplastic and infectious diseases [14]. While the early results have been limited, even disappointing to some observers, at the very least they evidence proof of principle. Direct vector infusion for correction of genetic disease has reached the phase of animal testing [15]. As new information about the molecular control of biologic processes emerges, the promise of dramatic advances in diagnosis and treatment has moved rapidly from bench to bedside. There is little doubt that this movement will continue if not accelerate.

References

1. Jacob F, Monod J. Genetic regulatory mechanisms in the synthesis of proteins. J Mol Biol 1961;3:318-56.
2. von Dungern E, Hirzfeld L. Ueber Vererbung gruppenspezifischer Strukturen des Blutes. Strukturen des Blutes. Z Immunitatsforsch 1910;6:284-92.
3. Hieter P, Boguski M. Functional genomics: it's all how you read it. Science 1997; 278:601-02.
4. Saiki RK, Scharf S, Faloona FA, et al. Enzymatic amplification of B-globin genomic sequences and restriction site analysis for diagnosis of sickle cell anemia. Science 1985;230:1350-54.
5. Duggan DJ, Bittner M, Chen Y, Meltzer P, Trent J. Expression profiling using cDNA microarrays. Nature Gen 1999;21(suppl):10-14.
6. Vyas GN, Yang G, Murphy EL. Transfusion-related transmissible diseases: Detection by polymerase chain reaction-amplified genes of the microbial agents. Transfus Med Rev 1994;8:253-66.

7. Erlich HA, Sheldon EL, Horn G. HLA typing using DNA probes. Biotechnology 1986;4:975-81.
8. Rufer N, Breuer-Vriesendorp BS, Tiercy JM et al. High-resolution histocompatibility testing of a group of sixteen B-44 positive, ABDR serologically matched unrelated donor-recipient pairs. Human Immunol 1993; 38:235-39.
9. Stroncek DF, Konz R, Clay ME, Houchins JP, McCullough J. Determination of ABO glycosyltransferase genotypes by use of polymerase chain reaction and restriction enzymes. Transfusion 1995;35:231-40.
10. Bray PF, Jin Y, Kickler T. Rapid genotyping of the five major platelet alloantigens by reverse dot blot hybridization. Blood 1994;84:4361-67.
11. Ory PA, Clark MR, Kwoh EE, Clarkson SB, Goldstein IM. Sequences of complementary DNAs that encode the NA1 and NA2 forms of Fc receptor III on human neutrophils. J Clin Invest 1989;84:1688-91.
12. Roddie PH, Ludlam CA. Recombinant coagulation factors. Blood Rev 1997;11:169-77.
13. Blaese RM, Culver KW, Miller AD et al. T-lymphocyte directed gene therapy for ADA-SCID: initial trial results after 4 years. Science 1995; 270:475-80.
14. Klein HG. Cellular gene therapy: a credible odyssey. Transfusion 1997; 37:675-70.
15. Greengard JS, Jolly DJ. Animal testing of a retroviral-mediated gene therapy for Factor VIII deficiency. Thromb Haemostas 1999;82:555-61.

ADVANCES IN BLOOD GROUPING –THE MOLECULAR BIOLOGY APPROACH[1]

M.E. Reid[2]

Introduction

Blood group antigens are inherited, polymorphic, structural characteristics located on proteins, glycoproteins or glycolipids on the outside surface of the red blood cell (RBC) membrane. Because the antigens are on the surface of RBCs, they can, if introduced into the circulation of a person lacking a particular antigen, illicit an immune response. Indeed, a blood group antigen is defined by an alloantibody that reacts with the antigen in an observable way. It is the antibody that causes clinical problems in maternal-foetal incompatibility, transfusion incompatibility, and autoimmune haemolytic anaemia (AIMA). In terms of blood transfusion, ABO is the most clinically significant blood group system due to the natural presence of anti-A and/or anti-B in the plasma of most people who lack the corresponding antigens. These antibodies rarely cause severe haemolytic disease of the newborn (HDN). In both transfusion and HDN, the next most clinically significant antibodies are those in the Rh, Kell, Duffy, and Kidd blood group systems. Alloantibodies to other antigens occur with an incidence of less than 1% of the population [1-4].

The majority of blood group antigens are located on proteins, which pass through the RBC membrane once, or multiple times, or are attached via a glycosylphosphatidylinositol anchor (Figure 1). Others are carbohydrate residues attached to proteins or to lipids in a precise way as defined by specific glycosyltransferases. Blood group antigens that are encoded by alternative forms of the same gene are clustered into a blood group system. More than 250 antigens have been placed into 25 blood group systems by the International Society for Blood Transfusion (ISBT) Working Party on Terminology for Red Cell Surface Antigens [5-7]. Antigens that do not belong in a blood group system have been placed in either the 200 series (antigens that are clustered together based on similar immunochemical properties), in the 700 series (independent low incidence antigens) or in the 901 series (independent high incidence antigens).

---

1. Supported in part by grant HL-54459 from the National Institutes of Health.
2. Director, Immunochemistry Laboratory, New York Blood Center, NY, USA.

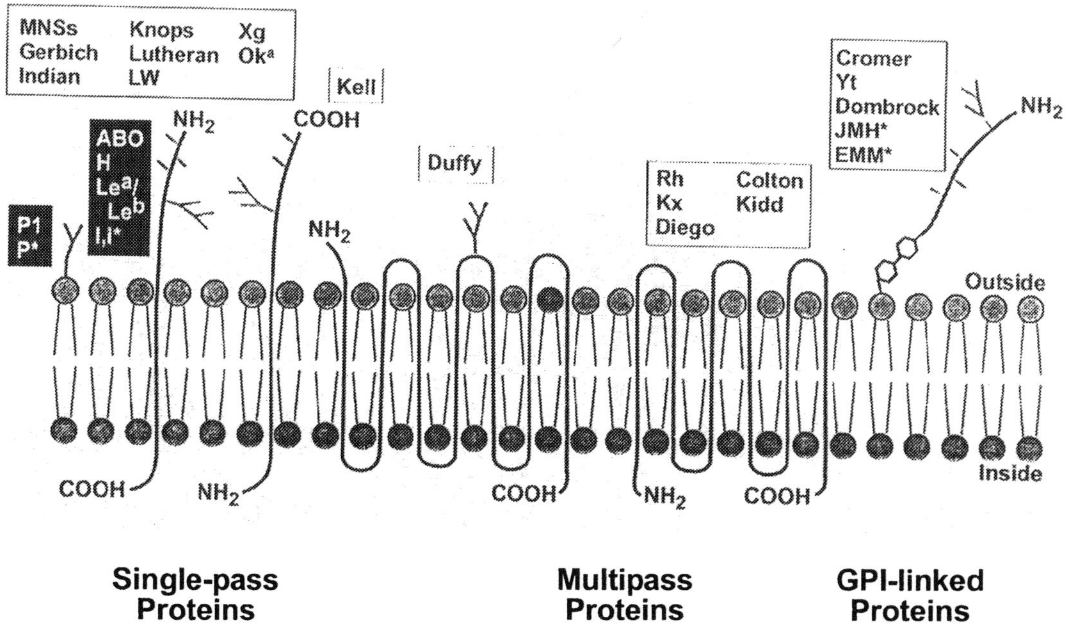

Figure 1. Schematic drawing of structures carrying blood group antigens on RBCs. Dark boxes represent antigens that are defined by carbohydrates and open boxes represent protein-based antigens. N-linked sugars are denoted by branched lines and O-linked sugars by straight lines.

* indicates antigens in collections (200 Series) or are high incidence antigens (901 Series).

Table 1. Molecular events and blood groups

– Point mutations	– Insertion
– Missense	– Exon
– Nonsense	– Nucleotide
– Alternative splicing	– Alternative initiation
– Deletion	– Chromosome translocation
– Gene	– Single crossover
– Exon	– Gene conversion
– Nucleotide	– Other gene rearrangements

Many years of meticulous haemagglutination studies have provided a vast knowledge base of the esoteric nature of blood group antigens. This information is valuable in selecting blood samples with unusual characteristics for molecular analysis. In turn, these samples have revealed valuable information about the genes that encode them. Since genes encoding all but three of the blood group systems have been cloned and sequenced, rapid progress has been made in elucidating the molecular events that generate diversity in blood group antigens and phenotypes [8-10][1] . While many molecular events give rise to blood group antigens and phenotypes (Table 1), missense mutations are by far the most common. To date, all the antigens in Kell, Diego and Lutheran blood group systems that have been studied are associated with a point mutation and single amino acid substitution [8-10]. Thus, allele-specific polymerase chain reaction (PCR) or PCR-restriction fragment length polymorphisms (RFLP) assays can readily be used in the clinical laboratory to predict the antigen profile of an individual. In addition to missense mutations, other molecular events have been described to give rise to various blood group antigens and phenotypes (Table 1). The molecular basis of antigens and phenotypes in MNS and Rh systems (which are encoded by homologous genes in tandem on a single chromosome) also involves various gene rearrangements. In some of the blood group systems further analysis is required to determine the molecular basis of variant forms of the gene, and in the majority, more extensive studies are required to establish the incidence of variants in different ethnic groups.

The molecular knowledge has opened the door to applying molecular genotyping in several areas of transfusion medicine providing answers to some clinically relevant problems that were not attainable by serological testing. These include; (1) identifying a foetus at risk for haemolytic disease of the newborn (HDN), (2) genotyping a patient who has been recently transfused (as a necessary part of antibody identification), (3) genotyping a patient whose RBCs are positive in the direct antiglobulin test (DAT) and no direct agglutinating antibody is available, (4) resolving a discrepancy in ABO or D typing (especially of weak antigens in donors), (5) determination of which phenotypically Fy(b−)

1. See data base: http:\\www.bioc.aecom.yu.edu/bgmut/index.htm.

patients can be transfused with Fy(b+) RBCs, (6) mass screening of donors when antibodies are not readily, available or are weakly reactive, (7) genotyping when an antigen is depressed, (8) mass screening of donors, (9) determination of zygosity of RhD (especially of D-positive prospective fathers), (10) determination of the origin of engrafted white blood cells in a stem cell recipient, (11) determination of the origin of lymphocytes in patients with graft-versus-host disease (GvHD) reactions, (12) tissue typing, (13) paternity testing, and (14) forensic testing.

This article will first review the molecular basis of the most clinically relevant blood group antigens, then consider some applications of molecular genotyping that are in current use and describe how this knowledge can be applied to advance blood grouping. The blood group systems are ordered according to their clinical significance.

Molecular Basis of Clinically Important Blood Group Systems

ABO Blood Group System

The gene that gives rise to the A and B antigens of the ABO blood group system occurs in several allelic forms, all of which contain seven exons that encode different glycosyltransferases of 354 amino acids. These enzymes add a specific sugar to a carbohydrate backbone whose terminal sugar is fucose (H antigen). The newly added terminal sugar gives rise to theA antigen (N-acetyl-D-galactosamine) or the B antigen (D-galactose). The H sugar, fucose, is added by α(1,2)fucosyltransferase encoded by the FUT 1 gene and point mutations in this gene give rise to H-deficient RBCs (including the null Bombay phenotype) that do not express A or B antigens [11-20]. The O phenotype results from a single nucleotide deletion that causes a frame shift and production of an inactive transferase, thereby leaving the H antigen unaltered. The difference between the genes encoding blood group A and B specific transferases is seven nucleotides, only four of which lead to amino acid substitutions (Arg179Gly; Gly235Ser; Leu266Met; Gly268Ala) [21]. There are many subgroups in the ABO system (e.g., A_2, A_x, cis AB, A_{el}, B_m) that are associated with point mutations and deletions in the transferase gene (in exons 6 and 7); the reduced enzymatic activity results in a weak expression of the A or B antigen [22-32]. Knowledge of the molecular basis of the A and B alleles makes it possible to determine the genotype of weak A and B antigens and to distinguish them from group O and H-deficient phenotypes. It is worthy of note that even with the sophisticated molecular techniques available today, the major source of patient mortality and morbidity due to ABO mismatched transfusions is clerical error.

Rh Blood Group System

The Rh blood group system antigens are encoded by variant forms of two highly homologous genes (*RHD*; *RHCE*), which are on chromosome 1p36.13-p34.3 in the probable order of *RHCE-RHD* [33]. Each gene consists of 10 exons and encodes a polypeptide of 417 amino acids (RHCE, RHD), which spans the RBC lipid bilayer 12 times and may have a transport function [34-36]. Depending on the Rh phenotype, 32 to 36 amino acid substitutions constitute the difference between the RhCE and RhD polypeptides [35].

The five most clinically significant Rh system antigens are D, C, c, E and e. C/c and E/e occur as pairs of antithetical antigens, encoded by the *RHCE* gene. The E/e antigens differ by a single amino acid at position 226, while a combination of at least three amino acid substitutions is associated with the C/c antigens. In sharp contrast, the D antigen, encoded by the *RHD* gene, has no antithetical counterpart. The D antigen consists of many different epitopes and people whose D-positive RBCs lack one or more of these epitopes (partial D) can make anti-D [37]. Besides the five common antigens, there are at least 40 other discrete Rh antigens [5-7]. Some of these antigens and the partial D antigens have been associated with a single point mutation or gene rearrangement (of *RHCE* into *RHD*, or *RHD* into *RHCE*) which occur because of the high degree of homology of the two genes [34-36].

The prevalence of D-negative people is as high as 20% in Caucasians and as low as 0.5% in some Asian populations. For the most part, the molecular basis for the D-negative phenotype is an absence of the entire *RHD* gene. The prevalence of the D-negative phenotype, together with the absence of the entire RhD polypeptide (rather than the presence of an altered protein) may account, in part, for the high immunogenicity of the D antigen.

One area of concern in the clinical laboratory is the differentiation of weak D (D^{U}) phenotypes (associated with an apparently normal *RHD*) [38-40] from weak partial D phenotypes. This is because people with the weak D phenotype (whose RBCs have a weak expression of all the D epitopes) do not make anti-D while people with the partial D^{VI} phenotype (whose RBCs express a qualitative altered partial D antigen) can make anti-D. D^{VI} can be differentiated from other weak D phenotypes by immunoblotting with a monoclonal antibody (LOR-15C9) [41,42].

D-negative people are not usually exposed to the D antigen by transfusion but can be by pregnancy. It is interesting to note that while D-negative, C-negative (dce/dce) people and some D-positive (with a partial D antigen) people make anti-D, certain D-negative, C-positive people do not appear to do so [Reid, unpublished observation]. The partial D, D^{IIIa}, differs from wild type RhD polypeptide by only three amino acid substitutions, which are identical to the amino acids at the same position in the RhCE polypeptide (Asn152Thr, Thr201Arg, and Phe223Val) [43]. The three types of partial D^{VI} result from different homologous exchanges of exons from *RHCE* into *RHD* (in type I exons 4 and 5, in type II exons 4 to 6, and in type III exons 3 to 6) [44-48]. While some D-negative phenotypes lack the *RHD* gene, others [d(Ce)ce^{S} and d(Ce)Ce] are associated with homologous exchanges of exons 3 to 7 [49], or exons 2 to 9 [50] of *RHCE* into *RHD*. Based on this knowledge, one would predict d(Ce)ce^{S} and d(Ce)Ce people, if challenged, would make anti-D more readily than the partial D^{IIIa} and D^{VI} phenotypes. However, the reverse appears to be true. If the presence of an altered gene precludes production of anti-D in some people, the knowledge of the molecular basis of Rh phenotypes may allow prediction of which phenotypically D-negative people can be transfused with D-positive

RBCs without risk of anti-D production. This would have practical implications especially for patients whose serum contains multiple antibodies.

The requirements for expression of Rh antigens are not fully understood. However, it is clear that interaction of RhD and RhCE polypeptides with at least one other protein is essential and that the Rh50 protein is critical because a lack of Rh antigens on RBCs from Rh_{null} and Rh_{mod} people is associated with an altered *RH50* gene [34,51]. Interestingly, *RH50* is homologous to *RHD* and *RHCE* but resides on chromosome 6. Several distinct molecular bases for the lack of Rh50 in RBCs have been described [34,36,51] and it is likely that there will be others as more people are tested.

Numerous reports describing approaches for determination of the molecular basis of various Rh antigens and phenotypes have been described. As reviewed by Huang [34], early methods taking advantage of differences in the 3'-regions of *RH*, and later methods using the size difference of intron 4 (*RHCE* is larger than *RHD*) encountered discrepancies. More recent approaches have concentrated on amplification of the majority of the differences between *RHD* and *RHCE* genes, [34,38,49,52-59]. However to date there is no completely reliable molecular methodology available to accurately genotype and predict Rh phenotypes. A better understanding of the molecular basis of Rh antigens and determination of factors involved in their expression is needed and it is possible that direct sequencing of PCR products, as used for HLA typing, will be required.

Kell Blood Group System

Antigens in the Kell blood group system are carried on variant forms of a protein consisting of 732 amino acids that are encoded by a gene with 19 exons located on chromosome 7q33. Kell is a type II membrane protein that is highly folded via disulfide bonds and, in the RBC membrane, is linked to a multi-pass protein, XK [60,61]. An absence of XK results in the McLeod syndrome and a weak expression of Kell system antigens (reviewed in Redman and Lee [62]; Øyen, *et al.*[63]). While there are no known polymorphisms associated with the XK protein, all Kell polymorphisms are associated with point mutations that are predicted to encode single amino acid substitutions [64]. The single amino acid substitution (Pro597Leu) associated with Js^a and Js^b antigens is situated between two cysteines (at residues 596 and 599) [64]. This most likely explains why these antigens are more sensitive to reagents that disrupt disulfide bonds than other Kell antigens [65].

Antibodies to antigens in the Kell blood group system can cause severe transfusion reactions and anaemia of the newborn. Recent studies suggest that with Kell blood group incompatibility, anaemia of a foetus may be associated with suppression of erythropoiesis as well as haemolysis [66-69]. Unlike anti-D, the strength of the antibody does not correlate with the severity of the infant's anaemia [70] and, therefore genotyping of the foetus is a valuable clinical test for determining if a foetus is at risk for severe anaemia. Since Kell polymorphisms are associated with point mutations and new or ablated restriction enzyme sites, it is possible to genotype for Kell variants using allele-specific primers with PCR, and PCR-RFLP assays [64]. Indeed, both approaches have been used successfully for genotyping DNA samples from amniocytes [71-75].

Kp(a+) RBCs have a weakened expression of high incidence Kell antigens [63]. Since the Kell protein carrying the Kp^a antigen has tryptophan (a nonpolar bulky residue) at residue 281, which has replaced arginine (a charged basic amino acid), it is conceivable that the conformation of the protein is affected so that it is no longer stable [76]. *In vitro* expression studies have revealed that most of the Kp^a form of Kell protein is retained in an intracellular compartment [77]. This is consistent with a model where the Kp^a isoform of Kell is not folded correctly so that less is present on the RBC membrane.

Duffy Blood Group System

The two polymorphic antigens in this system, Fy^a and Fy^b, are encoded by the Duffy gene and result from a single amino acid substitution of Gly44Asp [78-81]. The Duffy gene has two exons that generate a spliced transcript encoding a 336 amino acid protein which is predicted to span the membrane seven times. The protein is a receptor for the malarial parasites *Plasmodium vivax* and *Plasmodium knowlesi* and functions as a promiscuous chemokine receptor in the RBC (for review see Hadley and Peiper [82]). RBCs from 68% of African Americans and 100% of West Africans, which type as Fy(a-b-), are resistant to invasion by the malarial parasites. This phenotype is due to a point mutation within the promoter region of the Duffy gene, which to date has always occurred in the *FY*B* allele. The point mutation abolishes binding of the erythroid specific GATA-1 transcription factor resulting in the silencing of the *FY*B* gene in erythroid cells but not in other tissues [78,88-85]. This has an implication for transfusion medicine, since individuals with the Fy(b-) phenotype whom by genotyping are shown to have a non-functional GATA box, will be unlikely to make anti-Fy^b and, therefore, can be transfused with Fy(b+) RBCs without fear of eliciting an immune response to the Fy^b antigen. Until recently, the mutated GATA box was thought to be restricted to people of black ancestry, however, it also occurs in whites in Brazil [86], in non-Ashkenasi Jews [87], and in a woman with Swiss/Scottish ancestry (Chaudhuri, A; unpublished data). These studies further highlight the importance of genotyping different ethnic populations for identifying the prevalence of a given allele.

Allele-specific PCR and PCR-RFLP have been used to distinguish *FY*A* from *FY*B* alleles and to distinguish a functional from a non-functional GATA box [74,78,84,85,88,89]. GATA-box genotyping is required because there are rare people whose Fy(a-b-) phenotypes result from different molecular bases unrelated to a non-functional GATA box [81,88,90]. In rare individuals the Fy(a-b-) phenotype has resulted from molecular events unrelated to a non-functional GATA box [81,88,90] (unpublished data). These people can make potent anti-Fy3 (a high incidence antigen of the Duffy protein). Therefore unlike the Fy(a-b-) individuals who have mutated GATA box, these individuals should be transfused with Fy(a-b-) RBC products.

RBCs with the Fy^x phenotype have such a weak expression of Fy^b that it can go undetected by routine haemagglutination [87,91,92] The weak Fy^b expression is due to an approximately 10 fold reduction in the amount of Duffy protein present on RBC membranes [93-95]. Molecular genotyping studies have shown that two mutations, encoding the substitutions Arg89Cys and Ala100Thr in the coding region of the Duffy protein are associated with the Fy^x phenotype [93, 95,96] (see Pogo and Chaudhuri, this issue). By expression analysis in a heterologous system, it has been shown that the Arg89Cys substitution, which is predicted to be in the first intracellular loop is the critical mutation responsible for the weak Duffy expression [93,94]. It should be noted that since RBCs that are used for antibody identification are selected by serological testing, it is possible that they may be heterozygous for the Fy^x allele and weakly reactive antibodies may go undetected or unnecessarily complicate the antibody identification process. By performing PCR-RFLP analysis on samples from panel donors and selecting RBCs that do not have the Fy^x genotype, this potential source of error could be eliminated.

The prevalence of the Fy^x phenotype is approximately 3% in Caucasians [97], and a similar frequency has been estimated from genotype analysis [87,95,98] However, there are no reported cases of transfusion complications due to transfusion of Fy^x RBCs to Fy(b-) people.

Kidd Blood Group System

The Kidd system consists of two antithetical antigens, Jk^a and Jk^b and a high incidence antigen, Jk3. The gene, located on chromosome 18q12 encodes a multipass urea transporter protein consisting of 391 amino acids [99]. Jk^a and Jk^b alleles differ by a change of a single nucleotide 838G>A, which causes an Asp280Asn substitution and a *Mnl*I RFLP. Although the gene organisation has not yet been determined, this information makes it possible to genotype for Jk^a and Jk^b alleles in a PCR-RFLP assay [100].

Diego Blood Group System

The Diego antigens (Di^a and Di^b) are located on the anion exchanger (band 3), encoded by *SLC4A1* [101]. Another antithetical pair of antigens (Wr^a and Wr^b) and 10 low incidence antigens recently have been added to this system by virtue of point mutations in the *SLC4A1* gene [102-105]. Since these mutations produce a new or ablate an existing restriction enzyme site, genotyping can easily be performed by allele-specific or PCR-RFLP based methods. Many antibodies in this system, especially those to the low incidence antigens, have caused HDN. Thus, Diego genotyping of foetal RBCs is indicated when a mother has the antibody and the father is heterozygous for the gene encoding the low incidence antigen. It should be noted that *SLC4A1* has many alleles that are not associated with blood group antigens [106] but need to be considered when designing primers for genotyping in this system.

MNS Blood Group System

The MNS system is second only to Rh in terms of diversity and has over 40 antigens that are recognised by the ISBT [5-7]. Antigens in the MNS system are encoded by two highly homologous genes *GYPA* and *GYPB* that are adjacent on chromosome 4. The molecular bases of many of the antigens in this system are due to nucleotide substitutions, exon skipping, and gene recombination events that give rise to hybrid glycophorin A (GPA) and glycophorin B (GPB) proteins [10,90]. Although it is theoretically possible to detect a gene encoding any antigen for which the molecular basis is known, certain precautions are necessary when interpreting results in this blood group system. For instance, since *GYPA* and *GYPB* are homologous with a third gene in this family, *GYPE*, it is not always possible to design primers to specifically amplify only one of the genes. Several methods to genotype for M and N have been reported [107-110]. While anti-M and anti-N are not highly clinically significant, antibodies to low incidence antigens in this system have caused HDN.

A flow cytometric assay to detect M and N antigens has been used to monitor the level of exposure to radiation in atomic bomb survivors, after accidents involving radiation, or during radiation therapy [111-116]. This GPA-based somatic cell mutation assay enumerates peripheral blood erythrocytes that have lost the expression of M and N antigens in a dose dependent manner. Thus, this assay has practical applications in the evaluation of a person exposed to radiation.

Other blood group antigen systems

The molecular basis of many other blood group antigens has been determined. Single point mutations are associated with Lu^a/Lu^b, Au^a/Au^b [117,118], Yt^a/Yt^b [119], LW^a/LW^b [120], Co^a/Co^b [121] and In^a/In^b [122] polymorphisms, antigens in Cromer system [10], Ok^a [123] and three antigens in the Gerbich system [124]. This knowledge can be used to design PCR-based assays to perform genotyping.

The complement receptor 1 (CR1) or CD35 carries antigens in the Knops blood group system [125,126]. The molecular bases of the antigens have not been determined but soluble CR1 has been used to specifically inhibit antibodies to these antigens [127]. Antibodies to Knops system antigens are not clinically significant, availability of soluble CR1 would have practical value in the transfusion service since it could be used to inhibit Knops system antibodies in serum used for compatibility testing. Interestingly, RBCs with the form of CR1 that lacks Sl^a antigens have a reduced adhesion to the domain of the *Plasmodium falciparum* parasite-derived variant erythrocyte membrane protein that binds to normal RBCs. Since the Sl(a-) phenotype is common in Africans this finding suggests that the phenotype may protect against certain types of malaria [128].

Clinical Applications

The classical method of testing for blood group antigens and antibodies is haemagglutination. This technique is simple, inexpensive, does not require sophisticated equipment and, when done correctly, has a specificity and sensitivity that is appropriate for the clinical care of the vast majority of patients. However, there are a number of areas where haemagglutination may be limited, notably since it gives only an indirect measure of potential complications in an at risk pregnancy.

Identification of a foetus at risk for anaemia of the neonate

To identify a foetus at risk for HDN, DNA can be prepared from cells obtained by conventional invasive techniques such as amniocentesis, or chorionic villus sampling, and by non-invasive procedures from trophoblasts collected by transcervical sampling [129-131] or even from foetal erythroblasts from the maternal circulation [132,133]. Molecular genotyping should be considered when a mother's serum contains an alloantibody that has been associated with HDN and the father is heterozygous for the gene encoding the corresponding antigen, indeterminable, or he is not available for testing. For prenatal diagnosis of a foetus at risk of HDN, the approach to molecular genotyping should err on the side of caution. Thus, the DNA typing strategy should be to detect any gene (or part of a gene) even if the product is not expressed on the RBC membrane because then the mother will be monitored throughout pregnancy. In contrast, failure to detect a gene (false negative result) whose product is expressed could result in inadequate monitoring during the pregnancy. Since the incidence of certain genes varies in different populations it can be helpful to know the ethnicity of the parents. Further, in order to limit the gene pool, it is helpful to test DNA from the parents in parallel with tests on DNA from the foetus. It is also important to determine whether the mother has been the recipient of medical procedures such as, artificial insemination, *in vitro* fertilisation, or whether she is a surrogate mother.

Determination of blood group alleles in multi-transfused patients

Genotyping for blood group alleles can be performed on DNA prepared from white blood cells (WBCs), buccal epithelial cells or cells in urine sediment [134], all of which are obtained by non-invasive procedures. However, before interpreting results, it is important to obtain a medical history to ascertain whether the patient has been transplanted with allogeneic stem cells obtained from bone marrow, peripheral blood or placental blood. In these situations, the genotype result obtained with DNA prepared from peripheral blood leukocytes is likely to be different from DNA obtained from other tissues. Our data show that donor leukocytes, remaining after transfusion of filtered RBC components, do not interfere with the PCR reactions under the stringency conditions we employed [135]. However, it should be noted that contamination with allogeneic cells is not uncommon in samples exposed to clinical laboratory equipment [136] and that in trauma patients who received 8 to 10 units of relatively fresh blood, long term survival of leukocytes has been demonstrated when using sensitive assays [137].

Determination of blood group alleles when a patient's RBCs are coated with immunoglobulin

Molecular genotyping can be useful when the patient's RBCs are coated with IgG and appropriate direct agglutinating antibodies are not available and chemical treatment of the RBCs does not remove sufficient IgG to give confident results. This is especially true when only weakly reactive antibodies are available. This approach is also useful if the chemical treatment inactivates the antigen.

Resolution of weak A, B and D typing discrepancies

Using haemagglutination, interpretation of a person's ABO group is made after testing RBCs for the presence of antigens and serum for the presence of anti-A and anti-B. A proportion of blood donors and patients who historically have been typed as group O are now being typed as group A or group B because reagents contain monoclonal antibodies capable of detecting small amounts of the immunodominant carbohydrate responsible for A or B specificity. In the past, RBC products from such donors (i.e., having weak A or B antigens) have been transfused to group O patients without untoward sequelae. However, a typing result that differs from the historical record often results in time-consuming analyses, such as absorption and elution studies, which are needed to determine the true ABO group of the person. Since the molecular bases of many of the weak subgroups of A and B are associated with altered transferase genes, PCR-based techniques can be used to define the transferase gene and, thus, the ABO group [20,28,30,31,138].

A similar situation occurs with the D antigen of the Rh blood group system. A proportion of blood donors that historically have been typed as D-negative are now being typed as D-positive. Again, this is because reagents contain monoclonal antibodies that are capable of detecting small and specific parts of the D antigen. The molecular basis of numerous D variants has been determined and the knowledge can be used to identify the genes encoding altered RhD protein in these people.

Practical Considerations

Once the molecular basis of a blood group antigen is known, PCR-based assays can be designed to perform genotyping. However, interpretation of the results for use in clinical care, requires a combined expertise in molecular techniques, blood group antigens, and clinical outcomes, as outlined below.

Technical

Missense mutations associated with blood group antigens that give rise to a gain or loss of a restriction enzyme site are particularly useful in the clinical laboratory. This is especially true when they produce a unique banding pattern with both alleles because a distinct pattern is obtained thereby eliminating the need for an internal control to verify that each test has been performed correctly. A list of blood group antigens and informative restriction enzymes is given in Reid and Yazdanbakhsh [8].

Since the quantity and quality of DNA in clinical samples is often less than ideal, it is recommended to design primers that will generate relatively short PCR amplified products (smaller than 400 bp) prior to RFLP analysis. It is also recommended to test known homozygotes and heterozygotes in parallel and to perform the genotyping without knowledge of the phenotyping results or the expected answer. In the prenatal diagnosis setting, it is advisable to test DNA from the parents in parallel with DNA from the test amniocytes. This helps to reduce the size of the gene pool. For the same reason, it can be helpful to know the ethnicity of the patient under study. These and other technical considerations have been reviewed elsewhere [89].

Table 2. Examples of molecular events where genotype and phenotype do not agree.

Mechanism	Blood group system	Phenotype	Molecular basis/Outcome
Transcription	Duffy	Fy(a-b-)	GATA box –33 t→c
Splicing	MNS	S–s–	Partial/complete skipping exon 5
	Jk	Jk(a-b-)	Missense mutation
	Cromer	Dr(a-)	S165L and a 44 nucleotide deletion in exon 5; splicing small amount of normal length protein
Premature stop codon	Duffy	Fy(a-b-)	14 nucleotide deletion and +1 frameshift
			W136 stop codon
	Colton	Co(a-b-)	Frameshift
	Rh	r'	Q41 stop codon
	Rh	D–	4 nucleotide deletion and –1 frameshift
	GYPC	Ge:-2,-3,-4(LN)	–1 frameshift
	Rh	Rh_{null}	2 nucleotide deletion and +1 frameshift
Reduced amount of protein	Duffy	Fy^X	R89C
	Colton	Co(a-b-)	P38L
Unknown	Rh	Rh_{null}	Grossly normal gene, no RhCE or RhD
	Rh	D–	Grossly normal gene, no RhD
	Rh	D– – (Gou)	Grossly normal gene, no RhCE
	Kell	K_0	No Kell protein
Gene rearrange-ments	Rh	Various	Crossovers, gene conversions
	MNS	Various	Crossovers, gene conversions
Interacting proteins	Rh		For expression, Rh antigens require RhAG protein
	Kell		For expression, Kell antigens require XK protein
	Diego		For expression of Wr^b (a point mutation on band 3), amino acids 59 to 76 of glycophorin A are required
	Gerbich		For expression, Ge antigens require membrane skeletal protein, protein 4.1
Modifying genes	Kidd	Jk(a-b-)	*In(Jk)*
	Lutheran	Lu(a-b-)	*In(LU)*

Correlation of genotype with phenotype

Although the application of molecular genotyping to clinical situations is increasingly providing valuable information, it is important to remember that the genotype and phenotype of an individual may not correlate. Events leading to pseudo discrepant results can be broadly categorised into those that occur as a consequence of medical practices, those that occur from molecular events and those that occur because of a requirement of an interacting protein.

Medical Practice – Since certain medical practices can lead to the presence of non-self DNA , it is important to obtain an accurate medical history from the patient before interpreting genotyping results. For example:

- In patients who have had a transplant of allogeneic stem cells from cord blood, bone marrow, or peripheral blood, results obtained on DNA from white blood cells (WBCs) will differ from results obtained on DNA from buccal epithelial cells.
- In the prenatal setting, it is important to establish whether or not the mother has received artificial insemination from a sperm donor other than her husband, or is a surrogate mother. Although it is obvious that in such cases the genotype obtained from aminocytes will differ from the father/mother, this critical information is not always provided.
- In a patient who has received massive or chronic transfusions, the circulating RBCs are predominantly those of the donor. In such patients, it has been established that circulating WBCs are predominantly from the patient. Indeed, in the analysis of over 60 such patients, unexpectedly, we did not find any trace of microchimerism [135,139].

Molecular Events – Molecular events that can give discordant results between genotype and phenotype include a point mutation in a regulatory element of a gene thereby impairing transcription, a mutation in splicing sites of a gene, a premature stop codon, the absence of an interacting protein, and a decreased amount of protein in the RBC membrane. In the examples given in Table 2, PCR-based assays would predict a certain antigen profile and yet the gene product would not be expressed in the RBC membrane. There are other examples where a grossly normal gene is present and yet there is no RBC surface expression (e.g., D-negative, K_0, Jk(a-b-)). The molecular mechanism underlying these phenotypes has not been determined.

In other scenarios, a gene may not be detectable by specific prime and yet a product is present on the RBC membrane. This can occur with hybrid genes where, for instance, part of the *RHD* is replaced with *RHCE*: primers for the missing part of *RHD* will not amplify and yet a partial D antigen will be expressed on the RBC surface [36]. This can occur when two homologous genes are juxtaposed on the same chromosome, e.g., in MNS, Rh and Chido/Rogers blood group systems as a result of molecular events, such as crossovers, gene conversion, and other gene rearrangements. With hybrid genes, depending on their location, the primers may or may not anneal to the hybrid gene and thereby could give false negative or false positive genotyping results.

Interacting Protein(s) Required – The expression of some blood group antigens is dependant upon the product of modifying genes. Four well established cases have been reported where the required interacting protein has been described: Wr^b antigen is encoded by the wild type of band 3 [102], however, it is only expressed if amino acids 59 to 76 of glycophorin A (GPA) are present [140,141]. To be fully expressed, Kell system antigens require that the XK protein is present [63,142]. Rh antigens (D, C, c, E, e) require the presence of the Rh associated glycoprotein (RhAG) for expression [51,143]. In the absence of the membrane skeletal protein, protein 4.1, only a proportion (approximately 30%) of glycophorin C and glycophorin D is held in the RBC membrane and the cells appear to lack antigens in the Gerbich blood group system [144]. Antigens in the Kidd and Lutheran blood group systems are weakened by the presumed presence of, respectively, *In(JK)* and *In(LU)*. These modifying genes and their products have not been identified.

Other Considerations

Although, the molecular bases associated with a large number of antigens have been reported, in many cases the analysis has been restricted to a relatively small number of people with known antigen profiles. This information is being applied to molecular genotyping with the assumption that the molecular analysis will correlate with RBC antigen typing in all populations. However, it should be noted that we are in the infancy of our understanding of the gene polymorphisms in different ethnic groups and their significance to the expression of blood group antigens. A much larger number of people from a variety of ethnic backgrounds needs to be analysed in order to establish more firmly the correlation between genotype and the blood group phenotype. Until such a database is available, caution should be exercised when recommending clinical practice based on molecular genotyping of blood group antigens. Although molecular genotyping is clearly a valuable adjunct to haemagglutination, it is unlikely to replace this tried and true method any time soon.

Acknowledgements

The author is indebted to Christine Lomas-Francis for help in preparing the manuscript and thank Robert Ratner for typing the manuscript.

References

1. Giblett ER. A critique of the theoretical hazard of inter *vs.* intra-racial transfusion. Transfusion 1961;1:233-38.
2. Giblett ER. Blood group alloantibodies: An assessment of some laboratory practices. Transfusion 1977;4:299-308.
3. Hoeltge GA, Domen RE, Rybicki LA, Schaffer PA. Multiple red cell transfusions and alloimmunization: Experience with 6996 antibodies detected in a total of 159,262 patients from 1985 to 1993. Arch Pathol Lab Med 1995;119:42-45.
4. Geifman-Holtzman O, Wojtowycz M, Kosmas E, Artal R. Female alloimmunization with antibodies known to cause hemolytic disease. Obstet Gynecol 1997;89:272-75.

5. Daniels GL, Anstee DJ, Cartron J-P, et al. Blood group terminology 1995. ISBT working party on terminology for red cell surface antigens. Vox Sang 1995;69:265-79.
6. Daniels GL, Anstee DJ, Cartron JP, et al. Terminology for red cell surface antigens - Makuhari report. Vox Sang 1996;71:246-48.
7. Daniels GL, Anstee DJ, Cartron JP, et al. Terminology for red cell surface antigens - Oslo report. Vox Sang 1999;77:52-57.
8. Reid ME, Yazdanbakhsh K. Molecular insights into blood groups and implications for blood transfusions. Curr Opin Hematol 1998;5:93-102.
9. Avent ND. Human erythrocyte antigen expression: Its molecular bases. Br J Biomed Sci 1997;54:16-37.
10. Reid ME, Lomas-Francis C. The Blood Group Antigen FactsBook. San Diego: Academic Press, 1996.
11. Larsen RD, Ernst LK, Nair RP, Lowe JB. Molecular cloning, sequence, and expression of a human GDP-L- fucose:beta-D-galactoside 2-alpha-L-fucosyltransferase cDNA that can form the H blood group antigen. Proc Natl Acad Sci USA 1990;87:6674-78.
12. Wang BJ, Koda Y, Soejima M, Kimura H. Two missense mutations of H type α(1,2)fucosyltransferase gene (*FUT1*) responsible for para-Bombay phenotype. Vox Sang 1997;72:31-35.
13. Yu L-C, Yang Y-H, Broadberry RE, Chen Y-H, Lin M. Heterogeneity of the human *H* blood group α (1,2) fucosyltransferase gene among para-Bombay individuals. Vox Sang 1997;72:36-40.
14. Rouquier S, Lowe JB, Kelly RJ, Fertitta AL, Lennon GG, Giorgi D. Molecular cloning of a human genomic region containing the *H* blood group α(1,2)fucosyltransferase gene and two *H* locus-related DNA restriction fragments. J Biol Chem 1995; 270:4632-39.
15. Kelly RJ, Ernst LK, Larsen RD, Bryant JG, Robinson JS, Lowe JB. Molecular basis for H blood group deficiency in Bombay (Oh) and para-Bombay individuals. Proc Natl Acad Sci USA 1994;91:5843-47.
16. Kelly RJ, Rouquier S, Giorgi D, Lennon GG, Lowe JB. Sequence and expression of a candidate for the human *secretor* blood group α(1,2)fucosyltransferase gene (*FUT2*). J Biol Chem 1995;270:4640-49.
17. Chandrasekaran EV, Jain RK, Larsen RD, Wlasichuk K, Matta KL. Characterization of the specificities of human blood group H gene-specified α1,2-L-fucosyltransferase toward sulfated/sialylated/fucosylated acceptors: Evidence for an inverse relationship between α1,2-L-fucosylation of Gal and α1,6- L-fucosylation of asparagine-linked GlcNAc. Biochemistry 1996;35:8914-24.
18. Prieto PA, Larsen RD, Cho M, et al. Expression of human *H*-type α1,2-fucosyltransferase encoding for blood group H(O) antigen in Chinese hamster ovary cells - Evidence for preferential fucosylation and truncation of polylactosamine sequences. J Biol Chem 1997;272:2089-97.
19. Wagner FF, Flegel WA. Polymorphism of the *h* allele and the population frequency of sporadic nonfunctional alleles. Transfusion 1997; 37:284-90.
20. Olsson ML, Guerreiro JF, Zago MA, Chester MA. Molecular analysis of the *O* alleles at the blood group ABO locus in populations of different ethnic origin reveals novel crossing-over events and point mutations. Biochem Biophys Res Commun 1997;234:779-82.
21. Yamamoto F-I, Clausen H, White T, Marken J, Hakomori S. Molecular genetic basis of the histo-blood group ABO system. Nature 1990;345:229-35.
22. Gassner C, Schmarda A, Nussbaumer W, Schönitzer D. ABO glycosyltransferase genotyping by polymerase chain reaction using sequence-specific primers. Blood 1996;88:1852-56.

23. Grunnet N, Steffensen R, Bennett EP, Clausen H. Evaluation of histo-blood group ABO genotyping in a Danish population: frequency of a novel O allele defined as O^2. Vox Sang 1994;67:210-15.
24. Ogasawara K, Yabe R, Uchikawa M, et al. Molecular genetic analysis of variant phenotypes of the ABO blood group system. Blood 1996;88:2732-37.
25. Ogasawara K, Bannai M, Saitou N, et al. Extensive polymorphism of ABO blood group gene: Three major lineages of the alleles for the common ABO phenotypes. Hum Genet 1996;97:777-83.
26. Olsson ML, Chester MA. Frequent occurrence of a variant O^1 gene at the blood group ABO locus. Vox Sang 1996;70:26-30.
27. Olsson ML, Chester MA. Evidence for a new type of *O* allele at the ABO locus, due to a combination of the A^2 nucleotide deletion and the A^{el} nucleotide insertion. Vox Sang 1996;71:113-17.
28. Olsson ML, Chester MA. A rapid and simple ABO genotype screening method using a novel B/O^2 versus A/O^2 discriminating nucleotide substitution at the ABO locus. Vox Sang 1995;69:242-47.
29. Stroncek DF, Konz R, Clay ME, Houchins JP, McCullough J. Determination of *ABO* glycosyltransferase genotypes by use of polymerase chain reaction and restriction enzymes. Transfusion 1995;35:231-40.
30. Yamamoto F. Molecular genetics of the ABO histo-blood group system. Vox Sang 1995;69:1-7.
31. Watanabe G, Umetsu K, Yuasa I, Suzuki T. Amplified product length polymorphism (APLP): A novel strategy for genotyping the ABO blood group. Hum Genet 1997;99:34-37.
32. Suzuki K, Iwata M, Tsuji H, et al. A de novo recombination in the ABO blood group gene and evidence for the occurrence of recombination products. Hum Genet 1997;99:454-61.
33. Carritt B, Kemp TJ, Poulter M. Evolution of the human RH (rhesus) blood group genes: A 50 year old prediction (partially) fulfilled. Hum Mol Genet 1997;6:843-50.
34. Huang C-H. Molecular insights into the Rh protein family and associated antigens. Curr Opin Hematol 1997;4:94-103.
35. Cartron J-P. Defining the Rh blood group antigens: Biochemistry and molecular genetics. Blood Rev 1994;8:199-212.
36. Avent ND, Reid ME. The Rh blood group system: A review. Blood 2000;95:375-87.
37. Tippett P, Lomas-Francis C, Wallace M. The Rh antigen D: Partial D antigens and associated low incidence antigens. Vox Sang 1996;70:123-31.
38. Avent ND, Martin PG, Armstrong-Fisher SS, et al. Evidence of genetic diversity underlying Rh D^-, weak D (D^u), and partial D phenotypes as determined by multiplex polymerase chain reaction analysis of the *RHD* gene. Blood 1997; 89:2568-77.
39. Rouillac C, Gane P, Cartron J, Le Pennec PY, Cartron JP, Colin Y. Molecular basis of the altered antigenic expression of RhD in weak D (D^u) and RhC/e in R^N phenotypes. Blood 1996;87:4853-61.
40. Beckers EAM, Faas BH, Ligthart P, et al. Lower antigen site density and weak D immunogenicity cannot be explained by structural genomic abnormalities or regulatory defects of the *RHD* gene. Transfusion 1997;37:616-23.
41. Apoil PA, Reid ME, Halverson G, et al. A human monoclonal anti-D antibody which detects a nonconformation-dependent epitope on the RhD protein by immunoblotting. Br J Haematol 1997;98:365-74.
42. Reid ME, Halverson GR, Roubinet F, Apoil PA, Blancher A. Use of LOR-15C9 monoclonal anti-D to differentiate erythrocytes with the partial D^{VI} antigen from those with other partial D antigens or weak D antigens. Immunohematology 1998; 14:89-93.
43. Huang C-H, Chen Y, Reid M. Human D^{IIIa} erythrocytes: RhD protein is associated with multiple dispersed amino acid variations. Am J Hematol 1997;55:139-45.

44. Flegel WA, Müller TH, Schunter F, Gassner C, Schönitzer D, Wagner FF. *D category VI Type III:* A D-Ce(3-6)-D hybrid protein with normal RHD antigen density on red cells (abstract). Transfusion 1997;37(Suppl):101S.
45. Rouillac C, Colin Y, Hughes-Jones NC, et al. Transcript analysis of D category phenotypes predicts hybrid Rh D-CE-D proteins associated with alteration of D epitopes. Blood 1995;85:2937-44
46. Avent ND, Liu W, Jones JW, Scott ML, Voak D, Pisacka M et al. Molecular analysis of Rh transcripts and polypeptides from individuals expressing the D^{VI} variant phenotype: An *RHD* gene deletion event does not generate all D^{VI}ccEe phenotypes. Blood 1997;89:1779-86.
47. Huang C-H. Human D^{VI} category erythrocytes: Correlation of the phenotype with a novel hybrid RhD-CE-D gene but not an internally deleted RhD gene. Blood 1997;89:1834-35.
48. Matassi G, Chérif-Zahar B, Mouro I, Cartron JP. Characterization of the recombination hot spot involved in the genomic rearrangement leading to the hybrid *D-CE-D* gene in the D^{VI} phenotype. Am J Hum Genet 1997;60:808-17.
49. Blunt T, Daniels G, Carritt B. Serotype switching in a partially deleted RHD gene. Vox Sang 1994;67:397-401.
50. Huang C-H. Alteration of *RH* gene structure and expression in human dCCee and DC^{W}- red blood cells: Phenotypic homozygosity versus genotypic heterozygosity. Blood 1996;88:2326-33.
51. Chérif-Zahar B, Raynal V, Gane P, et al. Candidate gene acting as a suppressor of the *RH* locus in most cases of Rh-deficiency. Nat Genet 1996;12:168-73.
52. Okuda H, Kawano M, Iwamoto S, et al. The *RHD* gene is highly detectable in RhD-negative Japanese donors. J Clin Invest 1997;100:373-79.
53. Rose NC, Hurwitz C, Silberstein L, Andovalu R, Stoerker J. Prenatal analysis of rhesus CcDEe blood groups by heteroduplex generator. Am J Obstet Gynecol 1997; 176:1084-89.
54. Aubin JT, Kim CL, Mouro I, et al. Specificity and sensitivity of RHD genotyping methods by PCR-based DNA amplification. Br J Haematol 1997;98:356-64.
55. Steers F, Wallace M, Johnson P, Carritt B, Daniels G. Denaturing gradient gel electrophoresis: A novel method for determining Rh phenotype from genomic DNA. Br J Haematol 1996;94:417-21.
56. Gassner C, Schmarda A, Kilga-Nogler S, et al. *RHD/CE* typing by polymerase chain reaction using sequence-specific primers. Transfusion 1997;37:1020-26.
57. Van den Veyver IB, Moise KJ, Jr. Fetal RhD typing by polymerase chain reaction in pregnancies complicated by rhesus alloimmunization. Obstet Gynecol 1996; 88:1061-67.
58. Daniels G, Green C, Smart E. Differences between RhD-negative Africans and RhD-negative Europeans. Lancet 1997;350:862-63.
59. Faas BHW, Beckers EAM, Wildeboer P, et al. Molecular background of VS and weak C expression in blacks. Transfusion 1997;37:38-44.
60. Carbonnet F, Hattab C, Collec E, Le Van Kim C, Cartron JP, Bertrand O. Immunochemical analysis of the Kx protein from human red cells of different Kell phenotypes using antibodies raised against synthetic peptides. Br J Haematol 1997;96:857-63.
61. Russo D, Redman C, Lee S. Association of XK and Kell blood group proteins. J Biol Chem 1998;273:13950-56.
62. Redman CM, Lee S. The Kell blood group system. Transfus Clin Biol 1995;4: 242-49.
63 Oyen R, Halverson GR, Reid ME. Review: Conditions causing weak expression of Kell system antigens. Immunohematology 1997;13:75-79.
64. Lee S. Molecular basis of Kell blood group phenotypes. Vox Sang 1997;73:1-11.

65. Branch DR, Muensch HA, Sy Siok Hian AL, Petz LD. Disulfide bonds are a requirement for Kell and Cartwright (Yt[a]) blood group antigen integrity. Br J Haematol 1983;54:573-78.
66. Vaughan JI, Warwick R, Letsky E, Nicolini U, Rodeck CH, Fisk NM. Erythropoietic suppression in fetal anemia because of Kell alloimmunization. Amer J Obstet Gynec 1994;171:247-52.
67. Weiner CP, Widness JA. Decreased fetal erythropoiesis and hemolysis in Kell hemolytic anemia. Am J Obstet Gynecol 1996;174:547-51.
68. Parilla BV, Socol ML. Hydrops fetalis with Kell isoimmunization. Obstet Gynecol 1996;88:730.
69. Manning M, Warwick R, Vaughan J, Roberts IAG. Inhibition of erythroid progenitor cell growth by anti-Kell: A mechanism for fetal anaemia in Kell-immunized pregnancies (abstract). Br J Haematol 1996;93(Suppl 1):13.
70. Leggat HM, Gibson JM, Barron SL, Reid MM. Anti-Kell in pregnancy. Br J Obstet Gynaecol 1991;98:162-65.
71. Lee SH, Bennett PR, Overton T, Warwick R, Wu X, Redman CM. Prenatal diagnosis of Kell blood group genotypes: KEL1 and KEL2. Am J Obstet Gynecol 1996;175: 455-59.
72. Mifsud NA, Haddad AP, Sparrow RL, Condon JA. Use of the polymerase chain reaction sequence specific oligonucleotide technique for the detection of the K1/K2 polymorphism of the Kell blood group system. Blood 1997;89:4662-63.
73. Avent ND, Martin PG. Kell typing by allele-specific PCR (ASP). Br J Haematol 1996;93:728-30.
74. Murphy MT, Fraser RH. Detection of Kell blood groups: Molecular methods in the diagnostic laboratory. Blood Rev 1997;11:8-15.
75. Spence WC, Maddalena A, Demers DB, Bick DP. Prenatal determination of genotypes Kell and Cellano in at-risk pregnancies. J Reproduct Med 1997;42: 353-57.
76. Lee S, Wu X, Son S, et al. Point mutations characterize *KEL10*, the *KEL3, KEL4,* and *KEL21* alleles, and the *KEL17* and *KEL11* alleles. Transfusion 1996;36:490-94.
77. Yazdanbakhsh K, Lee S, Yu Q, Reid ME. Identification of a defect in the intracellular trafficking of a Kell blood group variant. Blood 1999;94:310-18.
78. Chaudhuri A, Polyakova J, Zbrzezna V, Pogo O. The coding sequence of Duffy blood group gene in humans and simians: Restriction fragment length polymorphism, antibody and malarial parasite specificities, and expression in nonerythroid tissues in Duffy-negative individuals. Blood 1995;85:615-21.
79. Tournamille C, Le Van Kim C, Gane P, Cartron J-P, Colin Y. Molecular basis and PCR-DNA typing of the Fy^a/Fy^b blood group polymorphism. Hum Genet 1995;95: 407-10.
80. Iwamoto S, Omi T, Kajii E, Ikemoto S. Genomic organization of the glycophorin D gene: Duffy blood group Fy^a/Fy^b alloantigen system is associated with a polymorphism at the 44-amino acid residue. Blood 1995;85:622-26.
81. Mallinson G, Soo KS, Schall TJ, Pisacka M, Anstee DJ. Mutations in the erythrocyte chemokine receptor (Duffy) gene: The molecular basis of the Fy^a/Fy^b antigens and identification of a deletion in the Duffy gene of an apparently healthy individual with the Fy(a-b-) phenotype. Br J Haematol 1995;90:823-29.
82. Hadley TJ, Peiper SC. From malaria to chemokine receptor: The emerging physiologic role of the Duffy blood group antigen. Blood 1997;89:3077-91.
83. Tournamille C, Colin Y, Cartron JP, Le Van Kim C. Disruption of a GATA motif in the *Duffy* gene promoter abolishes erythroid gene expression in Duffy-negative individuals. Nat Genet 1995;10:224-28.
84. Iwamoto S, Li J, Sugimoto N, Okuda H, Kajii E. Characterization of the Duffy gene promoter: Evidence for tissue-specific abolishment of expression in Fy(a-b-) of black individuals. Biochem Biophys Res Commun 1996;222:852-59.

85. Chaudhuri A, Nielsen S, Elkjaer ML, Zbrzezna V, Fang F, Pogo AO. Detection of Duffy antigen in the plasma membranes and caveolae of vascular endothelial and epithelial cells of nonerythroid organs. Blood 1997;89:701-12.
86. Castilho L, Pellegrino J, Jr., Saad STO, Hue-Roye K, Rios M. Duffy genotyping in Fy(b) Brazilians and the prevalence of the GATA box mutation (abstract). Transfusion 1998;38(Suppl):38S.
87. Parasol N, Reid M, Rios M, Castilho L, Harari I, Kosower NS. A novel mutation in the coding sequence of the FY*B allele of the Duffy chemokine receptor gene is associated with an altered erythrocyte phentoype. Blood 1998;92:2237-43.
88. Mallinson G, Soo KS, Anstee DJ. The molecular basis of Fy^a and Fy^b antigens (abstract). Transf Med 1994;4(Suppl):45.
89. Reid ME, Rios M. Applications of molecular genotyping to immunohaematology. Br J Biomed Sci 1999;56:145-52.
90. Reid ME. Molecular basis for blood groups and function of carrier proteins. In: Silberstein LE, ed. Molecular and Functional Aspects of Blood Group Antigens. Arlington, VA: American Association of Blood Banks, 1995:75-125.
91. Chown B, Lewis M, Kaita H. The Duffy blood group system in caucasians: Evidence for a new allele. Am J Hum Genet 1965;17:384-89.
92. Murphy MT, Templeton LJ, Fleming J, Ferguson M, Peterkin M, Fraser RH. Comparison of Fy(b) status as determined serologically and genetically. Transf Med 1997;7:135-41.
93. Tournamille C, Le Van Kim C, Gane P, et al. Arg89Cys substitution results in very low membrane expression of the Duffy antigen/receptor for chemokines in Fy^x individuals. Blood 1998;92:2147-56.
94. Yazdanbakhsh K, Yu Q, Powell VI. Reduced expression of the Duffy chemokine receptor in the weak Fy^b (Fy^x) phenotype is due to a mutation (Arg89Cys) of a conserved residue in the chemokine receptor superfamily (abstract). Blood 1998;92(Suppl 1):671a.
95. Olsson ML, Smythe JS, Hansson C, et al. The Fy^x phenotype is associated with a missense mutation in the Fy^b allele predicting Arg89Cys in the Duffy glycoprotein. Br J Haematol 1998;103:1184-91.
96. Reid ME, Rios M, Roye K, Chaudhuri A, Pogo O, Yazdanbakhsh K et al. Molecular basis of Fy^x (abstract). Transfusion 1998;38(Suppl):102S.
97. Race RR, Sanger R. Blood Groups in Man. 6th ed. Oxford, England: Blackwell Scientific Publications, 1975.
98. Castilho L, Pellegrino J, Jr., Saad STO, Reid ME, Rios M. Molecular events in FY*B silencing among Fy(b) Brazilians (abstract). Blood 1998;92(Suppl 1):563a.
99. Olivès B, Mattei M-G, Huet M, et al. Kidd blood group and urea transport function of human erythrocytes are carried by the same protein. J Biol Chem 1995;270: 15607-10.
100. Olivès B, Merriman M, Bailly P, et al. The molecular basis of the Kidd blood group polymorphism and its lack of association with type 1 diabetes susceptibility. Hum Mol Genet 1997;6:1017-20.
101. Bruce LJ, Anstee DJ, Spring FA, Tanner MJ. Band 3 Memphis variant II. Altered stilbene disulfonate binding and the Diego (Di^a) blood group antigen are associated with the human erythrocyte band 3 mutation Pro^{854}-->Leu. J Biol Chem 1994;269:16155-58.
102. Bruce LJ, Ring SM, Anstee DJ, Reid ME, Wilkinson S, Tanner MJA. Changes in the blood group Wright antigens are associated with a mutation at amino acid 658 in human erythrocyte band 3: A site of interaction between band 3 and glycophorin A under certain conditions. Blood 1995;85:541-47.
103. Jarolim P, Murray JL, Rubin HL, Coghlan G, Zelinski T. A Thr_{552}-->Ile substitution in erythroid band 3 gives rise to the Warrior blood group antigen. Transfusion 1997;37:398-405.

104. Jarolim P, Murray JL, Rubin HL, Smart E, Moulds JM. Blood group antigens Rb[a], Tr[a], and Wd[a] are located in the third ectoplasmic loop of erythroid band 3. Transfusion 1997;37:607-15.
105. Jarolim P, Rubin HL, Reid M. Characterization of six low incidence blood group antigens carried by erythrocyte band 3 (abstract). Transfusion 1996;36(Suppl): 49S.
106. Jarolim P, Murray JL, Rubin HL, et al. Characterization of 13 novel band 3 gene defects in hereditary spherocytosis with band 3 deficiency. Blood 1996;88:4366-74.
107. Corfield VA, Moolman JC, Martell R, Brink PA. Polymerase chain reaction-based detection of MN blood group- specific sequences in the human genome. Transfusion 1993;33:119-24.
108. Huang C-H, Guizzo ML, McCreary J, Leigh EM, Blumenfeld OO. Typing of MNSs blood group specific sequences in the human genome and characterization of a restriction fragment tightly linked to S-s- alleles. Blood 1991;77:381-86.
109. DuPont BR, Grant SG, Oto SH, Bigbee WL, Jensen RH, Langlois RG. Molecular characterization of glycophorin A transcripts in human erythroid cells using RT-PCR, allele-specific restriction, and sequencing. Vox Sang 1995;68:121-29.
110. Murray JC, Buetow KH, Ferrell RE, Sieberg PD, Fukuda M. An RFLP for glycoprotein A (MN) is in linkage disequilibrium with MN and Ss. Cytogenet Cell Genet 1988;47:149-51.
111. Grant SG, Jensen RH. Use of hematopoietic cells and markers for the detection and quantitation of human *in vivo* somatic mutation. In: Garratty G, ed. Immunobiology of Transfusion Medicine. New York: Marcel Dekker, Inc., 1994: 299-323.
112. Jensen RH, Langlois RG, Bigbee WL, et al. Elevated frequency of glycophorin A mutations in erythrocytes from Chernobyl accident victims. Radiat Res 1995;141: 129-35.
113. Schiwietz J, Lorenz R, Scheubeck M, Börner W, Hempel K. Improved determination of variant erythrocytes at the glycophorin A (GPA) locus and variant frequency in patients treated with radioiodine for thyroid cancer. Int J Radiat Biol 1996; 70:131-43.
114. Kyoizumi S, Akiyama M, Cologne JB, et al. Somatic cell mutations at the glycophorin a locus in erythrocytes of atomic bomb survivors: Implications for radiation carcinogenesis. Radiat Res 1996;146:43-52.
115. Jensen RH, Reynolds JC, Robbins J, et al. Glycophorin A as a biological dosimeter for radiation dose to the bone marrow from iodine-131. Radiat Res 1997;147:747-52.
116. Bigbee WL, Jensen RH, Veidebaum T, et al. Biodosimetry of Chernobyl cleanup workers from Estonia and Latvia using the glycophorin a *in vivo* somatic cell mutation assay. Radiat Res 1997;147:215-24.
117. Parsons SF, Mallinson G, Daniels GL, Green CA, Smythe JS, Anstee DJ. Use of domain-deletion mutants to locate Lutheran blood group antigens to each of the five immunoglobulin superfamily domains of the Lutheran glycoprotein: Elucidation of the molecular basis of the Lu^a/Lu^b and the Au^a/Au^b polymorphisms. Blood 1997;89:4219-25.
118. El Nemer W, Rahuel C, Colin Y, Gane P, Cartron JP, Le Van Kim C. Organization of the human LU gene and molecular basis of the Lu^a/Lu^b blood group polymorphism. Blood 1997;89:4608-16.
119. Bartels CF, Zelinski T, Lockridge O. Mutation at codon 322 in the human acetylcholinesterase (ACHE) gene accounts for YT blood group polymorphism. Am J Hum Genet 1993;52:928-36.
120. Hermand P, Le Pennec PY, Rouger P, Cartron JP, Bailly P. Characterization of the gene encoding the human LW blood group protein in LW^+ and LW^- phenotypes. Blood 1996;87:2962-67.
121. Smith BL, Preston GM, Spring FA, Anstee DJ, Agre P. Human red cell aquaporin CHIP. I. Molecular characterization of ABH and Colton blood group antigens. J Clin Invest 1994;94:1043-49.

122. Telen MJ, Udani M, Washington MK, Levesque MC, Lloyd E, Rao N. A blood group-related polymorphism of CD44 abolishes a hyaluronan-binding consensus sequence without preventing hyaluronan binding. J Biol Chem 1996;271:7147-53.
123. Spring FA, Holmes CH, Simpson KL, et al. The Ok[a] blood group antigen is a marker for the M6 leukocyte activation antigen, the human homolog of OX-47 antigen, basigin and neurothelin, an immunoglobulin superfamily molecule that is widely expressed in human cells and tissues. Eur J Immunol 1997;27:891-97.
124. Reid ME, Spring FA. Molecular basis of glycophorin C variants and their associated blood group antigens. Transf Med 1994;4:139-49.
125. Rao N, Ferguson DJ, Lee SF, Telen MJ. Identification of human erythrocyte blood group antigens on the C3b/C4b receptor. J Immunol 1991;146:3502-07.
126. Moulds JM, Nickells MW, Moulds JJ, Brown MC, Atkinson JP. The C3b/C4b receptor is recognized by the Knops, McCoy, Swain-Langley, and York blood group antisera. J Exp Med 1991;173:1159-63.
127. Moulds JM, Rowe KE. Neutralization of Knops system antibodies using soluble complement receptor 1. Transfusion 1996;36:517-20.
128. Rowe JA, Moulds JM, Newbold CI, Miller LH. *P-falciparum* rosetting mediated by a parasite-variant erythrocyte membrane protein and complement-receptor 1. Nature 1997;388:292-95.
129. Adinolfi M, Sherlock J, Kemp T, et al. Prenatal detection of fetal RhD DNA sequences in transcervical samples. Lancet 1995;345:318-19.
130. Bennett PR, Overton TG, Lighten AD, Fisk NM. Rhesus D typing. Lancet 1995; 345:661-62.
131. Kingdom J, Sherlock J, Rodeck C, Adinolfi M. Detection of trophoblast cells in transcervical samples collected by lavage or cytobrush. Obstet Gynecol 1995; 86:283-88.
132. Cheung M-C, Goldberg JD, Kan YW. Prenatal diagnosis of sickle cell anaemia and thalassaemia by analysis of fetal cells in maternal blood. Nat Genet 1996; 14:264-68.
133. Lo YMD, Corbetta N, Chamberlain PF, et al. Presence of fetal DNA in maternal plasma and serum. Lancet 1997;350:485-87.
134. Rios M, Cash K, Strupp A, Uehlinger J, Reid ME. DNA from urine sediment or buccal cells can be used for blood group molecular genotyping. Immunohematology 1999;15:61-65.
135. Reid ME, Rios M, Powell VI, Charles-Pierre D, Malavade V. DNA from blood samples can be used to genotype patients who have been recently transfused. Transfusion 2000;40:48-53.
136. Reed W, Lee T-H, Busch MP, Vichinsky EP. Sample suitability for the detection of minor leukocyte populations by polymerase chain reaction (PCR) (abstract). Transfusion 1997;37(Suppl):107S.
137. Lee T-H, Paglieroni T, Ohro H, Holland PV, Busch MP. Longterm multi-lineage chimerism of donor leukocytes in transfused trauma patients (abstract). Blood 1996;88(Suppl 1):265a.
138. Yamamoto F, McNeill PD, Hakomori S. Genomic organization of human histo-blood group ABO genes. Glycobiology 1995;5:51-58.
139. Wenk RE, Chiafari FA. DNA typing of recipient blood after massive transfusion. Transfusion 1997;37:1108-10.
140. Huang C-H, Reid ME, Xie SS, Blumenfeld OO. Human red blood cell Wright antigens: A genetic and evolutionary perspective on glycophorin A-band 3 interaction. Blood 1996;87:3942-47.
141. Reid ME. Contribution of MNS to the study of glycophorin A and glycophorin B. Immunohematology 1999;15:5-9.
142. Lee S, Russo D, Redman CM. The Kell blood group system: Kell and XK membrane proteins. Semin Hematol 2000;37:113-21.

143. Huang C-H, Liu PZ, Cheng JG. Molecular biology and genetics of the Rh blood group system. Semin Hematol 2000;37:150-65.
144. Reid ME, Takakuwa Y, Conboy J, Tchernia G, Mohandas N. Glycophorin C content of human erythrocyte membrane is regulated by protein 4.1. Blood 1990; 75:2229-34.

APPLICATION OF MOLECULAR BIOLOGY TO PARENTAGE TESTING AND FORENSIC MEDICINE[1]

Robert W. Allen[2]

Introduction

The last 20 years have witnessed a dramatic expansion in our understanding of the structure and function of the human genome. The expansion of our knowledge of chromosomal DNA is due, in part, to the development of molecular tools with which to analyze DNA. Their availability, coupled with the basic desire to understand the functional role played by each of the three billion nucleotides constituting the haploid human genome has fueled the intense research in molecular genetics.

As a consequence of the research into the genomes of a variety of species, and as a natural evolution of the biotechnology revolution of the 1970s and 1980s, the clinical laboratory has seen the widespread application of molecular genetics into diagnostic testing. To date, DNA based assays have been developed and, in some cases are marketed in the form of self-contained kits. In general, DNA based diagnostic tests employ the use of straightforward methods and yield reliable test results in areas as diverse as infectious disease testing [1-5], the diagnosis of inherited genetic disorders [6,7], blood group analysis [8,9], and human identification in the areas of parentage analysis and forensic medicine [10-12]. Diagnostic testing in many of these areas was possible before the introduction of molecular genetic techniques, typically using serological methods. However, serological methods were often technically diverse and sometimes were less than fully informative. It is for these and other reasons that molecular methods for DNA analysis have become a major part of clinical diagnostic testing in virtually every area of laboratory medicine.

Human Identification Testing: Methods and Markers

One field in which DNA analysis has had a dramatic impact is in human identification testing (human ID testing). Included in the general field of human ID testing are family relatedness testing (i.e. parentage testing), forensic medicine, the monitoring of bone marrow engraftment, the development of convicted felon

1. Supported in part by funds from the H.A.Chapman Charitable Trust and the Children's Medical Center Foundation
2. H.A. Chapman Institute of Medical Genetics, Tulsa, OK, USA

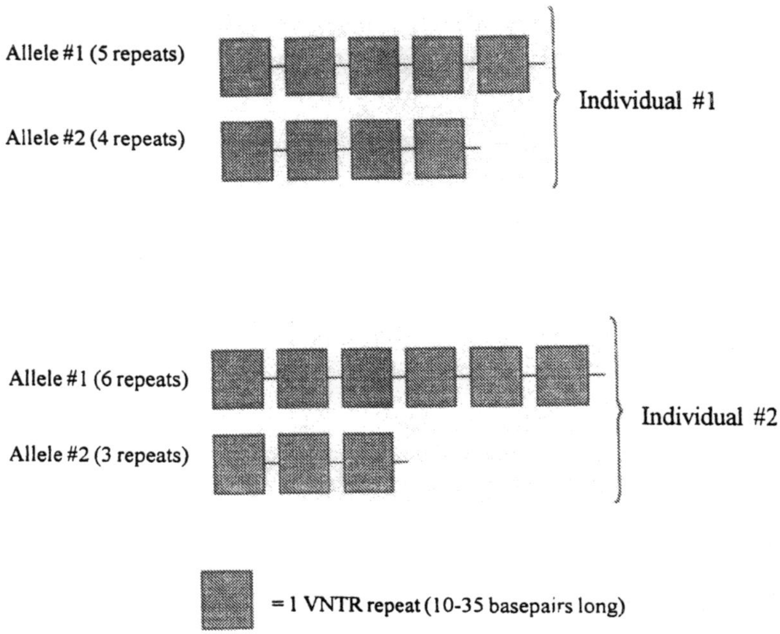

Figure 1. Structure of a VNTR locus. Shown in the figure is the basic structural organization of a typical VNTR locus in two individuals (#1 and #2). Each allele is composed of a short sequence of nucleotides that is tandemly repeated a variable number of times and is contained within the chromosome. The tandem repeat is flanked by single copy DNA (not shown) which contains the restriction enzyme sites used to liberate the VNTR from the chromosome for analysis. Digestion of chromosomal DNA with a restriction enzyme liberates the VNTR whose size is proportional to the number of repeats.

databases, and testing whether or not two individuals are identical or fraternal twins. The application of DNA analysis to human identification relies on the detection of genetic markers in chromosomal DNA that derive from differences in nucleotide sequence. Indeed, DNA sequencing is perhaps the best method to identify one individual to the exclusion of all others on the planet inasmuch as current dogma holds that everyone's DNA is different at the nucleotide sequence level, with the exception of identical twins. DNA sequencing, however, is neither efficient nor inexpensive as a means of DNA analysis. One consequence of differences in nucleotide sequence among members of a population is an alteration in the spatial arrangement of restriction enzyme sites within the chromosome. Restriction enzyme site polymorphisms can thus serve as genetic markers in the same way as blood groups or other more traditional markers. Subjecting chromosomal DNA to restriction enzyme digestion followed by agarose gel electrophoresis, Southern blotting, and hybridization to a DNA probe that is complementary to the region of interest yields a banding pattern of restriction fragments that can be visualized. This process, collectively known as restriction fragment length polymorphism mapping (RFLP mapping), was pioneered by Botstein and his collaborators [13] and provided one of the first methods for DNA analysis that was efficient. The RFLP mapping technique was also the principal development in molecular genetics that spawned the field of modern human ID testing.

RFLP mapping has revealed several types of genetic marker to exist within chromosomal DNA. One type consists of single nucleotide substitutions within the chromosome that either create or destroy a restriction enzyme site. These simple sequence markers are numerous in chromosomal DNA and typically consist of 2-3 alleles within the population [14,15]. Their detection using RFLP mapping is strictly dependent upon the use of the appropriate restriction enzyme. Due to their limited degree of polymorphism, markers of this type are no more informative for human ID testing than are more classical di- and tri-allelic genetic markers detected using serological methods.

In the mid to late 1980s a second type of genetic marker was discovered using RFLP technology that was found to be extremely informative for human identification [16-19]. These markers consist of small sequences of nucleotides that are tandemly repeated a variable number of times along a length of chromosome (Figure 1). Restriction enzyme digestion of chromosomal DNA isolated from different donors was found to produce restriction fragments containing differing numbers of repeats thereby yielding different banding patterns, or DNA profiles. In addition, these polymorphic DNA profiles could be visualized following digestion with any restriction enzyme that did not cut the DNA within the tandem repeat [20]. Population analyses of VNTR markers reveal the presence of large numbers of alleles that can be combined into literally thousands of possible phenotypes in all populations examined [12,21-23] (Figure 2). The very high degree of variability exhibited by VNTR markers make them especially informative for parentage testing and forensic investigations. Furthermore, since DNA is contained in most body fluids and tissues, and is relatively stable to environmental insult, DNA profiling has the capability to provide test results with a wide variety of samples; both freshly obtained and often with older forensic-type specimens retrieved from crime scenes.

The success of DNA profiling for resolving cases of disputed parentage and in the investigation of crime has fostered intense research and development into refining the technology. Developments occurring during the late 1980s and early 1990s included methods to label probes with non-radioactive tracers that are visualized colorimetrically [24] or through the use of chemiluminescence [25], as well as refining gel electrophoretic methods to maximize resolution of banding patterns.

In addition to VNTR markers composed of repeats of 10-35 basepairs detected with RFLP analysis, other VNTR markers have been described that are composed of tandem repeats of 2-4 nucleotides [11,26,27]. Markers of this type have been called microsatellites or small tandem repeats (STRs). These markers are not as polymorphic as are the VNTRs described above but they have the advantage of being able to be amplified using polymerase chain reaction (PCR). One of the limitations of RFLP mapping is the need for relatively large amounts of reasonably intact DNA in order to obtain a DNA profile. While this limitation

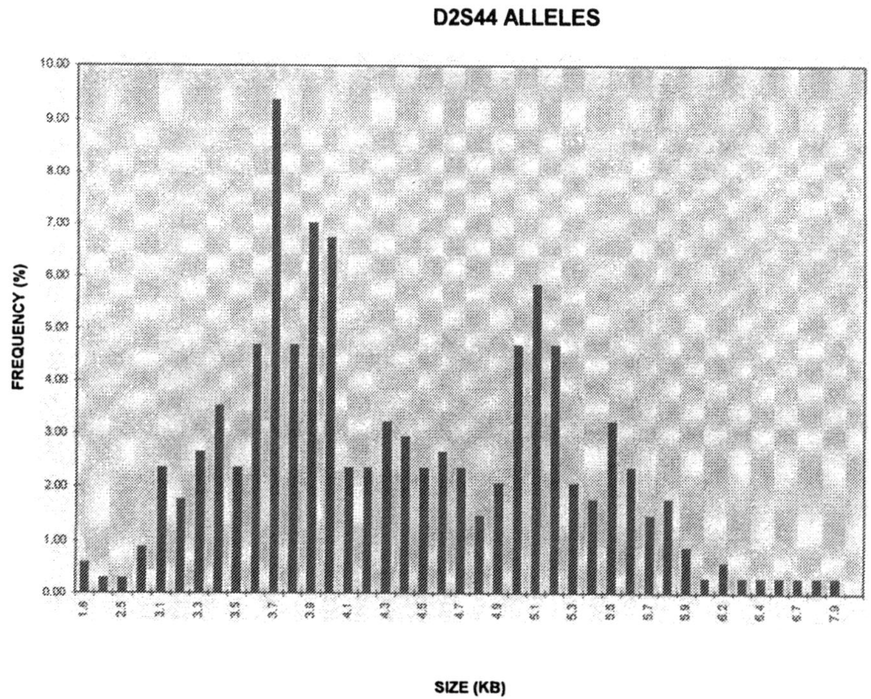

Figure 2. Variability of D2S44 alleles in the Caucasian population. The histogram depicts the approximate number of differently sized VNTR alleles at the D2S44 locus that have been detected in the Caucasian population from the US. Alleles were detected using RFLP mapping with Pvu II restriction endonuclease and the D2S44 probe YNH24 (18).

is not a significant problem for a lab performing parentage testing where fresh blood samples are readily available, it is a problem for a forensic DNA typing lab. Forensic labs often encounter crime scene samples that are limited in quantity, and furthermore, may have been subjected to environmental conditions that accelerate the partial or complete degradation of the DNA within the specimen. In such situations, PCR/STR typing is perhaps the only method able to produce a DNA profile. Whereas a typical RFLP marker may be composed of 30-50 alleles, the typical PCR/STR markers may exhibit 5-10 alleles, underscoring the reduction in discriminatory power afforded by PCR/STR technology.

Another PCR-based genetic marker developed during this period, especially for forensic DNA typing, was the DQα system of the major histocompatibility complex (MHC) [28,29]. The DQα system consists of eight alleles that together encode the different allelic forms of the DQα polypeptide chain of HLA-D region antigens, important in immune recognition and transplantation. Detection of alleles at the DQα locus is accomplished using dot blotting technology in which allele specific oligonucleotide probes are typically bound to a solid support which is hybridized to DQα product amplified from a specimen. The amplified product hybridizes to the corresponding allele specific oligonucleotide and the hybridized PCR product is subsequently visualized colorimetrically following hybridization to a second reporter probe. The locations of the resulting colored spots on the support define the phenotype of the donor of the sample. The DQα genetic marker is especially valuable for forensic testing because of the robust nature with which the locus is amplified using PCR. DQα typing is possible using saliva deposited on cigarette butts or stamps, single hair roots, and even single sperm as a source of genomic DNA template [30].

The dot-blotting technology used for the DQα locus has also been applied to the detection of alleles at five other genetic loci of limited polymorphism. These five loci (including LDLR, GYPA, HBGG, D7S8, and GC) are collectively known as the Polymarker™ system (Perkin Elmer) and are amplified together during PCR.

Human Identification Testing: Interpretation of DNA Typing Results in Various Clinical Situations

Parentage Testing

Perhaps the greatest number of DNA profiling tests performed are for parentage testing purposes, regardless of the DNA typing technology used. In 1998 in the US, approximately 250,000 parentage tests were performed, the majority of which employed the use of either RFLP or PCR/STR typing [31]. Typical results for a parentage analysis using RFLP technology are shown in figure 3. The results contain the DNA profiles of an alleged father, a child, a mother, and a mixture of DNA from the alleged father and the child (Figure 3). Analysis of results begins with identifying the allele transmitted to the child from the mother, thereby defining the obligate allele inherited from the biological father. The DNA profile of the alleged father is then examined to determine if the obli-

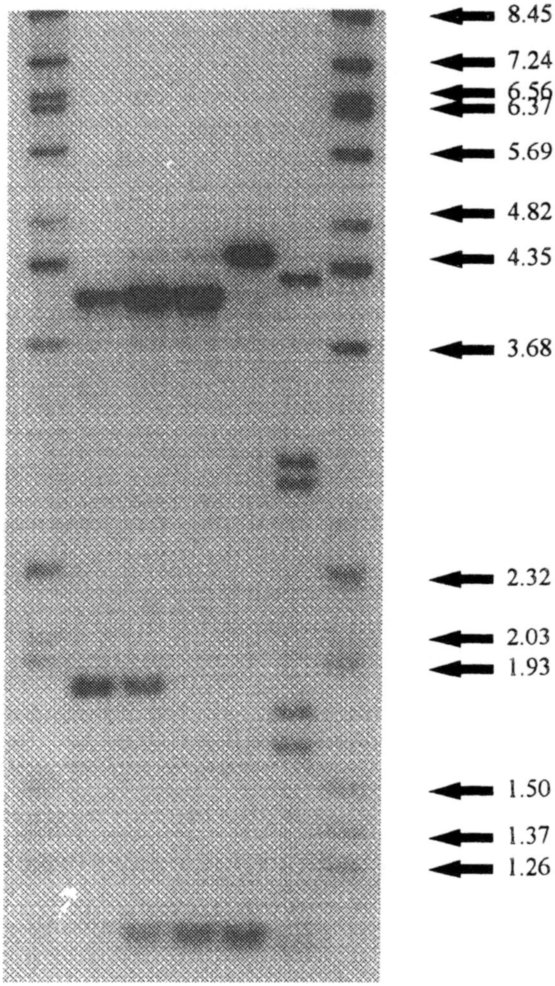

Figure 3. Application of RFLP analysis to a case of disputed parentage. Chromosomal DNA was extracted from blood samples obtained from an alleged father, a child, and the mother. DNA was subjected to RFLP analysis at the D16S85 locus using Pvu II restriction endonuclease. The alleged father is included as the father of the child based upon a sharing of the obligate gene. Lanes in the figure reflect: A) The DNA profile of the alleged father. X) The profile from a mixture of DNA of the alleged father and the child. C) The DNA profile of the child. M) The DNA profile of the mother. G) The DNA profile of a mixed genomic control sample from three unrelated donors. Lanes labeled S con-contain size ladders (sizes in the right margin are in kilobasepairs).

gate allele is present. If not, this constitutes evidence of non-paternity for the tested man, and comparable evidence at a second DNA locus confirms the exclusion. Evidence of non-paternity for two genetic systems is needed to exclude the tested man since RFLP markers have a reasonably high mutation rate during meiosis resulting in length changes for RFLP alleles that can lead to false exclusions.

If the tested man exhibits an allele matching the obligate allele in the child, he is included among a pool of men who also have the allele (and therefore could also be the father), and the question then becomes how much weight to assign to the match. Weighting inclusionary DNA typing results is accomplished in the traditional way by calculating a paternity index (PI) [32]. The paternity index (PI) is a genetic odds ratio, or likelihood ratio, comparing the chance the tested man contributed the allele to the child versus the chance an untested, unrelated man of the same ethnic background transmitted the allele. If both the tested man and the mother are heterozygous, each has a chance of passing the obligate allele equal to 0.5. The chance the mother and tested man will pass the maternal and paternal alleles visualized in the child is 0.5 X 0.5= 0.25. The alternate possibility is that the mother and a random man mated to produce the child, the odds of which are 0.5 for the maternal transmission probability times the frequency of the obligate allele in the appropriate ethnic population. If the obligate allele has a frequency of 5% (i.e. 0.05), then the combined probability of producing the child by a mating of the mother and a random man is 0.5 X 0.05= 0.025.

The traditional formula for paternity index is:

PI= X/Y

where X is the probability of producing the phenotype of the child in question through a mating of the mother and the tested man (i.e. 0.25 in the example above) and Y is the probability of producing the phenotype of the child in question through a mating of the mother with the random man (i.e. 0.025 in the example above). In this case the PI equals 10.0, which means that the likelihood the tested man is the father is favored over the likelihood a random man is the father by a factor of 10:1. The compilation of inclusionary test results for several statistically independent genetic systems is accomplished simply by multiplying the individual PI s together, producing the combined PI, or, cumulative PI [32]. This statistical logic is used for weighting inclusionary evidence in cases of disputed parentage regardless of the technology used.

One aspect of the statistical analysis of parentage tests performed using RFLP mapping that is unique to this technology is the estimation of allele frequency within the population. Alleles at RFLP loci are defined by an estimate of size relative to known size markers that are also present in the gel during electrophoresis. Complicating allelic designation however, is the fact that RFLP systems exhibit a continuous distribution of alleles whose size differs from nearest neighbors by single repeats [10,21,33,34]. Given the resolution possible with agarose gel electrophoresis, two distinct alleles that differ by few repeats may be indistinguishable with RFLP analysis thereby creating a level of uncer-

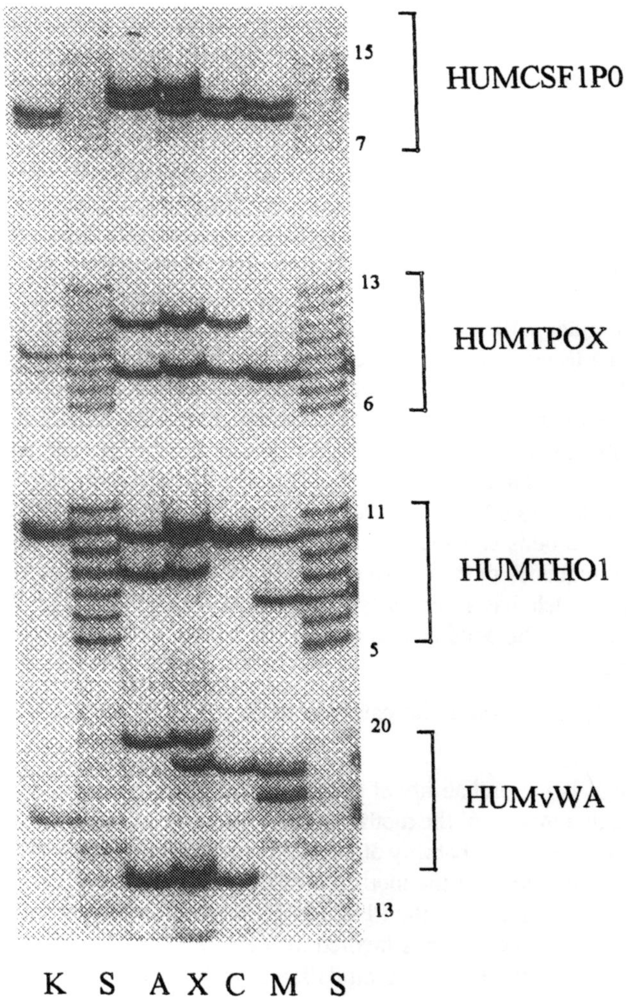

Figure 4. PCR/STR analysis of a case of disputed parentage. Chromosomal DNA was obtained from buccal swabs from an alleged father, a child and the mother. These samples were then subjected of PCR/STR analysis using a multiplex amplification reaction. PCR primers included in the multiplex were fluoresceinated and directed the amplification of STR alleles at the (from top to bottom) HUMCSF1P0, HUMTPOX, HUMTHO1, and HUMvWA31 loci. Amplification products were separated by size using polyacrylamide gel electrophoresis and alleles in each sample were visualized using fluorescence with a laser scanner (Molecular Dynamics, Sunnyvale, CA). Lanes shown contain: K) The DNA profile of the genomic control (K562 erythroleukemic cells). A) The DNA profile of the alleged father. X) The profile of a mixture of DNA from the alleged father and child. C) The DNA profile of the child. M) The DNA profile of the mother. Lanes labeled S represent the allelic ladders. The smallest and largest alleles for each locus contained within the ladders are labeled in the right hand margin and denote the number of tandem repeats within each allele.

tainty in designating alleles. In other words, it is not possible to identify discreet alleles in RFLP systems like it is with blood groups, HLA, or other markers used for parentage testing. Thus, while RFLP results demonstrate in a rather compelling fashion that the alleged father is included, their statistical analysis requires measures be taken to account for the lack of precision in identifying alleles when estimating the frequency of the shared gene from an allele frequency database. The steps taken to ensure a reasonable level of conservatism in assigning allele frequency is to create a size bin that surrounds the obligate allele and sums observations of all alleles that are of similar size into a single frequency. The width of the bin (and hence the number of allele observations summed) depends upon the known resolution of the gel system and the labs reproducibility in estimating allele size. In most laboratories, alleles that differ by 1-2% of their size can be reproducibly resolved as distinct. Likewise, reproducibility in estimating allele size averages ±1% for most RFLP laboratories. In the floating bin method, a defined percentage of the estimated size of the obligate allele is added to and subtracted from the observed size estimate for the obligate allele, thereby defining the limits of the frequency bin [10,21,33,34]. All allele counts in a database that are contained within the bin are then summed and this value divided by the total alleles observed for the locus becomes the frequency of the obligate allele used in the calculation of paternity index.

Another common method used to compensate for the imprecision in allele designation for RFLP markers is the use of a fixed bin. The fixed bin method [35] utilizes the space between size markers within a size ladder as allele frequency bins. All alleles falling between two flanking size markers within the ladder are assigned a single frequency for calculations. Thus, if 10 electrophoretically distinct alleles migrate during electrophoresis to positions that fall within the space flanked by the same two size markers, all 10 alleles are considered a single allele group and are divided by the total number of alleles observed for the locus to arrive at a frequency estimate. One frequency value is thus assigned to each of the 10 alleles. Alleles migrating at the boundary of two fixed bins are assigned the frequency of the bin with the higher frequency.

The use of floating versus fixed bins among labs using RFLP methods varies somewhat. In general, most parentage testing labs use the floating bin method for allele frequency determination whereas most forensic labs use the fixed bin method. Both methods yield statistical results that are fairly comparable.

The statistical interpretation of DNA typing results obtained using PCR/STR methods are more direct and are similar to serological markers in that the alleles detected are distinct [11]. Figure 4 demonstrates typical PCR/STR results from a paternity test. It is seen in the figure that the tested man is included and the obligate allele is defined by comparison to the allelic ladder flanking the DNA profiles of the trio. Identification of alleles can either be accomplished manually, or through the use of computer software capable of measuring the distance migrated by an unknown during electrophoresis and comparing that value to the distances migrated by individual alleles within the allelic ladder. The fact that the size of the tandem repeat is typically 4 basepairs, and the resolution of the polyacrylamide gel systems used is 1-2 basepairs, allows for discreet allele designations to be made. In other words, the gel system is usually capable of discriminating all alleles at the locus. The establishment of allele frequency databases is a straightforward task as is their use for calculations.

Regardless of the technology used, the goal of the parentage testing laboratory is to either exclude the alleged parent, or to include him or her with a combined parentage index of at least 99:1. This corresponds to a 99% probability of parentage and has become the threshold level for the courts to conclude that a tested alleged parent is indeed the true parent of the child. This threshold has been established, in part, through the setting of minimum standards for parentage testing laboratories by the American Association of Blood Banks (AABB), the organization providing guidelines, standards, and accreditation for parentage testing laboratories in the USA [32,36].

It was stated earlier that RFLP systems are more powerful in their discriminatory capabilities than are PCR/STR systems. In order to achieve a comparable level of discriminatory power, laboratories employing the use of PCR/STR technology analyze a greater number of genetic systems as the routine test battery. Whereas an RFLP laboratory may incorporate the analysis of 4-6 systems as a routine test battery, a PCR/STR lab will typically analyze 8-12 systems. Although this may seem at first glance to constitute double the work for a technologist, multiple PCR/STR systems can be co-amplified in a single multiplex reaction, producing allelic products from 3-4 loci whose size range does not overlap (Figure 4). Moreover, if a laboratory has the capability of detecting PCR/STR alleles using multicolor fluorescent detection methods, the different systems can be labeled with different colored fluorescent primers enabling alleles from 8-10 systems to be simultaneously amplified and analyzed.

PCR/STR technology is gaining in popularity among parentage testing laboratories worldwide. One reason for the popularity is the ability to use a greater variety of specimens as a source of DNA. For example, the use of buccal swabs has greatly expanded within the parentage testing community in the USA [31,37]. Swabs yield ample chromosomal DNA template for PCR/STR analysis and are easily collected and transported thereby reducing collection costs. Likewise, the use of PCR/STR analysis creates the possibility of performing tests using uncultured amniocytes, blood spot cards, or coroner specimens that may be partially degraded as sources of DNA for analysis. Most importantly, PCR/STR analysis allows for rapid turn-around times, a feature that is often attractive to potential clients.

Forensic Testing

Many of the methods and interpretations of DNA typing results discussed above for parentage testing also apply to the investigation of crime using genetic testing. Significant differences include the nature of the specimens often used as a source of DNA for analysis and how the analysis of results is accomplished. DNA typing has had an impact on forensic medicine that generally exceeds that experienced by the field of parentage testing

Samples used for DNA typing by forensic laboratories typically include tissues or body fluids deposited on various substrates. Forensic samples are often retrieved by investigators only after a significant period of exposure to environmental insults that accelerate decomposition of all biological materials. Nonetheless, DNA often survives such exposure in a form suitable for analysis, especially when PCR based testing methods are utilized. Methods for extraction of DNA from forensic specimens do not differ significantly from fresh blood or buccal swabs. One common forensic analysis, namely the investigation of sexual assault, however, does incorporate a variation of the common extraction procedure. Most sexual assault evidence consists of a mixed specimen with contributions of DNA from both the victim and the assailant(s) (Figure 5). The mixed sample will produce a DNA profile containing alleles from two individuals which can sometimes lead to ambiguity in tracing which alleles in the pattern belong to the knowns being tested. The variation of the typical extraction method strives to separate the victim's contribution of DNA present in epithelial cells in a stain or swab from the spermatozoa thereby simplifying the DNA profiles visualized and hence the analysis of results. The differential extraction procedure [38,39] involves the use of a mild extraction procedure to dissolve epithelial cells originating from the victim in evidentiary specimens, thereby liberating her DNA. The spermatozoa from the assailant are unaffected by the mild pre-extraction method and can be recovered by centrifugation. Sperm are then lysed to liberate their DNA using more rigorous extraction methods involving proteinase K and SDS with the addition of a disulfide reducing agent to break the disulfide links in the sperm capsid.

One challenge often associated with the retrieval of forensic specimens for analysis is the actual collection of the biological material from the substrate. The use of swabs, swatches of cloth, and scraping instruments often will transfer a biological sample from the original substrate onto one from which DNA is more easily extracted.

Another significant difference between the application of DNA typing for parentage and forensic purposes is the statistical interpretation of results. In a parentage analysis, the only match considered in weighting the evidence in favor of paternity is the match between the one allele in the alleged parent's sample and the obligate parental allele in the child. Statistically, this means the frequency of only the matching allele is incorporated into the statement of odds. In contrast, in forensic DNA typing where the donor of a specimen retrieved from a crime scene is in question, both alleles for each genetic marker evaluated must match between evidentiary and known sample (Figure 5). When both alleles do match, assigning weight to the match incorporates the frequencies of both alleles in the calculation of a phenotype frequency. The frequency of the alleles in question are extracted from a database using binning if RFLP profiles are involved. This is performed in the same manner as in parentage testing. Once allele frequencies have been determined however, the laboratory must go one step further and calculate a phenotype frequency for the donor of the specimen. The Hardy-Weinberg equation is used for this calculation, and if the matching DNA profiles are heterozygous, the 2pq portion of the equation is applied, substituting the frequencies of the two alleles for the p and q elements in the formula. If the DNA profile exhibits a single band, the laboratory could calculate a phenotype frequency as p^2, the predicted phenotype frequency of the homozygote. However, the limitation in resolution of alleles associated with RFLP

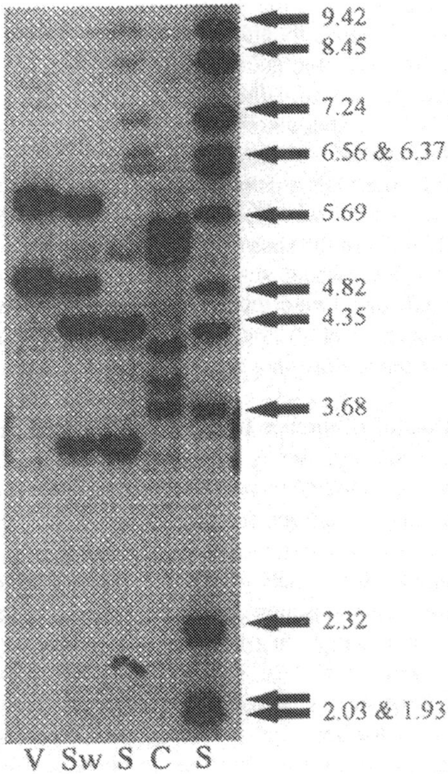

Figure 5. Application of RFLP analysis to the investigation of sexual assault. RFLP analysis was performed at the D2S44 locus as described in Figure 3. DNA used for the analysis was extracted from known blood samples from the victim (lane V), a vaginal swab (lane Sw), and a suspect (lane S). Other lanes in the figure include the size markers (lanes S) and the genomic control (lane C). Size markers shown in the right hand margin are in kilobasepairs.

methods, and the prevailing tendency for forensic labs to be conservative has resulted in the calculation of phenotype frequencies for single banded patterns as 2p, or as $p^2 + 2p(1-p)\theta$, where θ represents a coefficient of inbreeding within the population. This conservative approach is recommended in a report on forensic DNA typing prepared by the National Research Council of the National Academy of Sciences [40]. Once phenotype frequencies have been calculated for the various DNA markers analyzed, they can be multiplied together to yield a composite phenotype frequency which then becomes the denominator in the likelihood ratio calculation as was discussed earlier for producing a paternity index. The numerator of the ratio is 1 which reflects the fact that the suspect's DNA profile matches that of the evidence. Although this method of calculating match probabilities may sound straightforward, there has been much heated discussion among population geneticists around the world as to the accuracy of the numbers produced [23,41-47]. Questions of Hardy Weinberg equilibrium and statistical independence of loci were raised among population geneticists throughout the early to mid 1990s, casting doubt on the validity of forensic calculations. These arguments have all but disappeared now from both the courtroom and the literature due primarily to extensive research on populations around the world and the development of better methods with which to validate calculations [23,35,43,45,48,49].

Other applications

While parentage testing and forensic analyses represent the traditional application of DNA typing for human identification, the high degree of polymorphism associated with DNA markers makes their analysis useful for resolving other situations of questioned family relationships or identifying the source of biological materials. For example, questioned family relationships other than alleged parent and child can be evaluated technically and statistically using DNA typing. Wenk et.al. [50] have published methods to incorporate DNA typing results into a statistical statement of the likelihood of sibship among two or more individuals and Traver [51] have extended this logic to evaluate other questioned family relationships. DNA sequencing of the mitochondrial genome or analysis of DNA markers on the Y chromosome allow for investigations of the maternal and paternal lineages respectively within families. In fact, polymorphisms in mitochondrial DNA and in the Y chromosome have allowed investigators to study the historical origins and migratory routes of early human populations around the world [52-55].

Another area for the application of DNA typing is in monitoring engraftment following bone marrow transplantation [56]. Donors and recipients for bone marrow transplantation are normally matched precisely for blood groups and tissue antigens making these markers of little value in post-transplant monitoring of engraftment. PCR/STR or RFLP markers however, have no known function related to rejection of a transplant, thereby making them ideal markers for distinguishing between the donor and the recipient as the source of hemopoietic cells in the bloodstream. Furthermore, since these genetic markers are so polymorphic, there is a very good chance that even if donor and recipient are closely related, there will be differences in the DNA profiles that can be used for post-transplant analysis. Thus, donor and recipient are subjected to DNA typing prior to transplantation to identify systems in which their DNA profiles differ. Following transplantation, it is these systems that are analyzed to determine the source of cells circulating in the bloodstream. Both RFLP mapping and PCR/STR analysis exhibit a sensitivity of about 10% admixture of cell types (unpublished observation).

Finally, DNA typing is being used widely in the US and elsewhere in the world to establish DNA databanks composed of profiles from felons convicted of violent crime. Samples are collected from felons typically at the time of parole and subjected to DNA profiling and the results are entered into a centralized databank. If the parolee commits a crime and leaves biological evidence at the scene, the DNA profile can be produced from that sample and matched against known profiles in the databank. Those that match thus become suspects in the investigation and more extensive DNA profiling can be repeated with the remains of the evidentiary sample and freshly obtained samples from suspects. Early in the development of the combined DNA indexing system (CODIS) program, RFLP markers were used for data banking. However, recently, PCR/STR loci have replaced RFLP markers for the program since discreet alleles are identified making matching against databank entries more straightforward. At the end of 1999, it is estimated that over 750,000 samples will have been collected from convicted felons nationwide, and of those, almost 500,000 will have been analyzed and the results entered into the databank [57].

Human Identification Testing: Present Status of the Field

The field of human ID testing has grown to the point of no longer being highly specialized. The availability of DNA typing kits from several manufacturers for both RFLP and PCR/STR typing and the widespread availability of potential clients for parentage testing and other forms of human ID services have spawned the establishment of small testing laboratories. In addition, human ID services have been added to test menus of larger laboratories providing other forms of clinical testing. In the field of parentage testing, the American Association of Blood Banks tracks developments and prepares an Annual Report Summary that dicusses changes and developments in the field. The most recent report [31] indicates that for 1998, there were approximately 250,000 cases of disputed parentage resolved by laboratories in the USA. Of these, 94% were resolved with the use of DNA analysis methods, a trend that has continued to grow since the late 1980s. Clearly, DNA related technologies are the mainstay of parentage testing. Other developments in the field of parentage testing are the increasing use of robotic workstations for DNA extraction, restriction enzyme digestions, PCR setup, etc. The use of such workstations is likely to increase as pressure to further reduce costs is maintained.

The forensic use of DNA typing methods has also grown. The Bureau of Justice Statistics estimates that DNA typing methods will be applied to almost 50,000 criminal investigations in the USA by the end of 1999 [57], which is almost double the number of analyses performed in 1998. As in the case of parentage testing, a number of smaller laboratories have begun to offer forensic

DNA typing for private clients. Services requested typically involve repeat testing in criminal investigations in which a police laboratory as incriminated a suspect and a defense attorney wants the test confirmed with what remains of the evidence. Insurance companies sometimes request forensic DNA typing to investigate auto accidents using bloodstains or tissue samples retrieved from the car as a means to place occupants in the vehicle and determine who was driving. As has been re-iterated several times, DNA typing is able to provide compelling information in a myriad of situations requiring sample identification.

Human Identification Testing: Future Directions

Future directions for the field of human ID testing are likely to revolve predominately around technologic issues, particularly new or improved methods for examining a greater number of genetic loci more efficiently, with less sample, and at a more economic cost. Among the technologies presently in use, further developments in PCR/STR technology using fluorescent detection methods are likely to continue as are developments in automation. One new method for the detection of PCR/STR systems makes use of mass spectrometry to analyze amplified STR alleles. This method, known as time-of-flight mass spectrometry (or MALDI-TOF-MS) employs mass spectrometry to ionize STR alleles with pulses from a laser [58]. The ionized amplification products travel through a vacuum and impact a detector which records the time lapse from ionization to detection. The time-of-flight is directly related to allele size which provides a means of distinguishing alleles. Advantages of MALDI-TOF are sample throughput and a high degree of reproducibility in allele designation. The principal disadvantage of this technology at present is the 100-150 basepair upper limit of allele size that can be ionized and hence analyzed using the technology. Many of the PCR/STR loci commercially available have allele sizes that exceed this limit. As further research and development efforts are applied to mass spectrometry, ways will likely be found to extend the size range of alleles, and hence loci, that can be analyzed using this technology.

Another method that has been developed for human identification is known as Genetic Bit Analysis technology (GBA) [59]. GBA is a PCR based technology tailored to detect simple sequence polymorphisms in DNA. GBA is actually a collection of several well established molecular methods that begin with PCR amplification of a DNA template known to contain a single nucleotide polymorphism. Amplification is initiated from a 5'-phosphorothiolated primer. The phosphorothiolate end group protects the amplified product from the degradative action of T7 gene 6 exonuclease which degrades the hybridized unphosphorothiolated template. The resulting single stranded phosphorothiolated amplicon is then hybridized to an oligonucleotide probe immobilized onto a solid support (typically a 96 well polystyrene plate). The probe is designed to be complementary to both polymorphic forms of the diallelic DNA system and to terminate at the base just 5' to the polymorphic nucleotide. A single oligonucleotide probe thus hybridizes to both allelic forms of amplified product and furthermore serves as a primer for elongation by a single dideoxynucleotide that is either biotinylated or fluoreseinated. The dideoxynucleotide triphosphate coupled to biotin is complementary to one allelic form of the diallelic marker and the fluoresceinated dideoxynucleotide is complementary to the other allele. Once amplicons are hybridized to the probe, DNA polymerase elongates the probe sequence one nucleotide, incorporating either the biotinylated or fluoresceinated dideoxynucleotide, depending upon the allelic form of the amplicon hybridized to the probe. The final step in the procedure is to develop color in the 96 wells to identify which alleles are present in each. This is accomplished either by using peroxidase or alkaline phosphatase coupled to anti-biotin or anti-fluorescein antibodies respectively. The enzymes covert their respective substrates to blue and yellow colored products which can be easily detected and quantitated using ELISA based spectrophotometric detection methods. In practice, plates are incubated with one substrate, the color documented, the plates are washed and incubated with the second substrate. In this way amplified product from a DNA donor can be analyzed for the presence of either or both alleles in each well. Homozygotes for one allele will yield only one color whereas heterozygotes will yield both blue and yellow colors indicating the presence of both alleles.

The question arises as to why this methodology is under development for human ID testing given the low level of discriminatory power that is characteristic for diallelic genetic markers. As is the case for PCR/STR systems versus RFLP systems, the lower discriminatory power can be compensated for by adding additional systems to the panel. Thus, systems incorporated into GBA panels have been selected to be especially robust and reliable and PCR multiplexes of 8-10 loci have been developed. One sample of DNA can ultimately be amplified in 3-4 multiplex PCR amplifications and alleles at a total of 35-40 different diallelic loci produced [59]. Using this number of systems, it is possible to achieve a level of statistical power suitable for parentage testing and other situations requiring human identification. The one attractive feature of GBA technology is the rather simplistic and reliable nature of the chemistry needed to achieve a result and the suitability of such liquid handling tests for automation from sample preparation through data analysis. It is conceivable that GBA could be completely automated once blood samples were manually loaded onto a robotic workstation.

Closing Comments

Molecular biology has had dramatic impact on many areas of science. It could be argued however, that in no other discipline have developments in molecular science had such a dramatic effect on our daily lives as in the field of human identification. The popular press and the media continually convey stories involving DNA analysis methods and their use in solving everything from high visibility criminal activity to answering age-old questions such as whether or not bones found in a mass grave belong to Czar Nicholas and his family, or whether or not Thomas Jefferson sired children with one of his slaves. On a more practical note, DNA analysis technologies have greatly simplified the resolution of cases of disputed parentage and criminal investigation and a level of standardization is now being realized worldwide. Such standardization allows laboratories to share information leading to further understanding and improvements in the statistical interpretation of DNA typing information and also helps law enforcement agencies everywhere to investigate crime in a coordinated manner.

For all the practical benefit realized by society from the efforts of so many over the years in the development of human ID testing, there is another more basic gain being realized by the scientific community. We are gaining a better understanding of who we are and how we have evolved to our present condition. Ways in which this has occurred include studies of mitochondrial DNA and Y chromosome polymorphisms from human populations around the world yielding clues to relationships and migratory patterns of ancient wanderers. Likewise, each of the 250,000 parentage analyses performed in 1998 by AABB accredited labs represent family studies and offer the opportunity to encounter a mutation within a family group. These events occurring during the formation of sperm and egg allow us to witness evolutionary changes at the DNA level that underlie the high degree of polymorphism observed for the genetic systems we investigate. Given that the frequencies and characteristics of these events differ among loci, one can only conclude that there is some feature of the different VNTR loci that either promotes or protects against mutations. As answers to these and other questions become more apparent, our understanding of mutations that lead to polymorphism will undoubtedly increase, as will our understanding of mutations that lead to disease.

References

1. Persing DR, Smithe TF, Tenover FC, White TJ (eds). Diagnostic molecular microbiology: Principles and application. Washington, DC; American Society for Microbiology Rochester, MN; Mayo Foundation, 1993.
2. Wolcott M. Advances in nucleic acid based detection methods. Clin Microbiol Res 1992;5:370-86.
3. Macario AJL, Macario EC (eds). Gene probes for bacteria. New York, NY, Academic Press, 1990.
4. Busch MP. Application of molecular biology to infectious disease screening of blood donors. In: Allen RW, Aubuchon JP (eds). Molecular genetics in diagnosis and research. Bethesda, MD, AABB Press, 1995:111-38.
5. Busch MP, Stramer SL, Kleinman SH. Evolving applications of nucleic acid amplification assays for prevention of virus transmission by blood components and derivatives. In: Garratty G (ed). Application of molecular biology to blood transfusion medicine. Bethesda, MD, AABB Press, 1997:123-76.
6. Drocopoli NC, Haines JL, Korf BR, et al. (eds). Current protocols in human genetics. Vols. 1 and 2, New York, NY, John Wiley and Sons, 1998.
7. Ely P. Uses of molecular genetic technologies in diagnosis and prognosis. In: Allen RW, Aubuchon JP (eds). Molecular genetics in diagnosis and research. Bethesda, MD, AABB Press, 1995:65-80.
8. Telen MJ Structural aspects of blood group antigens. In: Garratty G (ed.). Application of molecular biology to blood transfusion medicine. Bethesda, MD, AABB Press, 1997:1-16.
9. Allen RW. Pre-natal blood grouping using the polymerase chain reaction. In: Garratty G (ed). Application of molecular biology to blood transfusion medicine. Bethesda, MD, AABB Press, 1997:103-22.
10. Allen RW, Wallhermfechtel M, Miller WV. The application of restriction fragment length polymorphism mapping to paternity testing. Transfusion 1990;30:552-64.

11. Hammond HA, Jin L, Zhong Y, Caskey CT, Chakraborty R. Evaluation of 13 short tandem repeat loci in personal identification applications. Am J Human Genet 1994;55:175-89.
12. Federal Bureau of Investigation. VNTR population data: A worldwide study. Vol. 5, Quantico, VA, FBI Academy, 1993.
13. Botstein K, White RL, Skolnick M, Davis RW. Construction of a genetic linkage map in man using restriction fragment length polymorphisms. Am J Hum Genet 1980;32:314-20.
14. Skolnick MH, Willard HF, Menlove LA. Report of the committee on human gene mapping by recombinant DNA techniques. Cytogenet Cell Genet 1984;37:210-73.
15. Cuticchia AJ, Pearson PL, and Klinger HP, eds. Chromosome coordinating meeting. Volume 1, Genome Priority News, Basel, Karger, 1993.
16. Jeffreys AJ, Wilson V, Thein SL. Hypervariable minisatellite regions in human DNA. Nature 1985;314:67-73.
17. Jeffreys AJ, Wilson V, Thein SL. Individual specific "fingerprints" of human DNA. Nature 1985;316:76-79.
18. Nakamura Y, Leppert M, O'Connell P, et al. Variable number of tandem repeat (VNTR) markers for human gene mapping. Science 1987;235:1616-22.
19. Wong Z, Wilson V, Patel I, et.al. Characterization of a panel of highly variable mini-satellites cloned from human DNA. Ann Hum Genet 1987;51:269-88.
20. Allen, RW, Hoover BA, Santiago L, Miller J. Compatibility of several HVR-type probes with Pvu II restriction endonuclease and advantages of Pvu II over other commonly used enzymes. Adv Forensic Hemogenet 1990;3:83-87.
21. Balasz I, Baird M, Clyne M, Meade E. Human population genetic studies of five hypervariable DNA loci. Am J Hum Genet 1989;44:182-90.
22. Odelberg SJ, Plaetke R, Eldridge JR, et al. Characterization of eight VNTR loci by agarose gel electrophoresis. Genomics 1989;5:915-24.
23. Krane DE, Allen, RW, Sawyer SA, Petrov DA, Hardtl DL. Genetic differences at four DNA typing loci in Finnish, Italian, and mixed Caucasian populations. Proc Natl Acad Sci (USA) 1992;89:10583-87.
24. Dykes DD. The use of biotinylated DNA probes in parentage testing: Non-isotopic labeling and non-toxic extraction. Electrophoresis 1988;9:359-68.
25. Beck S, O'Keefe T, Corell JM, Koster H. Chemiluminescent detection of DNA: Application for DNA sequencing and hybridization. Nucl Acids Res 1989;17:5115-23.
26. Weber JL, May PE. Abundant class of human DNA polymorphisms which can be typed using the polymerase chain reaction. Am J Hum Genet 1989;44:388-96.
27. Litt M, Luty JA. A hypervariable microsatellite revealed by in vitro amplification of a dinucleotide repeat within the cardiac muscle actin gene. Am J Hum Genet 1989;44:397-401.
28. Saiki RK, Bugawan TL, Horn GT, et al. Analysis of enzymatically amplified beta-globin and HLA-DQ alpha DNA with allele specific oligonucleotide probes. Nature 1986;324:163-66.
29. Horn GT, Bugawan TL, Long C, Erlich HA. Allelic sequence variation of the HLA-DQ loci: Relationship of serology and insulin dependent diabetes susceptibility. Proc Natl Acad Sci (USA) 1988;85:6012-16.
30. Comey CT, Budowle B, Adam DE, Baumstark AL, Lindsey JA, Presley LA. PCR amplification and typing of the HLA-DQα gene in forensic samples. J Forensic Sci 1993;38:239-49.
31. Parentage Testing Standards Committee, American Association of Blood Banks. Annual Report Summary for 1998. Bethesda, MD. AABB Press, 1999.
32. Walker RH, ed. Inclusion probabilities in parentage testing. Arlington, VA: AABB Press, 1983.

33. Brenner C, Morris JW. Paternity index calculations in single locus hypervariable DNA probes: Validation and other studies. In: Data acquisition and statistical analysis for DNA typing laboratories. Proc. Intern. Symp. Human Identification, Madison, WI, Promega Corp. 1989;21-54.
34. Endean D. RFLP analysis for paternity testing: Observations and caveats. In: Data acquisition and statistical analysis for DNA typing laboratories. Proc Intern Symp Human Identification, Madison, WI, Promega Corp., 1989;55-70.
35. Budowle B, Giusti AM, Waye JS, et al. Fixed bin analysis for statistical evaluation of continuous distribution of allelic data from VNTR loci for use in forensic comparisons. Ann Hum Genet 1991;48:841-55.
36. Allen RW. Parentage testing in the United States: The role of the American Association of Blood Banks. Profiles in DNA 1998;2:7-8.
37. Parentage Testing Standards Committee, American Association of Blood Banks. Annual Report Summary for 1997. Bethesda, MD: AABB Press, 1998.
38. Gill P, Jeffreys AJ, Werrett DJ. Forensic application of DNA "fingerprints". Nature 1985;318:577-79.
39. Giusti A, Baird M, Pasquale S, Balasz I, Glassberg J. Application of deoxyribonucleic acid (DNA) polymorphisms to the analysis of DNA recovered from sperm. J Forensic Sci 1986;31: 409-17.
40. Committee on DNA forensic science: The evaluation of forensic DNA evidence. Washington, DC, National Academy Press, 1996.
41. Roberts L. Science in Court: A culture Clash. Science 1992;257:732-36.
42. Lewontin RC, Hardtl DL. Population genetics in forensic DNA typing. Science 1991;254:1745-50.
43. Budowle B, Monson KL, Giusti AM. A reassessment of frequency estimates of PvuII generated VNTR profiles in a Finnish, and Italian, and a General U.S. Caucasian database: No evidence for ethnic subgrups affecting forensic estimates. Am J Hum Genet 1994;55:533-39.
44. Lander ES. Research on DNA typing catching up with courtroom application. Am J Hum Genet 1991;48:819-23.
45. Chakraborty R, Srinivasan MR, Daiger S. Evaluation of standard error and confidence interval of estimated genotype probabilities and their implications in DNA forensics. Am J Hum Genet 1993;52:60-70.
46. Devlin B, Risch N, Roeder K. Statistical evaluation of DNA fingerprinting: A critique of the NRC's report. Science 1993;259:748-50.
47. Weir BS. Population genetics in the forensic DNA debate. Proc Natl Acad Sci (USA) 1992;89:11654-59.
48. Federal Bureau of Investigation. The application of forensic DNA testing to solve violent crimes. Washington, DC: US Dept of Justice, 1993.
49. Weir BS. Independence of VNTR alleles defined as floating bins. Am J Hum Genet 1992;51:992-97.
50. Wenk RE, Traver M, Chiafari FA. Determination of sibship in any two persons. Transfusion 1996;36:259-62.
51. Traver M. Paternity calculations: Special situations. In: Statistics Workshop. Madison, WI, Promega Corp., 1996: Chapter 4.
52. Cooper G, Amos W, Hoffman D, Rubinsztein D. Network analysis of human Y microsatellite haplotypes. Hum Mol Genet 1996;5:1759-66.
53. Cann RL, Stoneking M, Wilson AC. Mitochondrial DNA and human evolution. Nature 1987;325:31-36.
54. Deka R, Jin L, Shriver MD, et al. Disperson of human Y chromosome haplotypes based on five microsatellites in global populations. Genome Res 1996;6:1177-84.
55. Hammer M, Zegura SL. The role of the Y chromosome in human evolutionary studies. Evol Anthropol 1996;5:116-34.

56. Gardiner N, Lauler M, O'Riordan J, DeArce M, McCann SR. Donor chimerism is a strong indicator of disease free survival following bone marrow transplantation for chronic myeloid leukemia. Leukemia 1997;11:512-15.
57. Bureau of Justice Statistics. web site address: WWW.OJP.USDOJ.GOV/BJS, February 1997.
58. Butler JM. STR analysis by time-of-flight mass spectrometry. Profiles in DNA 1999;2:3-6.
59. Nikiforov TT, Rindle RB, Goclet P, et al. Genetic bit analysis: A solid phase method for typing single nuleotide polymorphisms. Nucl Acids Res 1994;22: 4167-75.

TOWARDS A UNIVERSAL BLOOD GROUP – MANIPULATING THE MOLECULAR STRUCTURE OF BLOOD GROUP ANTIGENS

D J Anstee[1]

Introduction

The human red cell is a relatively simple highly specialised vehicle for facilitating the transport of oxygen from the lungs to the tissues and carbon dioxide from the tissues to the lungs. By encapsulating haemoglobin within a membrane it is possible to maintain much higher concentrations of haemoglobin in the circulation than would be possible if haemoglobin were free in solution. However, placing haemoglobin within a membrane requires that the membrane displays some key properties. It must be flexible so that the red cell can traverse capillaries. It must be permeable to allow respiration and essential maintenance of the metabolism of the cell and it must be strong enough to resist fragmentation as a result of the shear forces it is exposed to during circulation. The framework upon which the membrane is built is a lipid bilayer. Inserted into this lipid bilayer are proteins with a wide variety of functions (reviewed in Anstee and Cartron [1]). The most important of these proteins is the anion transport protein, band 3, which is responsible for the 'one-for-one' exchange of chloride and bicarbonate ions across the membrane and thereby facilitates carbon dioxide transport from the tissues to the lungs. Most, but not all, membrane proteins accessible on the extracellular face of the red cell are glycosylated. The glycosylation is either in the form of N-glycans or O-glycans. N-glycans can be large, highly branched structures frequently carrying the critical sugars for A,B,H activity at the non-reducing ends. O-glycans are generally smaller sialylated structures which are mainly carried on Glycophorins. The surface negative charge of the red cell is mainly attributable to sialylated O-glycans on Glycophorins A and B (reviewed in Anstee [2]). Glycolipids also contribute to the surface carbohydrate of the red cell. The most abundant glycolipid is globoside 1 (P glycolipid) which is present at about 14 million molecules/ red cell [3]. The surface carbohydrate or glycocalyx extends out from the surface of the red cell about 12.5nM and is thought to provide a protective barrier minimising membrane damage and microbial invasion [4,5]. The flexibility of the membrane is controlled by an elaborate network of proteins on the cytosolic face known collectively as the red cell skeleton [6]. The skeleton is attached to the lipid bilayer through protein:protein interactions at two main sites. The first, and most

---

1. Director, Bristol Institute for Transfusion Sciences, Bristol, UK.

important, site of interaction involves the N-terminal cytoplasmic domain of band 3 and the skeletal proteins, ankyrin, spectrin and pallidin (syn protein 4.2). The second site involves the cytoplasmic domain of Glycophorins C and D and the skeletal proteins protein 4.1, actin and protein p55. Inherited abnormalities in proteins involved in the first site are associated with hereditary spherocytosis. Inherited abnormalities affecting proteins involved in the second site of interaction are associated with hereditary elliptocytosis.

The foregoing discussion portrays the red cell as an elegant bag containing haemoglobin but does not allude to the immense structural diversity of the molecules that make up this bag when the structures are compared between individuals. Genetic variability which gives rise to the structural diversity found in red cell proteins can, in many cases, be linked to mans constant battle to survive malaria. A key stage in the life cycle of malarial parasites is development within red cells and it is well established that many inherited variations of globin genes give some protection from malaria at the expense of normal oxygen carrying capacity. Inherited anaemias resulting from globin gene abnormalities are extremely common. The WHO estimates that in the year 2000 approximately 7% of the world's population will be carriers of inherited disorders of haemoglobin [7]. Protection against malaria is afforded to the heterozygote. Individuals who are homozygous for the sickle gene or α- or β–thalassaemia genes have very severe disease. Such protection in the heterozygote is not confined to the oxygen carrying pigment. It has been known for many years that glucose-6-phosphate dehydrogenase deficiency provides protection against malaria and more recently it has become clear that inheritance of South East Asian Ovalocytosis (SAO) affords protection against severe cerebral malaria [8]. Individuals with SAO cells are heterozygous for a mutant band 3 protein which cannot transport anions and alters the linkage between the N-terminal domain of band 3 and the underlying skeleton to make this cell ovalocytic and less deformable. In this case, ability to transport carbon dioxide (as bicarbonate) is reduced by 50% but protection from the most serious consequences of malaria is obtained. Given the enormous impact that malaria has had on the structure of the red cell as evidenced by the widespread occurrence of haemoglobinopathies, glucose-6-phosphate dehydrogenase deficiency and SAO in areas of the world where the disease is endemic the argument that much of the structural diversity manifested at the surface of the red cell as blood group polymorphisms is also a consequence of selection against malaria is difficult to refute. Indeed, there is experimental evidence for involvement of the Fy protein, Glycophorin A and the CR1 protein in the disease process [9-11].

Several of the blood group antigens found at the surface of the red cell, particularly those defined by carbohydrate structures, are widely expressed at mucosal surfaces and at these sites selection is likely to operate to avoid colonisation by bacteria and viruses (reviewed in Moulds et al. [12], Ilver et al. [13]).

A Universal Blood Group - some experimental approaches

Whatever the causes, it is the structural diversity at the surface of the red cell which provides the main barrier to transfusion of red cells between individuals. For the most part such barriers are easily overcome by typing for the major antigens and screening for antibodies against other antigens or by compatibility testing. In populations of European origin, but not necessarily in other populations, the major antigens are those of the ABO system and the RhD antigen [14]. Despite these relatively simple procedures for providing compatible blood for transfusion there are circumstances when transfusion without the need for testing would be advantageous, for example, military use in a battlefield. In situations where the distribution of blood groups in the donor population does not match that of the patient population shortages of blood of suitable group can arise and blood of a universal group would be useful. Blood of a universal group would also find application for transfusion to multitransfused patients who have developed a mixture of blood group antibodies which makes provision of compatible blood almost impossible.

One approach which has received a lot of attention over many years is the development of red cell substitutes. Essentially, the challenge is to provide oxygen carrying capacity as an haemoglobin product or using perfluorocarbons and dispense with the red cell membrane altogether. Some haemoglobin products have reached the stage of clinical trial (reviewed in Prowse [15]).

Another approach with an equally long gestation involves the conversion of group A and group B blood to group O by incubation with purified exoglycosidases (reviewed in Goldstein [16]). That group B antigen can be converted to group O by incubation with the α–galactosidase derived from coffee beans was demonstrated 40 years ago [17]. Since then the coffee bean enzyme has been used in purified form and as an expressed recombinant protein to convert group B red cells to group O for use in phase 1 trials in human volunteers. The treated units were tolerated well although in some group O recipients a slight elevation of anti-B titre was observed suggesting that not all B antigen sites were removed under the conditions of treatment employed [16]. Conversion of group B cells to group O is of limited value in European populations where group B individuals are a minority (9% in the UK) and the real challenge for this approach is to effect the efficient conversion of group A to group O (41% of the population are group A in the UK). Unfortunately, α-N-acetylgalactosaminidase is not readily available. Goldstein and co-workers have used an enzyme from chicken liver to effect a substantial reduction of A antigen expression but could not eliminate A activity completely. They speculate that this inability to remove A antigen is due to the presence of cryptic A antigens in Type 3 oligosaccharides found on group A_1 red cells and discuss the possibility of using endoglycosidases in conjunction with the chicken liver enzyme to produce a therapeutic product.

Recently, several research groups have attempted to render red cells antigenically inert by coating the red cell surface with methoxy-polyethylene glycol (mPEG), producing what one group refers to as stealth cells. Polyethylene glycol has been used extensively to mask the surface of individual therapeutic proteins to prolong their half-life by reducing clearance through normal peptide and carbohydrate recognition systems and conferring resistance to proteolysis. There is some evidence that pegylated proteins are also less antigenic and therefore less likely to stimulate antibodies which compromise the therapeutic efficacy of the products. Nevertheless, a recent clinical trial of pegylated thrombopoietin had to be terminated when recipients developed antibodies to the product so reduced immunogenicity cannot be regarded as a guaranteed outcome of pegylation [15]. Pegylation is effected through covalent linkage to amino groups on the target proteins so if amino groups are critical for the function of the target proteins the process of pegylation may result in the inactivation of the protein. Other problems may result from the chemistry used to effect coupling (severity of treatment inactivates protein, use of a linker sequence which is itself immunogenic).

Jeong and Byun [18] appear to be the first to use mPEG to reduce the antigenicity of red cells. These authors found that pegylation (up to 3mM mPEG) effected a progressive increase in the amount of band 3 and Glycophorin A at the origin of gels after SDS-PAGE suggesting that these molecules were crosslinked by pegylation. They also showed a marked decrease in the agglutinability of pegylated red cells by anti A, anti-B and anti-D with a maximum decrease at 4mM mPEG. The morphology of pegylated cells remained unchanged and the oxygen affinity was similar to controls. Several other groups have extended these studies to show that other blood group antigens can be masked effectively without compromising the function of the red cells with respect to oxygen and anion transport, deformability and survival *in vivo* in a mouse model [19-22]. There are however major concerns that the amount of pegylation required to completely mask A,B antigens is such that the structure and function of the red cell may be adversely effected. The main therapeutic niche for this product at the present time appears to be in minimising the production of blood group antibodies after transfusion of ABO compatible red cells in the chronically transfused.

Conclusions

The red cell is a highly specialised cell, honed to perfection over millions of years. The genetic diversity of its constituent parts is largely a consequence of mans struggle to survive infectious disease in general and malaria in particular. When placed in this context the problems caused by blood group incompatibility seem small. For the vast majority of transfusions, pre-transfusion typing and screening is entirely satisfactory. A universal blood group is a laudable aspiration that is difficult to achieve and approaches which rely on modified haemoglobin, glycosidase-treatment or pegylation of blood have yet to yield therapeutic products of proven value. Perhaps we are taking the wrong approach. Rather than trying to modify nature should we seek to reproduce it in the laboratory [23]? If we could grow normal red cells of a required phenotype by culture *in vitro* from pluripotent stem cells we could have *designer* red cells to meet the needs of patients with difficult transfusion requirements.

References

1. Anstee D J, Cartron J-P. Toward an understanding of the red cell surface. In: Garratty G, ed. Applications of molecular biology to Blood Transfusion Medicine, American Association of Blood Banks, Bethesda, MD, USA 1997:17-49.
2. Anstee DJ. The blood group MNSs-active sialoglycoproteins. Semin Hematol 1981; 18:13-31.
3. Fletcher KS, Bremer EG, Schwarting GA. P blood group regulation of glycophingo-lipid levels in human erythrocytes. J Biol Chem 1979;254:11196-98
4. Tanner MJA. Erythrocyte glycoproteins. Current Top Membrane Transport. 1978;11: 279-325.
5. Viitala J, Järnfelt J. The red cell surface revisited. Trends Biochem Sci 1985;10:392-95.
6. Mohandas N, Gascard P. What do mouse gene knockouts tell us about the structure and function of the red cell membrane. In: Tanner MJA, Anstee DJ eds. Red Cell Membrane and its Disorders. Baillieres Clinical Haematology 2000;12(4):605-20.
7. Weatherall DJ. Science and the quiet art. Oxford University Press, UK 1995.
8. Genton B, al-Yaman F, Mgone CS, et al. Ovalocytosis and cerebral malaria (letter). Nature 1995;378:564-65.
9. Miller LH. Impact of malaria on genetic polymorphism and genetic diseases in Africans and African Americans. Proc Natl Acad Sci, USA 1994;91:2415-19.
10. Sim BKL, Chitris CE, Wasniowska K, et al. Receptor and ligand domains for invasion of erythrocytes by *Plasmodium falciparum*. Science 1994;264:1941-44.
11. Rowe JA, Moulds JM, Newbold CI, Miller LH. *P falciparum* rosetting mediated by a parasite-variant erythrocyte membrane protein and complement receptor 1. Nature 1997;388:292-95.
12. Moulds JM, Nowicki S, Moulds JJ, Nowicki BJ. Human blood groups: incidental receptors for viruses and bacteria. Transfusion 1996;36:362-74.
13. Ilver D, Arnqvist A, Ogren J, Frick I-M, et al. *Helicobacter pylori* adhesion binding fucosylated histo-blood group antigens revealed by retagging. Science 1998;373-77.
14. Anstee DJ. The functional importance of red cell antigens in transfusion medicine. In: Pamphilon DH, eds. Modern Transfusion Medicine. CRC Press Inc, Florida, USA 1995:217-28.
15. Prowse CV. Alternatives to standard blood transfusion: availability and promise. Transfus Med 1999;9:287-99.
16. Goldstein J. Conversion of ABO blood groups. Transfus Med Rev 1989;3:206-12.
17. Zarnitz ME, Kabat EA. The action of coffee bean α-galactosidase on blood group B and BP1 substance. J Amer Chem Soc 1960;82:3953.
18. Jeong ST, Byun SM. Decreased agglutinability of methoxy-polyethylene glycol attached red blood cells: significance as a blood substitute. Art Cells Blood Subs Immob Biotech 1996;24:503-11.
19. Hortin GL, Lok HT, Huang ST. Progress toward preparation of universal donor red cells. Artif Cells Blood Substit Immobil Biotechnol 1997;25:487-91.
20. Armstrong JK, Meiselman HJ, Fisher TC. Covalent binding of poly(ethylene glycol) (PEG) to the surface of red blood cells inhibits aggregation and reduces low shear blood viscosity. Am J Hematol 1997;56:26-28.
21. Scott MD, Murad KL, Koumpouras F, et al. Chemical camouflage of antigenic determinants: "Stealth" erythrocytes. Proc Natl Acad Sci USA 1997;94:7566-71.
22. Murad KL, Mahany KL, Kuypers FA, et al. Structural and functional consequences of antigenic modulation of red cells with methoxypoly(ethylene glycol). Blood 1999; 93:2121-27.

23. Mantalaris A, Keng P, Bourne P, Chang AYC, Wu JHD. Engineering a human bone marrow model: a case study on *ex vivo* erythropoiesis. Biotechnol Prog 1998;14: 126-33.

HLA AND HPA TYPING - ADVANCES IN THE APPLICATION OF MOLECULAR BIOLOGY

Chr. E.M. Voorter , E. M. van den Berg-Loonen[1]

HLA

One of the most polymorphic multigene systems in man is the Major Histocompatibility Complex (MHC). The MHC region is located on the short arm of chromosome 6, housing among others the HLA (human leucocyte antigen) class I and class II genes [1].

The classical HLA class I molecules A, B and C are composed of 2 different polypeptide chains, the α or heavy chain, which is encoded by the HLA gene and the β2 microglobulin chain, encoded by a gene located on chromosome 15. HLA class I molecules are expressed on virtually all nucleated cells of the body and play an important role in antigen presentation to cytotoxic T lymphocytes. In contrast the expression of class II molecules is in general restricted to B lymphocytes, macrophages, monocytes, dendritic cells and activated T lymphocytes. The class II molecules DR, DQ and DP are composed of an α and a β chain, which are encoded by the corresponding HLA class II A and B genes [2-4].

Beside these HLA genes that show normal expression of their gene product several pseudogenes have been identified in the HLA class I and II region. The high homology of these pseudogenes with classical HLA genes at the nucleotide level has complicated the development of DNA based typing techniques, and has made molecular HLA typing a major challenge for molecular biologists.

The polymorphism of HLA class I and II genes is largely concentrated in certain areas of the gene, namely the region encoding the peptide binding groove. For class I the groove is formed by the α1 and α2 domains which are encoded by exons 2 and 3 of the gene. For class II both the α1 and β1 domain, encoded by exon 2 of the A and B gene, are involved in groove formation. The extreme polymorphism of HLA is illustrated by the huge number of different alleles identified for each locus. In july 1999 the number of different alleles identified were 136 for HLA A, 283 for HLA B, 78 for HLA C, 224 for DRB, 42 for DQB and 84 for DPB [5,6, and published on internet]. These numbers are growing continuously.

1. Tissue Typing Laboratory, University Hospital Maastricht, Maastricht, NL.

Molecular Methods

The serological methods traditionally used to detect the HLA polymorphisms are limited by the lack of available monospecific alloantisera and monoclonal antibodies and the requirement of viable cells. The discovery of DNA amplification techniques has brought a revolution in HLA typing by enabling detection of polymorphisms at the genetic level. Several molecular methods were developed to type for the HLA genes; the 3 major techniques in current use being PCR-SSOP (sequence-specific probes), PCR-SSP (sequence specific primers) and SBT (sequence based typing) [7].

PCR-SSOP

The method first developed is SSOP, it utilizes the PCR reaction to specifically amplify target HLA genes and hybridization with oligonucleotide probes to detect polymorphic sequences in the amplified products. The principle of the method is illustrated in Figure 1. The target DNA molecule is amplified by locus or group-specific PCR. The PCR product is hybridized with a labelled oligonucleotide probe. Each probe is complementary to a different motif within the hypervariable HLA sequences. The specific patterns of probe hybridization can be used to identify individual alleles and combinations of alleles. Two different hybridization methods have been used. With the dotblot method the amplified DNA is immobilized on a membrane and subsequently hybridized with a labelled probe [8]. Most procedures use a 96 well format for the blot, enabling immobilization of 96 different DNA samples on the membrane. Given enough blots and probes an HLA typing can be deduced. This method is convenient for large scale through-put, but less ideal for fast typing. Therefore a reversed dot blot technique was developed, the different probes are immobilized on the membrane and subsequently hybridized with the labelled PCR products [9]. One membrane can contain sufficient probes to assign the HLA type of an individual. Both methods have been used to type for HLA class I and class II genes [10-16].

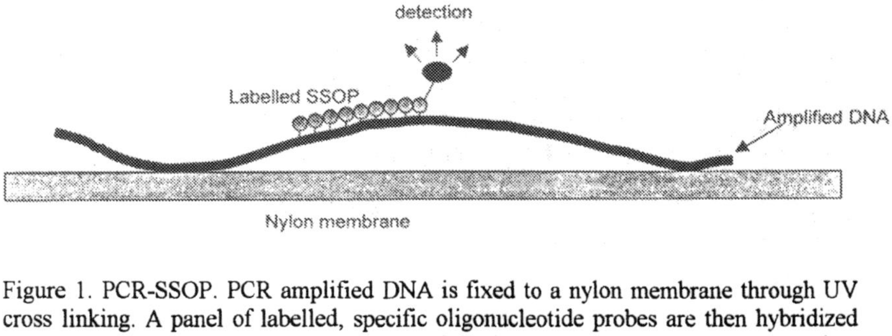

Figure 1. PCR-SSOP. PCR amplified DNA is fixed to a nylon membrane through UV cross linking. A panel of labelled, specific oligonucleotide probes are then hybridized with the template DNA on the membrane. Membranes are washed to an appropriate stringency to remove any false hybridization. Detection of hybridization is through the labelled probe. Both radio-labelled and non-isotopic probes can be used in association with appropriate methods of detection. Figure adapted from reference 7.

PCR-SSP

The second method developed was PCR-SSP (sequence specific primers), also known as ASP (allele-specific PCR) or ARMS-PCR (amplification refractory mutation system) [17–23]. In this method highly specific PCR reactions are used to discriminate between the different HLA types. This method, as illustrated in Figure 2, is based on a perfect match between primers and the DNA present. For the PCR reaction 2 primers are required and both primers will anneal and extend only if the 3' end of the primer matches completely with the DNA present, as shown in Figure 2A. Figure 2B shows that a mismatch between primer and DNA will prevent extension of the DNA by the polymerase and no PCR product will be formed. By performing many PCR reactions with different primer combinations specific for certain HLA types an HLA typing can be generated. After the PCR reactions the presence or absence of PCR products can be checked by agarose gel electrophoresis. An internal control is present in each PCR reaction to prevent false negative results due to technical problems.

Both SSP and SSOP will detect only those polymorphisms located in the probe or primer regions. This implicates that regular updating of primers and probes is necessary due to a constant increase of newly identified alleles and polymorphisms.

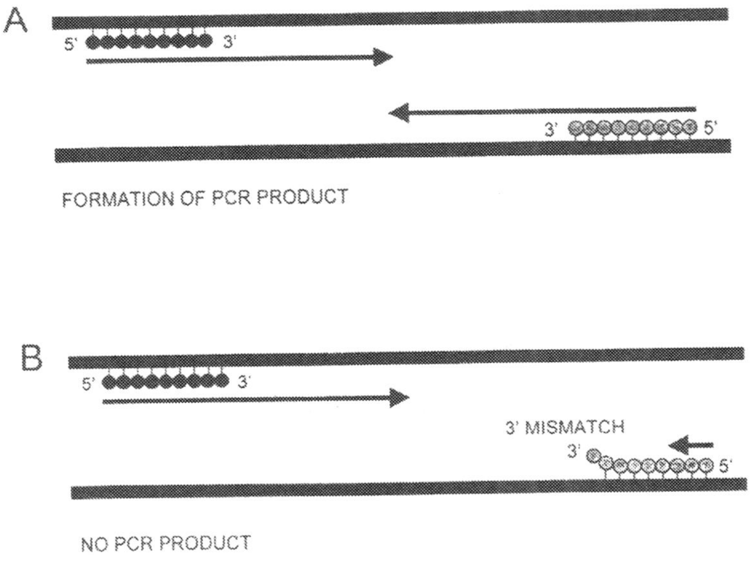

Figure 2. PCR-SSP. Highly specific PCR reactions are used to discriminate between different HLA types. PCR primers are designed to match with certain HLA types but to mismatch others. For amplification to take place under the appropriate PCR stringency, both primers need to be matched at their 3' residues as shown in Figure A. In Figure B the primer is matched to the sense-strand DNA and extension proceeds on this strand. However, a 3' mismatch on the primer binding to the antisense-strand DNA inhibits extension, thus causing the PCR reaction to fail. Figure adapted from reference 7.

SBT

To overcome the problem of regular updating the method of sequence based typing was introduced a couple of years ago and has since then become the ultimate high resolution typing technique for HLA [24-30]. Direct DNA sequencing as typing strategy has the advantage of unequivocal allele assignment as well as direct identification of new polymorphisms, since complete nucleotide information is obtained for the gene region of interest.

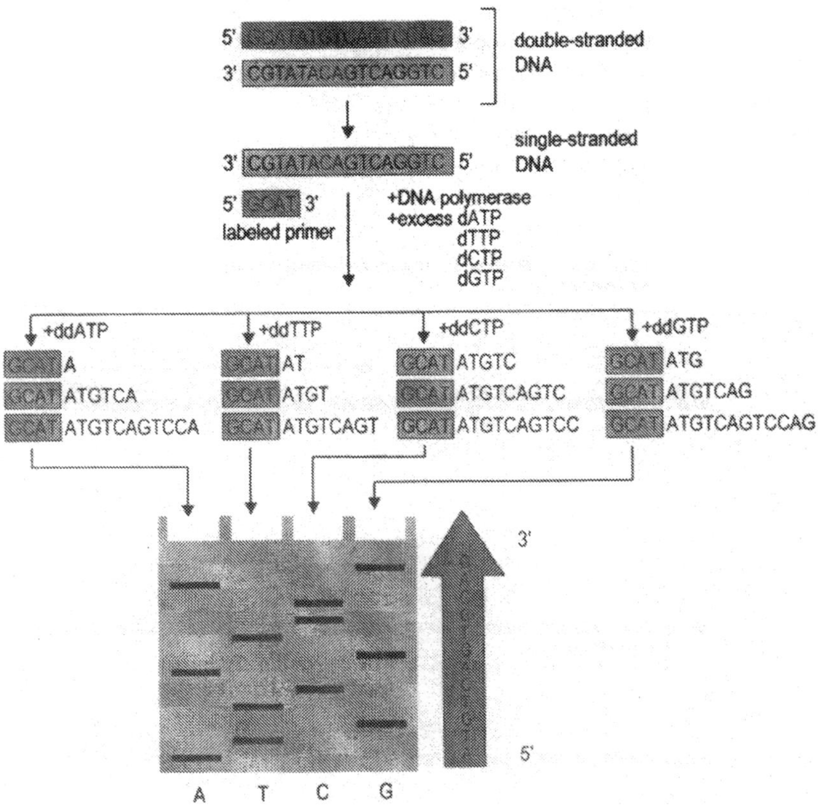

Figure 3. Schematic illustration of the DNA sequencing method. Double stranded DNA is denatured to single stranded template DNA. A labelled sequencing primer anneals to the single stranded DNA. Four different chain-terminating nucleoside triphosphates (red) are used in separate DNA elongation reactions on the same template DNA. When the products are analyzed by electrophoresis in four parallel lanes of a polyacrylamide-urea gel, the DNA sequence can be derived as illustrated.
Figure reproduced from reference 31.

The principle of DNA sequencing is explained in Figure 3. Double stranded DNA, usually a PCR product, is turned into single stranded DNA by denaturation. A labelled sequencing primer is added and anneals to the single stranded template DNA. Extension is realized by adding DNA polymerase e.g. Taq or T7 polymerase, in the presence of excess nucleotides. The mixture is divided into 4 parts and to each part one specific dideoxyribonucleoside triphosphate is added, i.e. ddATP, ddCTP, ddGTP, ddTTP. When the dideoxynucleotide is incorporated into the DNA chain, it blocks elongation by the next nucleotide. In each reaction mixture the DNA fragments formed will all end with the same nucleotide. In this way a set of partial replicas is built, all starting at the same position but terminating at different points along the DNA chain. By choosing the right conditions the different products are generated in a random fashion.

The products of the 4 reactions are analyzed by electrophoresis in 4 parallel lanes of a polyacrylamid-urea gel and the sequence can be read by combining the results of the four corresponding lanes. The reaction products on the gel are detected by the use of fluorescent dyes.

The picture obtained after gel electrophoresis on an ALFexpress automated DNA sequencer is illustrated in Figure 4. The nucleotides are represented by 4 different colors and the sequence is indicated underneath. In homozygous individuals one nucleotide is present at each position in the sequence. However the individual analyzed is heterozygous and possesses two different HLA alleles. This is clearly illustrated by the simultaneous presence of 2 different nucleotides at some of the polymorphic positions. Heterozygous peaks are indicated according to the IUB Nomenclature (W=A+T, Y=C+T, R=A+G). After careful checking of the sequence, the HLA type of the individual is determined by means of dedicated software. We use the HLA SequiTyper software (Pharmacia) and Figure 5 illustrates the identification by this programme of the HLA type DQB1*0202, DQB1*03032 for this individual. All conserved and non-conserved positions completely match the master sequence as can be read from the alignment information. The best subtype assignment as well as the next best assignments with their respective mismatches are shown in the Subtype Assignment part. The software program enables easy checking of all polymorphic positions. Many different approaches for SBT of HLA class I and class II are available now and can routinely be used in the laboratory [32-41].

HPA

Since HLA shows such a high degree of polymorphism sophisticated molecular methods like SBT are needed to obtain high resolution typing results. In contrast the human platelet antigen system (HPA) shows only few different polymorphisms as indicated in Table 1. The antigens are traditionally detected by antibody reactions directed against immunogenic epitopes located on the platelet membrane glycoproteins. The HPA system is a biallelic system, indicating that for each HPA determinant 2 different alleles can be present, the more prevalent allele is called a, the less prevalent one is called b. All known platelet specific antigens can be deduced to a single nucleotide difference leading to a single amino acid substitution at the protein level [42–44].

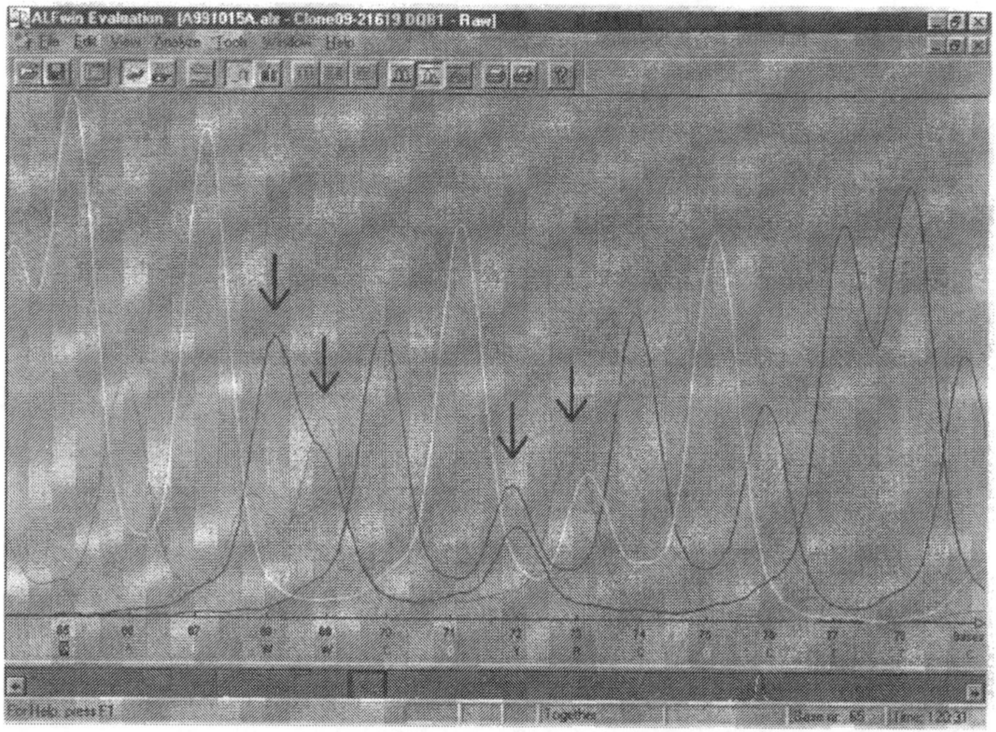

Figure 4. Partial HLA DQB1 sequence of a heterozygous sample. The four nucleotides are represented in different colours. Heterozygous peaks, i.e. simultaneous presence of two nucleotides, are indicated by arrows.

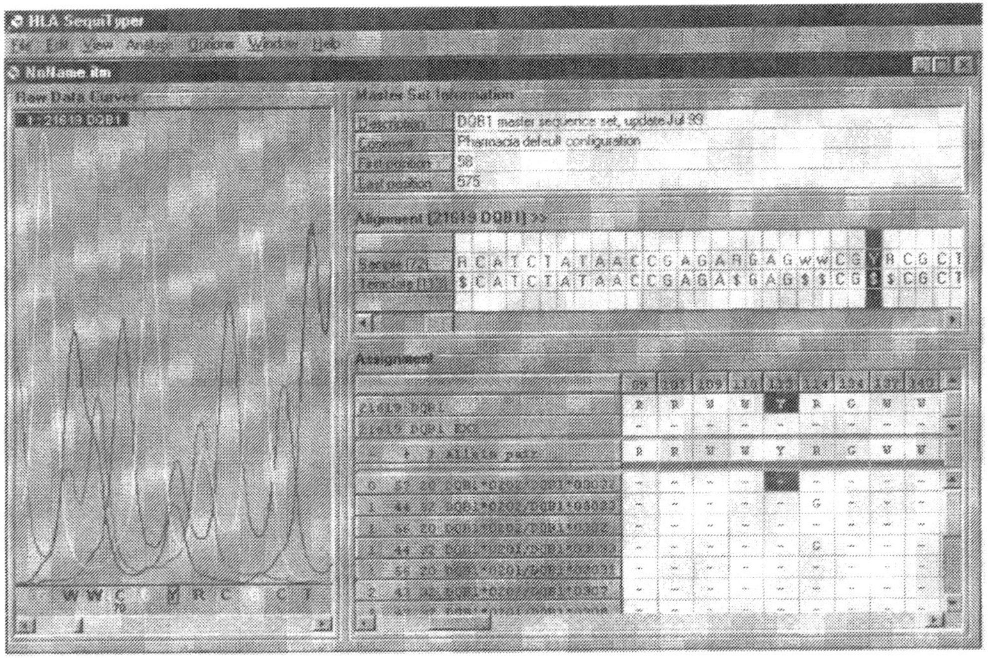

Figure 5. Typing assignment window indicating the best assignment and the number of mismatches. The master set allignment shows the concordance of the sample with the master sequence of DQB1. Polymorphic positions are indicated by dollar signs. The subtype assignment shows the best assignment without any mismatches and illustrates where mismatches are located in near best assignments.

Table 1. The 5 major HPA antigens with the original name in brackets, the localization and the polymorphism present in allele a (most prevalent allele) and b (less prevalent allele).

Antigen	Glycoprotein localization	Polymorphism	
		a	b
HPA-1 (Zw)	GPIIIa	T	196 C
HPA-2 (Ko)	GPIb	T	524 C
HPA-3 (Bak)	GPIIb	T	2622 G
HPA-4 (Yuk)	GPIIIa	A	526 G
HPA-5 (Br)	GPIa	A	1648 G

HPA

Since HLA shows such a high degree of polymorphism sophisticated molecular methods like SBT are needed to obtain high resolution typing results. In contrast the human platelet antigen system (HPA) shows only few different polymorphisms as indicated in Table 1. The antigens are traditionally detected by antibody reactions directed against immunogenic epitopes located on the platelet membrane glycoproteins. The HPA system is a biallelic system, indicating that for each HPA determinant 2 different alleles can be present, the more prevalent allele is called a, the less prevalent one is called b. All known platelet specific antigens can be deduced to a single nucleotide difference leading to a single amino acid substitution at the protein level [42–44].

The 2 most widely used molecular methods to determine HPA variants according to the literature are SSP and RFLP [45–52]. The method of SSP has already been described for HLA. It is a very convenient way for typing HPA since each HPA variant shows only one nucleotide difference at the DNA level, implicating that for each HPA determinant only 2 primer sets have to be used.

For RFLP DNA is incubated with a restriction enzyme resulting in cleavage of DNA if the correct sequence for the restriction enzyme is present. One nucleotide difference will prevent cleavage. Therefore the presence or absence of HPA variants can be deduced from the cleavage pattern with specific restriction enzymes. To use this method the polymorphism must involve a restriction enzyme site.

Conclusion

When comparing serological methods for HLA and HPA with the described molecular techniques, the advantages of the latter are obvious. The DNA methods give an increased quality of typing with the possibility of high resolution typing results. Furthermore, test material is easily available, which is a clear advantage in case of thrombocytopenia, and an unlimited supply of reagents can be produced, in contrast with the monospecific sera necessary for serological typing.

Acknowledgements

The authors thank Diana van Bakel for excellent secretarial assistance.

References

1. Campbell RD, Trowsdale J. Map of the human MHC. Immunol Today 1993;14: 349-52.
2. Trowsdale J, Young JAT, Kelly AP, et al. Structure, sequence and polymorphism in the HLA-D region. Immunol Rev 1985;85:5-43.
3. Strominger JL. Structure of class I and class II HLA antigens. Brit Med Bull 1987; 43:81-93.
4. Bjorkman PJ, Parham P. Structure, function and diversity of class I major histocompatibility complex molecules. Ann Rev Biochem 1990;59:253-88.
5. Mason PM, Parham P. HLA class I region sequences, 1998. Tissue Antigens 1998; 51:417-66.
6. Marsh SGE. HLA class II region sequences, 1998. Tissue Antigens 1998;51:467-507.
7. Krausa P and Browning M. Detection of HLA gene polymorphism. In: Browning M, McMichael A, eds. HLA and MHC: genes, molecules and function. Oxford: Bios Scientific Publishers Ltd, 1996:113-37.
8. Saiki RK, Bugawan TL, Horn GT, Mullis KB, Erlich HA. Analysis of enzymatically amplified beta-globin and HLA-DQ-α DNA with allele-specific oligonucleotide probes. Nature 1986;324:163-66.
9. Saiki RK, Walsh PS, Levenson CH, Erlich HA. Genetic analysis of amplified DNA with immobilized sequence-specific oligonucleotide probes. Proc Natl Acad Sci USA 1989;86:6230-34.
10. Erlich H, Bugawan T, Begovich AB, et al. HLA-DR, DQ and DP typing using PCR amplification and immobilized probes. Eur J Immunogenet 1991;18:33-55.
11. Tiercy JM, Morel C, Freidel AC, et al. Selection of unrelated donors for bone marrow transplantation is improved by HLA class II genotyping with oligonucleotide hybridization. Proc Natl Acad Sci USA 1991;88:7121-25.
12. Ng J, Hurley CK, Baxter-Lowe LA, et al. Large scale oligonucleotide typing for HLA-DRB1/3/4 and HLA-DQB1 is highly accurate, specific and reliable. Tissue Antigens 1993;42:473-79.
13. Bugawan TL, Apple R, Erlich HA. A method for typing polymorphism at the HLA A locus using PCR amplification and immobilized oligonucleotide probes. Tissue Antigens 1994;44:137-47.
14. Chen DF, Endres W, Meyer SA, Stangel W. A polymerase chain reaction-sequence-specific oligonucleotide procedure for HLA class II typing using biotin- and digoxigenin-labeled probes simultaneously in hybridization. Hum Immunol 1994;39:25-30.
15. Middleton D, Williams F, Cullen C, Mallon E. Modification of an HLA-B PCR-SSOP typing system leading to improved allele determination. Tissue Antigens 1995;45:232-36.
16. Kennedy LJ, Poulton KV, Dyer PA, Ollier WER, Thomson W. Definition of HLA-C alleles using sequence-specific oligonucleotide probes (PCR-SSOP). Tissue Antigens 1995;46:187-95.
17. Olerup O, Zetterquist H. HLA-DR typing by PCR amplification with sequence-specific primers (PCR-SSP) in 2 hours: an alternative to serological DR typing in clinical practice including donor-recipient matching in cadaveric transplantation. Tissue Antigens 1992;39:225-35.
18. Krausa P, Moses J, Bodmer W, Bodmer J, Browning M. HLA-A locus alleles identified by sequence specific PCR. Lancet 1993;341:121-22.

19. Sadler AM, Petronzelli F, Krausa P, et al. Low Resolution DNA typing for HLA-B using sequence-specific primers in allele- or group-specific ARMS/PCR. Tissue Antigens 1994;44:148-54.
20. Bunce M, Welsh KI. Rapid DNA typing for HLA-C using sequence-specific primers (PCR-SSP): identification of serological and non-serologically defined HLA-C alleles including several new alleles. Tissue Antigens 1994;43:7-17.
21. Bunce M, O'Neill CM, Barnardo MCNM, et al. Phototyping: comprehensive DNA typing for HLA-A, B, C, DRB1, DRB3, DRB4 DRB5 and DQB1 by PCR with 144 primer mixes utilizing sequence-specific primers (PCR-SSP). Tissue Antigens 1995; 46:355-67.
22. Savelkoul PHM, de Bruyn-Geraets DP, van den Berg-Loonen EM. High resolution HLA-DRB1 SSP typing for cadaveric donor transplantation. Tissue Antigens 1995; 45:41-48.
23. Tonks S, Marsh SGE, Bunce M, Bodmer JG. Molecular typing for HLA class I using ARMS-PCR: further developments following the 12th International Histocompatibility Workshop. Tissue Antigens 1999;53:175-83.
24. Santamaria P, Boyce-Jacino MT, Lindstrom AL, Barbosa JJ, Faras AJ, Rich SS. HLA class II "typing": direct sequencing of DRB, DQB and DQA genes. Hum Immunol 1992;33:69-81.
25. Santamaria P, Lindstrom AL, Boyce-Jacino MT, et al. HLA class I sequence-based typing. Hum Immunol 1993;37:39-50.
26. Spurkland A, Knutsen I, Markussen G, Vartdal F, Egeland T, Thorsby E. HLA matching of unrelated bone marrow transplant pairs: direct sequencing of in vitro amplified HLA-DRB1 and -DQB1 using magnetic beads as solid support. Tissue Antigens 1993;41:155-64.
27. Versluis LF, Rozemuller E, Tonks S et al. High resolution HLA-DPB typing based upon computerized analysis of data obtained by fluorescent sequencing of the amplified polymorphic exon 2. Hum Immunol 1993;38:277-83.
28. Knipper AJ, Hinney A, Schuch B, Enczmann J, Uhrberg M, Wernet P. Selection of unrelated bone marrow donors by PCR-SSP typing and subsequent non-radioactive sequence-based typing for HLA DRB1/3/4/5, DQB1 and DPB1 alleles. Tissue Antigens 1994;44:275-84.
29. McGinnis MD, Conrad MP, Bouwens AGM, Tilanus MGJ, Kronick MN. Automated, solid-phase sequencing of DRB region genes using T7 sequencing chemistry and dye-labeled primers. Tissue Antigens 1995;46:173-79.
30. Petersdorf EW, Hansen JA. A comprehensive approach for typing the alleles of the HLA B locus by automated sequencing. Tissue Antigens 1995;46:73-85.
31. Alberts B, Bray D, Lewis J, Raff M, Roberts K, Watson JD. Molecular Biology of the Cell. New York: Garland Publishing Inc, 1994.
32. Voorter CEM, Rozemuller EH, de Bruyn-Geraets D, van der Zwan A-W, Tilanus MGJ, van den Berg-Loonen EM. Comparison of DRB sequence-based typing using different strategies. Tissue Antigens 1997;49:471-76.
33. Voorter CEM, de Bruyn-Geraets D, van den Berg-Loonen EM. High resolution HLA typing for the DRB3/4/5 genes by sequence-based typing. Tissue Antigens 1997; 50:283-90.
34. Kotsch K, Wehling J, Kohler S, Blasczyk R. Sequencing of HLA class I genes based on the conserved diversity of the noncoding regions: sequencing-based typing of the HLA A gene. Tissue Antigens 1997;50:178-91.
35. Bettinotti MP, Mitsuishi Y, Bibee K, Lau M, Terasaki PI. Comprehensive method for the typing of HLA-A, B, and C alleles by direct sequencing of PCR products obtained from genomic DNA. J Immunotherapy 1997;20:425-30.
36. Eberle M, Knapp LA, Iwanaga KK, Domanico MJ, Aiyer K, Watkins DI. HLA-B typing by allele separation followed by direct sequencing. Tissue Antigens 1997; 49:365-75.

37. Norgaard L, Fugger L, Jakobsen BK, Svejgaard A. Sequencing-based typing of HLA-A locus using mRNA and a single locus-specific PCR followed by cycle-sequencing with AmpliTaq DNA polymerase, FS. Tissue Antigens 1997;49:455-65.
38. Voorter CEM, Kik MC, van den Berg-Loonen EM. High resolution HLA typing for the DQB1 gene by sequence-based typing. Tissue Antigens 1998;51:80-87.
39. van der Vlies SA, Voorter CEM, van den Berg-Loonen EM. A reliable and efficient high resolution typing method for HLA-C using sequence-based typing. Tissue Antigens 1998;52:558-68.
40. Pozzi S, Longo A, Ferrara GB. HLA-B locus sequence-based typing. Tissue Antigens 1999;53:275-81.
41. Kurz B, Steiert I, Heuchert G, Muller CA. New high resolution typing strategy for HLA-A locus alleles based on dye terminator sequencing of haplotypic group-specific PCR-amplicons of exon 2 and exon 3. Tissue Antigens 1999;53:81-96.
42. Simsek S, von dem Borne AE. Molecular genetics of human platelet antigens. Infusionsther Transfusionsmed 1994;21(suppl 3):29-33.
43. Mueller-Eckhardt C, Santoso S, Kiefel V. Platelet alloantigens - molecular, genetic and clinical aspects. Vox Sang 1994;67(suppl 3):89-93.
44. Newman PJ. Nomenclature of human platelet alloantigens: a problem with the HPA system? Blood 1994;83:1447-51.
45. Metcalfe P, Waters AH. HPA-1 typing by PCR amplification with sequence-specific primers (PCR-SSP): a rapid and simple technique. Br J Haematol 1993; 85:227-29.
46. Simsek S, Bleeker PM, Heeremans J, Von dem Borne AE. Human platelet antigen 5 (Br) genotyping by ASPA: allele-specific primer amplification (PCR-SSP). Br J Haematol 1994;88:659-61.
47. Skogen B, Bellissimo DB, Hessner MJ et al. Rapid determination of platelet alloantigen genotypes by polymerase chain reaction using allele-specific primers. Transfusion 1994;34:955-60.
48. Tanaka S, Taniue A, Nagao N, Ohnoki S, Okubo Y, Yamaguchi H. Simultaneous DNA typing of human platelet antigens 2, 3 and 4 by allele-specific PCR method. Vox Sang 1995;68:225-30.
49. Kluter H, Fehlau K, Panzer S, Kirchner H, Bein G. Rapid typing for human platelet antigen systems -1, -2, -3 and -5 by PCR amplification with sequence specific primers. Vox Sang 1996;71:121-25.
50. Matsuo K, Reid DM. Allele-specific restriction analysis of human platelet antigen system 4. Transfusion 1996;36:809-12.
51. Tanaka S, Taniue A, Nagao N, et al. Genotype frequencies of the human platelet antigen, Ca/Tu, in Japanese, determined by a PCR-RFLP method. Vox Sang 1996; 70:40-44.
52. Panzer S, Kaplan C. Report on the 1997 International Society of Blood Transfusion Workshop for genotyping of platelet alloantigens. Platelet and granulocyte workshop ISBT. Transfus Med 1998;8:125-28.

DISCUSSION

Moderators: H.G. Klein and C.Th. Smit Sibinga

R. Atamian (Beirut, Lebanon): Dr. Reid, is the D^u variant a different type, either quantitative or qualitative? And what is the relation of the Landsteiner-Wiener and the W-antigen and the Rhesus antigens based on the molecular studies?

M. Reid (New York, NY, USA): There are three separate genes. The D^u is a term which we now call weak D. There are several molecular bases; there are point mutations, there are rearranged D-genes and the CE gene. In all those cases the Rhesus D antigen gene has been normal. As far as I am aware the W^a and W^b gene require Rhesus antigen, but it is not related to expression of the weak D.

T. Smith (Ayrshire, UK): Dr. Voorter, can you tell us why the HLA system should have such polymorphism and why the platelet system has not; do you have a theory about it. What is the big difference?

C.E.M. Voorter (Maastricht, NL): The HLA system is of course taking care of the immune response in human beings. The more polymorphism there is the better the immune response can be taken care of. Since the HPA system is not taking care of the immune response in human beings diversity is not an issue.

W.H. Dzik (Boston, MA, USA): Dr. Anstee, your cartoons were very popular. Looking at the Kidd system Jk^a and Jk^b there seemed to be a third loop, but the loop was barely outside the membrane. Why the immune system would notice this tiny little bump on the cell surface. I wonder if you mean to imply that the epitope at the side of the antibody binding is really at the side of the polymorphism or is it possible that the polymorphic change which is in that spot caused that second big loop to somehow wiggle and in different directions change physical shape, and that is what the antibodies are recognising.

D.J. Anstee (Bristol, UK): I cannot rule out that there is a conformational change in the protein induced by that sequence change. But equally, the idea of there being an hydrophobic site in the lipid bi-layer which is antigenic is not a problem for me. The D antigen would be another example of a very hydrophobic protein close to the membrane, which generates an immune response.

M. Reid: The S antigen, which is clearly detectable in antibodies, is the predicted apology of the protein. I would say that position 245 which is a conformational change is embedded within the membrane.

H.G. Klein (Bethesda, MD, USA): Dr. Voorter, you showed an example of subtyping for HLA and explained the use of sequences. We are getting increasing requests for subtyping, so I have a question; How important clinically do you think the subtypes are? What about the clinical aspects involved? Are we going to end up using automated sequencing to do HLA-typing and what shall we do with all the polymorphisms we find in terms of transplantation compatibility?

C.E.M. Voorter: I was afraid of that question, because it takes a very long answer, but I will make it as short as possible. Of course HLA-typing is important for solid organ transplantation but also for bone marrow transplantation. It is especially the latter where the subtyping seems to be rather important. There have been cases reported where donor and recipient were fully matched by serology, but still there was a cytotoxic T-lymphocyte reaction from the recipient to the donor. So it resulted in rejection of the allograft. The reason was that the donor had B*4403 whereas the recipient had a B*4402 subtype of HLA. It was only one aminoacid difference that seemed to be enough to give rejection of the allograft. So, therefore it might be important in certain cases to do a subtyping and also for the solid organ transplantation it might be important. We are still working out when it will be important and in what cases we need the subtyping or not. When we do the subtyping of course the donor potential will be less because there will be fewer donors having that particular subtype. So you must be sure that it is of importance before we go in that direction.

A.H. Lazarus (Toronto, C): Dr. Allen, you said that the AABB labs had to be 99% sure of paternity testing and you did about 25,000 individuals; so does that mean that in 250 individuals the typing falls short?

R.W. Allen (Tulsa, OK, USA): Of course, when you do a test like this you look at a number of different genetic markers. So, statistically there is a possibility for each locus that perhaps 25,000 other men would fit the role that you found for the true father of the child. But of course if you look at another locus a 125,000 perhaps you have 25 that now fit the roles for both genetic markers you look at. If you look at the third marker you produce the number correspondingly again. Although there are a very large number of alleles present for each of these markers, you still have to look at a large enough proportion of them to eliminate and reduce the field to where you literally select one man out of a 100. But still in a city of 100,000 I believe there is a number of possible fathers. Of course there is a lot of non-genetic information that fits into the evaluation for parental test results. If the man that tested and matched is the father of the child although he was in another part of the world during the period of conception it is not very likely regardless what appears to be an instrumental genetic information that he really is the father. Than in fact he is not and he has somebody who is highly related to him genetically. The AABB standards are minimum standards; most laboratories actually test to a higher level, but at least at that 99% you know that falsely accused men are not going to occur with high frequency.

A.H. Lazarus: I agree with you that many other factors likely come into play. However, the laboratory would not be aware of these other factors and you probably wish to give a correct paternity test in virtually all instances, if the standard were raised to 99.999% and you assess 25,000 individuals per year, would that not mean that you will only have an 'unreliable paternity test' in less than 1 individual per year?

R.W. Allen: The experience is not so great to set the standard to a level. The AABB also has to cater to the peer group that it serves namely the laboratories. Now all laboratories are equipped with the methods, reagents and machines necessary to do the typing. So, you still have to make a provision for high-quality serology laboratories. That looks like only a handful of blood group and HLA facilities for example to still achieve an accredited status, which in the US at least is a prerequisite for doing large government subsidised parentage testing. So, you try and achieve a balance where you provide the try out of facts with good signs and compelling information and yet not may finalise. The laboratory does high quality work but achieves a lower level of resolution for just paying off in the process. There is in fact a preventive question that is existing in parentage testing built for years and years, but it just occurred in the last year. So how much testing is enough. Do we really need to get a 99.99% probability of paternity in each and every test that we do. One outcome or spin-off of the parentage testing methods is the non-traditional paternity test. Now we can get to a reconstruction of deceased alleged fathers testing their relatives, testing with the mothers not available. These kinds of scenarios carry with them necessarily a lower level of certainty and more ambiguity in interpretation of the test result. Sometimes it is not possible when you are doing the reconstruction for example for a child who inherits part of the deceased father's state were to gain a status, an immigration status or whatever, because of the number of people you have to test, to obtain samples from. It may not be possible to achieve a 99.99% probability of paternity. So you have to make the rules and regulations if you will credit test results low enough towards where you still have a level of certainty that is achievable in each and every case.

A. Fletcher (Sydney, AUS): Dr. Reid, you explained very nicely how molecular biology is a very useful supplement to serological testing in the context of a diagnostic referral. You showed how that could be useful in finding compatible blood for example, for different patients, but you also made the point that molecular biology is quite intensive and can take some time. Could you share with us how you go about making the decision about which cases you look at with molecular biology; what factors you consider to in making the decision: and for which blood systems.

M. Reid: I chose to do molecular biology on those samples that do not give me the answers based on the serological testing, the history I managed to get which is not always accurate and if we know that patients have a lot of transfusions. The potential of knowing the genotype might help us in the antibody identification we will do.

A. Fletcher: Is it really a case by case decision?

M. Reid: Absolutely.

H.P. Olthuis (Leiden, NL): Prof. Anstee, from your lecture I picked up that the antigenicity of the red cells when transfused is influenced by PEG treatment. Would this mechanism also work in platelets for transfusion especially for HLA antigens?

D.J. Anstee: Well, I think it comes back to the question of the functions that you want the product to perform. I suspect that you could coat platelets with polyethylene glycol and mask the antigens. But whether they function as platelets afterwards I doubt.

C.Th. Smit Sibinga (Groningen, NL): Could I come back to the red cell issue. Is there anything known at this point in time whether the mechanism of the regulation is truly a binding effect or is it a wrapping effect, an enveloping effect. If so, is there anything known about the persistence of the effect during further circulating life of transfused red cells.

D.J. Anstee: The chemistry couples the PEG co-valently to amino groups on the surface of the cell. Available evidence indicates that conditions of PEGylation can be found which permit a normal life span of red cells in mice.

R. Atamian: Dr. Voorter, you were saying that because of the immunity we have diversity, polymorphism. Now, could it be the other way around, that because of the immune response throughout our lives exposure to so many different microorganisms and activation and multiply different kinds of cytokines and so on; could it be that we are producing a milieu for auto-metagenicity unlikely for our own cells which are not multiplying frequently or more likely for the multiplying cells such as sperms and other. Therefore, our generation is helping some kind of different, diverse production of the HLA antigens.

C.E.M. Voorter: It is very difficult to speculate on that; to me it looks like the question what was first, the egg or the chicken. So it can be either the way you are suggesting or that the diversity was first or, because of the multiplicity of immune responses that we need nowadays, that the polymorphism has extended to that.

H. Hambley (Pontyclune, UK): Dr. Anstee, just to come back to the question of sickle cell disease and blood group expression. Is this acquired in sickle cell disease and what about some of the other phenotypes like SC and sickle cell thalassaemia. What is the expectation in these situations.

D.J. Anstee: I do not think it is an acquired phenomenon, I think it is a consequence. There are enhanced levels of Lutheran protein on sickled red cells. The reason why it is enhanced is not clear but one would speculate the abnormality of erythropoiesis is an indirect consequence of the altered haemoglobin. Abnormal haemoglobin is affecting the assembly of the Lutheran protein on the surface. So you have this strange situation where we have more Lutheran on the surface of the red cells. You also have conditions where there is damage to endothelium and laminin which is the receptor for Lutheran is exposed more than it would be in a normal circulation, so you have this coincidence of two properties that are conducive to such an adhesive interaction. So, you asked if it was acquired. I do not think it is acquired and I think it is a consequence of the pathology of the disease that you get this. It may transfer to other diseases like thalassaemia for example where adhesive occlusions of small capillaries are also reported. It might be a general phenomenon in these globin gene disorders. But I was picking one example of blood group research impacting on a disease. There are many other adhesive interactions that are implicated in the process of occlusion of capillaries and this is just one of them. As to which is the most important, we do not know yet. But the Lutheran laminin interaction is remarkably high affinity and it is very hard to ignore; it is of the order of antigen-antibody reaction; it is not a loose affinity that you often get in cell adhesion it is a very powerful interaction.

H. Hambley: Do you have any information about Lutheran expression in sickle cell trait or SC disease.

D.J. Anstee: Yes, that has been looked at, but not by us. The figures I was showing related to SS, not to heterozygotes.

G. Mertens (Antwerp, B): I have a question to Dr. Allen and to Dr. Reid. A traditional question in paternity testing when you use blood as the sample is to ask whether any person of the paternity trio (alleged father, mother, child) did receive a blood transfusion during the previous three months. Now, is it correct to state that when you only use Tandem Repeats (STR's) amplified by PCR, it is not relevant anymore to ask that question. Dr. Reid you still do RFLP tests even in multi-transfused individuals. When you perform an RFLP test do you only type the white blood cells of the individual itself without problems due to the transfused blood? But, is that also true do you think for PCR of STR's?

R.W. Allen: Well, with most of the blood products that are available today there is not a tremendous contamination of the transfused product by contaminating white cells as I understand. We still maintain the question because so many laboratories continue to use blood and so many laboratories continue to use serology in mixed fields and that sort of thing. In terms of DNA typing my experience from offering this service for recipients of bone marrow transplantation to look at congruity in the circulating cells both with RFLP methods as well as with PCR, suggest that there really is not too much difference in terms of the sensitivity. With either RFLP or PCR you can detect a 5-10% circulating cells reappearing from the recipient so in terms of sensitivity of the method there is really not a tremendous difference between RFLP and PCR. So, the question about blood transfusion is really not as relevant as perhaps it once was. The more important question and the new question that has appeared on is informed consent from individuals undergoing parentage testing and bone marrow transplantation. I once had a physician from New York who asked me whether or not bone marrow transplantation would eliminate him as a father of a child and he was asking purely theoretically. But I can guarantee you that he was taking copious notes of our conversation and was at least considering having a bone marrow transplantation prior to a paternity test in which he stood to be liable for support of a child for perhaps 18 years. So the bone marrow transplantation issue is still a valid one. In whatever advice I give physicians as alleged fathers, I always take swabs nowadays.

II. VIROLOGY AND QC ASPECTS

NUCLEIC ACID AMPLIFICATION TESTING OF BLOOD DONATIONS

M. P. Busch,[1,2,3] S.H. Kleinman,[4,5] S.L. Stramer[6]

Overview of Present Risks of Major Transfusion-Transmitted Viruses

There has been a dramatic reduction in risk of viral transmission by transfusions over the last two decades due in part to improvements in laboratory testing of donated blood (Figure 1) [1]. Remaining present risks range from 1 in 50,000 to 1 in 700,000 per unit for the major viral agents [2]. There are four sources of this remaining risk: pre-seroconversion (window phase) donations, viral variants, atypical seroconversion, and laboratory error (Table 1). Viral variants such as HIV-1, group O, unusual subtypes of HCV, or HBsAg mutants occur; however, surveillance data indicate that these viral variants are very rare in the US, particularly among blood donors [3,4]. Atypical seroconversion or immunosilent infection has generally been discounted. Several studies in the late 1980s reported high rates of HIV culture or polymerase chain reaction (PCR) positivity among seronegative persons [5,6] and in the early 1990s it was suggested that up to 10 percent of persons with HCV infections may be immunosilent (PCR-positive, antibody-negative) carriers [7]. These results were subsequently proven to be artifactual due to PCR contamination and false positive culture results [8,9], and continued study of high risk populations and early data from the international screening of blood and plasma donors by nucleic acid amplification tests (NAT) have failed to substantiate these alarming reports of frequent immunosilent infections [10,11]. Laboratory testing errors occur at low frequency (~0.1 percent) [12], but these errors only become significant if the test error occurs on a seropositive unit [13]. The rate of test error multiplied by the rate of seropositive units indicates that test errors contribute only a small fraction (5-10 percent) of the total risk for each agent [12].

1. Blood Centers of the Pacific, San Francisco, CA, USA.
2. Blood Systems, Inc., Scottsdale, AZ, USA.
3. University of California, San Francisco, CA, USA.
4. Westat, Inc., Rockville, MD, USA.
5. University of British Columbia, Victoria, BC, Canada.
6. American Red Cross National Confirmatory Testing Laboratory, Gaithersburg, MD, USA.

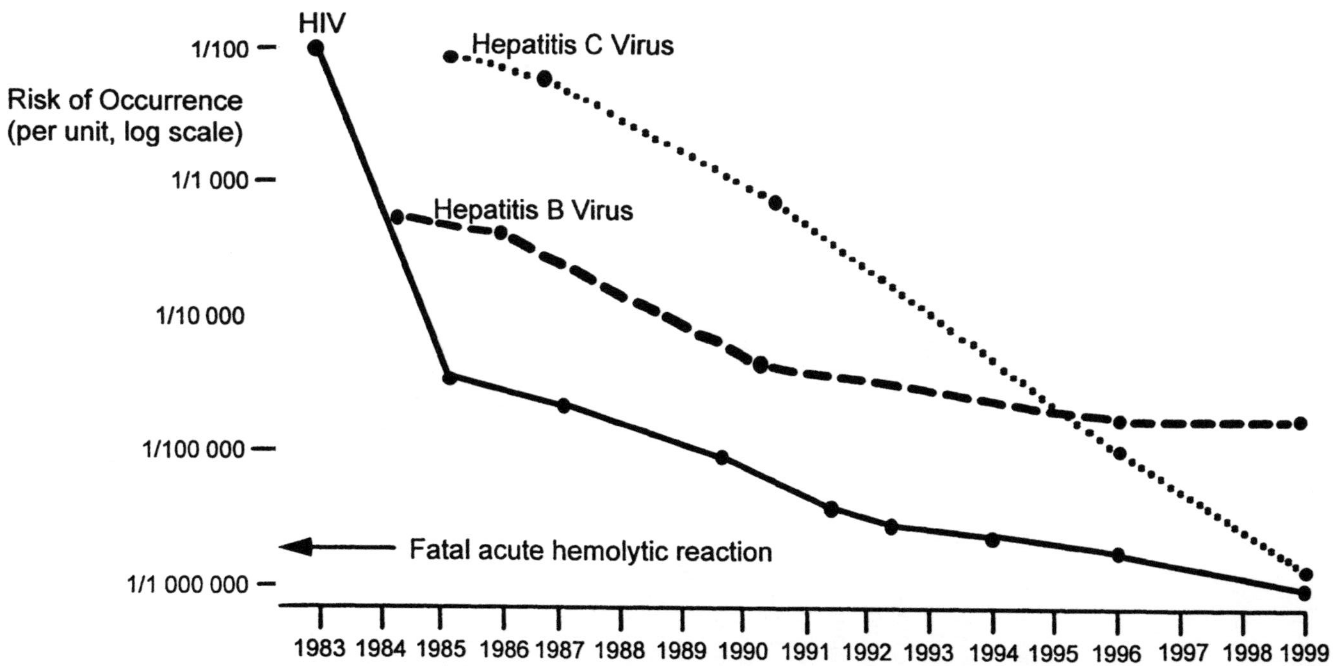

Figure 1. Transfusion conplication risk estimates. Updated from Ann Int Med 1997;127:904-09.

Table 1. Estimated risk of viral infections per 10^7 screened donations.

	Window phase	**Variants**	**Abnormal SC**	**Test error**	**Total**
HIV	15 (93.7%)	<0.6 (<3.7 %)	<0.1 (<0.6 %)	0.4 (2.5 %)	16
HCV	80 (72.1–87.9 %)	<1 (<1.1–<0.9 %)	0-20 (0–21.9 %)	11.2 (0 – 21.9 %)	91–111
HBV	63–150 (41.2–98.0%)	<0 (0 %)	1 (0.6–1.5 %)	1–3 (0.6–4.5 %)	66–153
HTLV	15 (93.7 %)	<1 (<6.2 %)	1 (6.2 %)	0.8 (5 %)	16
Total	183–260 (61.8–96.8 %)	<1 (<0.3–<0.5 %)	0–20 (0–10.6 %)	15.4 (5.2–8.1 %)	189–296

Updated from Reference 28. Stramer SL, Kleinman SH. Evolving Applications of Nucleic Acid Amplification Assays for Prevention of Viral Transmission by Blood Components and Derivatives. Chapter 6 in Garratty G, ed. Applications of Molecular Biology to Blood Transfusion Medicine (1997 Workshop book). American Association of Blood Banks, Bethesda, MD. 1997:123-76.

Donations during the window period related thus contribute the predominant risk for each of the major viral agents (Table 1). For example, in HIV and HBV at least 90 percent of the risk is attributable to donations of window phase units, and for HCV window period donations account for at least 75 percent of risk, with the remaining 25 percent reflecting the potential contributions of rare immunosilent HCV infections and test error (due to the relatively high HCV prevalence). Therefore a detailed understanding of viral dynamics during the window phase is critical to development of NAT screening systems.

Window Period and NAT

A variety of sources of data on primary infections can be used to form an overall model of the window period for each agent (reviewed in reference 14). When people present with symptoms of HIV, HBV or HCV, they are already well into the incubation phase and are often in the early stages of seroconversion; indeed, the symptoms (hepatitis, flu-like symptoms) are reflective of the immune response against the virus and consequent tissue damage rather than viral replication or viremia per se. Therefore, these symptomatic cases do not offer much information on the earlier pre-antibody (window) stage of primary infection. A second source of data involves nosocomial cases, such as needle stick accidents or transfusions. These cases are valuable sources of data because the time from exposure to viremia and seroconversion can be documented. Analysis of such cases leads to a distinction between the exposure to seroconversion and the viremia to seroconversion window periods [14-16]. Similarly, in animal models involving HIV, HCV and HBV, it is clear that there is a period of time following exposure during which viral replication is restricted to tissue sites and there is not an infection or detectable viremia [17-19]. During the first few days or weeks post-inoculation, viral replication can be demonstrated in lymph nodes or liver but not in the circulation [14,17]. These data form the basis for the concept of an *eclipse* phase of the window period, during which there is no evidence of infection in blood by laboratory methods including NAT. This eclipse phase is important when considering the potential yield of NAT [20]. Current NATs can first detect viral nucleic acids in blood several weeks following exposure. This has led to studies looking at infectivity during the eclipse phase preceding appearance of detectable viremia, to try to better understand when individuals become infectious, relative to exposure and to detection of viremia by NAT [20]. It is likely that no matter how sensitive these NATs become, we will never completely close the exposure to seroconversion window period, because of this non-viremic eclipse phase. However, NAT may be able to detect virtually all donations given during the viremic phase of primary infection that is relevant to transfusion risk.

HIV Window Period

With respect to the exposure to seroconversion window period in HIV infection, an analysis of data from health care workers who acquired HIV following nosocomial exposures, yielded an estimated median time to seroconversion of approximately 40 days [14]. In about five percent of these cases, however, seroconversion did not occur for over six months following exposure. In these prolonged window phase cases, specimens collected prior to seroconversion consistently tested negative by PCR and antigen assays until several weeks prior to seroconversion. Thus, the existence of delayed seroconversions is probably not a concern for blood transfusions because the delayed seroconversions appear to be due to a prolonged eclipse phase with delayed virus dissemination, and the development of viremia is always associated with rapid subsequent seroconversion.

Another important source of data on the HIV window period comes from lookback studies of recipients of seroconverting donors [15]. Overall, 20 percent of prior donations by seroconverted donors transmitted HIV. The longer the time period between the negative and positive donations, the less likely that prior donations transmitted the virus. An analysis of these data by Petersen, et al. [15] yielded an estimate for the infectious window period prior to seroconversion of approximately 20 days in the context of improved sensitivity HIV antibody tests.

The third source of relevant data on HIV primary infection involves studies of viral markers in plasma donor seroconversion panels [14,21,22]. Characterization of these panels has allowed definition of a series of stages of evolution of markers during primary infection. The first stage is the eclipse phase, which typically lasts ten days but may be as long as several months. This is followed by a period during which HIV RNA (but not p24 antigen) is detected by NAT; this RNA-only phase is estimated to last three to five days. This is followed by a five-day period when both RNA and p24 antigen are detectable but the presently licensed antibody screening tests are negative. During the early RNA-only phase of infection, the levels of the virus in plasma ranges from <100 copies to approximately 10^5 copies per mL. Once p24 antigen is detected, over 10,000 copies of RNA are consistently detected. By analyzing the slope of RNA levels, the rate at which virus levels increase over time in plasma during primary viremia can be estimated [22]. This rate of viral ramp-up in blood is termed the viral doubling time, and has been estimated to be slightly under one day; this rate of

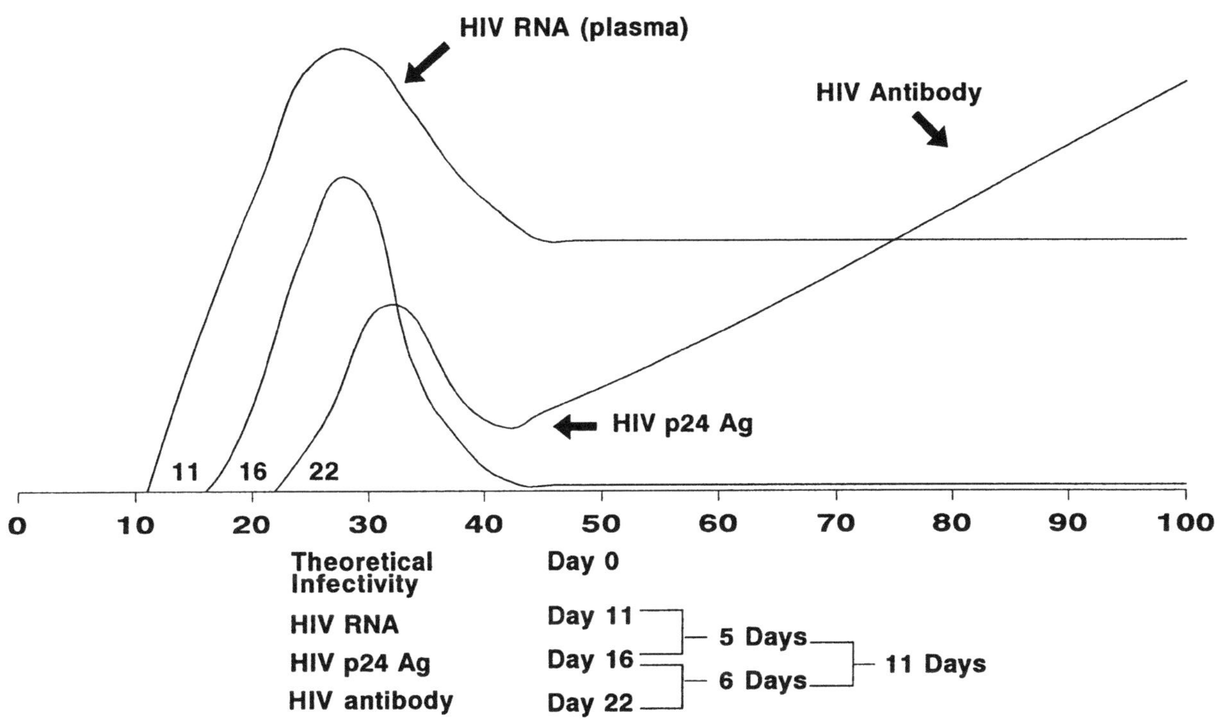

Figure 2. HIV markers during early infection.

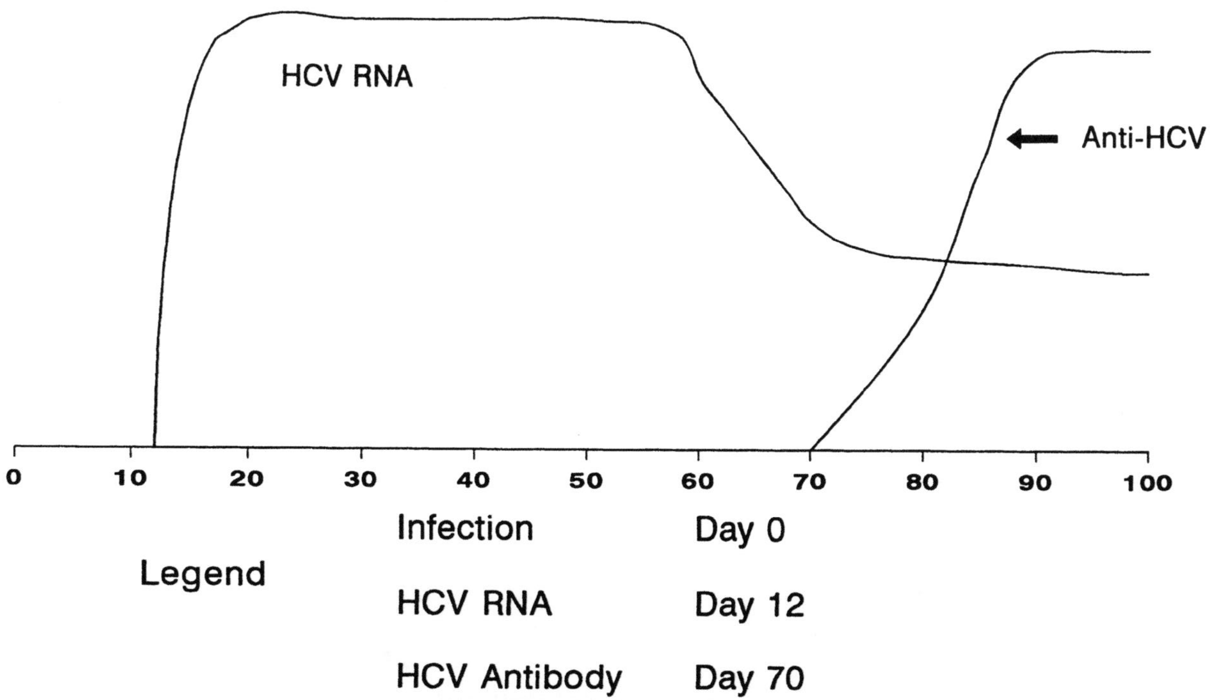

Figure 3. HCV markers during early infection.

viral dissemination is consistent with the estimated 20-day infectious window period, if one assumes that one copy per mL of HIV is infectious and that peak viremia coincides with seroconversion. Figure 2 summarizes the current model for HIV primary infection based on the above data.

HCV Window Period

Three sources of data have been used to understand the HCV window period: HCV seroconverting plasma donor panels, cases of transfusion-transmitted HCV infections from post-transfusion hepatitis studies from the 1970s and early 1980s, and chimpanzee infection studies. As with HIV, when a plasma donor seroconverts, the preceding serial donations can be studied to determine the duration and levels of HCV RNA prior to liver enzyme elevation and antigen formation. Unlike HIV, in which the RNA-positive/antibody-negative stage is brief, with HCV there is a prolonged high-titer viremic phase prior to an increase in ALT and antibody seroconversion [23,24]. In addition, the transition from a negative to a positive RNA NAT result is extremely abrupt with HCV, with an extraordinarily rapid viral doubling time estimated at ~0.1 day. In one study, the duration of HCV RNA positivity prior to seroconversion averaged 41 days [23]. The RNA copy number during this period averaged over 100,000 copies per mL. With this high titer viremia phase, HCV is very amenable to detection by pooled specimen NAT, even with pools as large as 100 to 500 donations [11].

The second source of data on the HCV window period involves post-transfusion hepatitis cases from the late 1970s [24]. The findings from analysis of these cases were very similar to those based on SC plasma donors, with a prolonged, high titer viremia preceding seroconversion. Because the date of exposure (i.e., transfusion) is known, it can be shown that most recipients develop high level viremia within ten days of infection. This is followed by an average 40 day period of high level viremia prior to ALT elevation and, approximately ten days later, antibody seroconversion.

Finally, chimpanzee infectivity studies are consistent with these observations in humans [18]. RNA is detected by NAT an average of nine days post-inoculation, with antibody detected on average of 73 days post-inoculation. Figure 3 summarizes these data in a model of HCV primary infection.

HBV Window Period

The HBV window period can likewise be studied using panels of samples from seroconverting plasma donors or archived samples from subjects in post-transfusion hepatitis studies. Data from post-transfusion cases indicate that HBsAg is first detected about 50 to 60 days following transfusion [25]. As with HIV, virus can be detected at low level in plasma for several weeks prior to HBsAg detection, based on testing serial samples collected prior to HBsAg positivity from plasma seroconversion panels [26,27]. Analyses of the HBV DNA concentrations in serial plasma donations yielded an average four-day

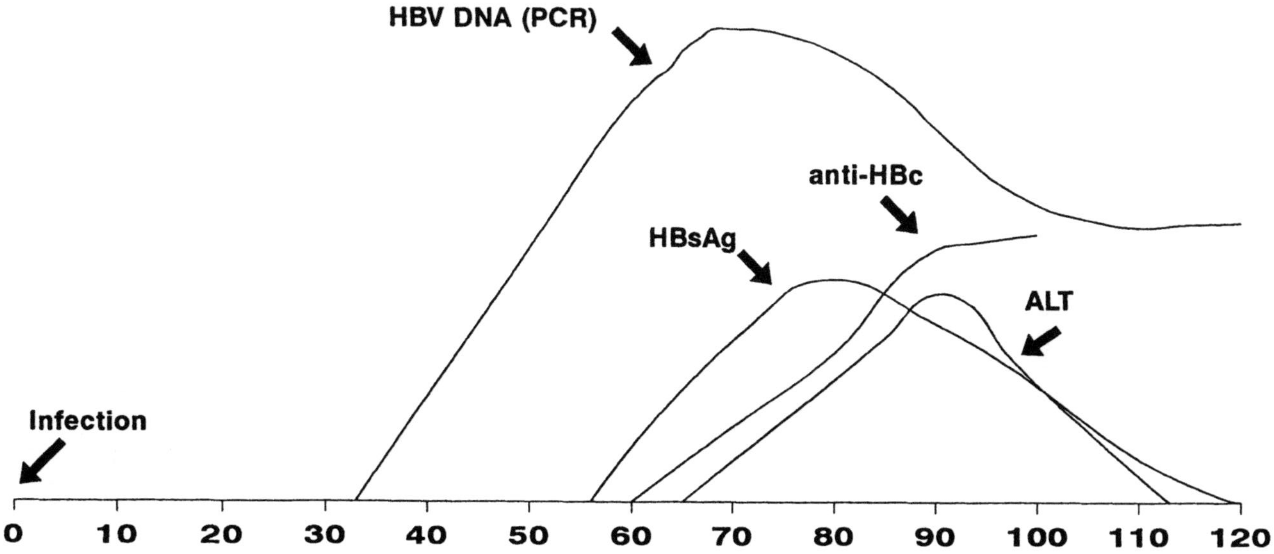

	Infection	Day 0
Legend	HBV DNA	Variable up to 23 days * prior to HBsAg *
	HBsAg	Day 56; disappears Day 120

* Average: 6 - 15 Days

Figure 4. HBV markers during early infection.

doubling time for HBV DNA during this pre-seroconversion phase. Thus, whereas HCV doubles every few hours and HIV doubles every day, HBV levels increase relatively slowly in plasma (Figure 4).

For HBV the levels of DNA in plasma have been correlated with infectivity in chimpanzees. Replicate dilutions of plasma that had previously been pedigreed with regard to chimp infectivity, were tested by PCR assays to determine the dilution when the assays became negative. Based on these data, it has been estimated that ten to twenty HBV particles or DNA equivalents are equivalent to one chimp infectious dose. These results address the issue of what sensitivity of NAT is needed to interdict infectivity. Based on the level of HBV DNA at HBsAg seroconversion, the doubling time, and the chimp infectivity data, it can be estimated that infectivity may exist as much as 40 days prior to detectable HbsAg [26,27]. However, due to the relatively good sensitivity of HBsAg assays, the concentration of HBV DNA are relatively low (average of less than 1,000 copies per mL, range 1-2,400 copies/mL) during this prolonged phase of HBV infectivity prior to surface antigen. Therefore, NAT pooling strategies will only allow detection of a small portion of the pre-HBsAg infectious window period. It is for that reason that most companies are viewing HBV as a less feasible target for NAT of pooled donation samples. Figure 4 summarizes the characteristics of HBV primary infection, Table 2 contrasts the key parameters for HIV, HCV, and HBV.

Table 2. Window period detection by NAT.

Window	HIV	HCV	HBV
Infection to antibody	22 d	70 d	59 d
Reduced by single donation NAT	10–15 d	41–60 d	6–15 d
Doubling time	1 d	<0.1 d	4 d
Viral load (g eq/mL)	10^2–10^7	10^5–10^7	10^2–10^4

HIV NAT vs p24: 3–8 days; RNA <10 g eq/mL.

Overview of Gene Amplification / Nucleic Acid Amplification and Detection Systems

The various gene amplification/ nucleic acid amplification technologies (NAT) have three steps in common: sample preparation or extraction, amplification of the target DNA or RNA and detection of the amplified product. The authors have previously reviewed these in reference 28. There have been many advances in the amplification and detection steps, with the solution of many associated problems and the development of automation capable of rapid sample processing. Until recently, commercially available sample preparation methods were largely manual, labor-intensive and time-consuming [29,30]. Most protocols involved a cell purification or viral particle centrifugation or capture step, followed by an extraction step to liberate the target nucleic acid from cells or virion particles. These steps were required to concentrate virus and remove or degrade proteins and other potential inhibitors of the amplification reaction. Over the last several years manufacturers have developed new nucleic acid concentration and purification techniques based on either target sequence specific probe capture methods or nonspecific binding of nucleic acids to silica particles (see below). Stability of viral nucleic acids during specimen storage and processing is a critical issue. Although for some viruses the nucleic acid targets are relatively stable (e.g. HIV and HBV) [31], for other viruses (e.g. HCV) nucleic acids are relatively labile [32-34], sensitive detection of these agents requires more rapid extraction and more care in subsequent storage and handling steps. Special plasma preparation tubes (PPT) may be required to maintain viral nucleic acid stability, particularly if samples are shipped to regional laboratories for extraction, pooling, and analysis. Development of front-end automation is currently a major focus by manufacturers involved in development of NAT for mass screening.

The nucleic acid extracts are then subjected to an amplification step, which generally results in almost a billion-fold increase in the concentration of the target sequences. The development of thermostable enzymes for amplifying DNA or RNA and of automated thermocycling instruments has considerably simplified the amplification step of PCR and related technologies (see below). Isothermal methods have also been developed that obviate the need for heat-stable enzymes and thermal cyclers. Some newer methods even allow the direct readout of test results from the vessel in which the amplification was performed. Many of the new amplification instruments use microplate formats and are therefore compatible with automated pipetting instruments that can also be used with sample pooling, extraction, and detection systems.

In research and some commercial reference laboratories, amplified products are generally detected through the use of radioactively labeled probes. For commercial applications, a variety of non-isotopic methods have been developed [28-30]. Detection methods can be either sequence-specific or nonspecific. One widely used sequence-specific approach involves capturing amplified products in microplate wells coated with nucleic acid probes, followed by detection of the bound product via EIA or chemiluminescence technology. This approach can take advantage of equipment routinely used in blood centers today for detecting viral antigens and antibodies. Over the past several years, novel nonspecific detection methods have also been developed that have the potential to increase the throughput and sensitivity of the detection step even further (see below). Integrated systems are thus being developed that will maintain sample identification throughout processing and analysis and will markedly increase throughput and decrease labor costs.

Several problems associated with early amplification assay systems that have since been solved are worthy of note. For example, some early studies using PCR were plagued by false-positive results because of the carry-over of amplified product from positive to negative samples [35]. Initial efforts to avoid this cross-contamination problem involved the dedication of separate laboratories for pre- and post-amplification steps, the use of disposable positive-displacement pipette tips and rigorous procedural controls [36]. More recently, new biochemical approaches have been developed that inactivate previously amplified product (reviewed in Persing, et al.[29]). One example of note is for

PCR carry-over control by inactivation of previously amplified DNA by selective action of uracil-N-glycosylase (UNG). All amplification reactions are performed using dUTP instead of dTTP. UNG is added to reactions prior to amplification. Selective cleavage of uracil residues from any previously amplified DNA present as a contaminant results in elimination of the template activity of contaminants. Other manufacturers have focused on the containment of amplified product in sealed amplification or detection cassettes [37]. Use of these approaches has virtually eliminated the carry-over problem.

Another problem relates to the inherent genetic variability of viruses such as HIV and HCV, with consequent false-negative results because of the failure of primers or probes to bind to target sequences or because of inefficient amplification of the target [38-40]. This problem has been solved by focusing amplification reactions at highly conserved sequences (e.g., the 5'-nontranslated region of HCV, and gag and pol regions of HIV) and by the use of batteries of primers and probes directed at known variants and designed to tolerate minor sequence mismatches [37-39].

A third key advance relates to the detection of multiple viruses in a single reaction. This is accomplished by adding primers (and probes) directed toward each target agent in each reaction tube. This approach requires the identification of optimized reaction conditions compatible with efficient and sensitive amplification and detection of each target sequence. The feasibility of co-amplifying and detecting more than one infectious agent in a given sample has been demonstrated by a number of groups using a variety of amplification technologies [37,41-43]. Co-amplification is also widely used for the quantification of viral burden through incorporation of spiked internal quantification standards that amplify in parallel with target viral sequences [43-47].

This approach should not only allow for development of a single assay for the viral agents of current greatest interest (i.e., HIV RNA, HBV DNA and HCV RNA in plasma), but it should also allow for addition of other agents to the assay, as warranted. Thus, in the future, a multiplex, nucleic acid amplification-based screening system might include primers and probes for HCV, HIV and HBV, as well as for additional viruses (e.g. HHV-8, CMV, parvo B19) [48-50], bacteria [51] and a variety of parasites (e.g., Chagas' disease, malaria, babesiosis, etc) [52]. When a donor sample is reactive by such a multiplex screening assay, the specific infection involved could then be determined by re-testing the specimen using virus-specific amplification assays. For obvious logistic and cost-effectiveness reasons, such a combination screening approach holds great promise for the future of infectious disease screening.

Major Methods for Amplification and Detection of Viral Nucleic Acids

Four major nucleic amplification methods have been commercialized for the detection of blood-borne pathogens, and will be briefly reviewed below.

Polymerase Chain Reaction

PCR uses pairs of primers that anneal to conserved regions flanking the target sequence of interest, thereby focusing the amplification reaction to the target area [53,54]. Through the use of repeated cycles of heating to denature (separate) previously amplified DNA sequences (so-called amplicons) into single-stranded templates, and then cooling to allow annealing (binding) of the primers to the target sequences, DNA synthesis is repeated numerous times. Each cycle theoretically doubles the copy number of target sequences, such that 30 cycles of amplification can produce a billion copies of the target. The amplified sequences can then be readily detected by virtue of their specific size following electrophoretic separation, or by the use of radiolabeled or enzyme-tagged probes that hybridize to internal, amplified sequences.

Roche Molecular Systems (Somerville, New Jersey, USA), which purchased the rights to PCR from Cetus in 1991, has developed commercial assays for the detection of HIV-1, HTLV-I/HTLV-II, and CMV DNA in peripheral blood mononuclear cells, as well as assays for HIV-1 RNA, HBV DNA, HCV RNA, and CMV DNA in plasma or other fluids [40,45-47,50]. Each assay is capable of detecting as few as one to 100 molecules of target DNA or RNA in the processed sample volume, and each uses primers and probes to conserved sequences of one or more viral genes to ensure the detection of most known variants. The currently available commercial assays from Roche use either a 96-well format for thermal cycling and detection or an instrument (COBAS) that automatically processes batches of 24 specimens through amplification and detection steps. The detection system is based on the use of biotinylated primers such that amplified products are biotinylated. Aliquots from PCR tubes are denatured and transferred to microwells or tubes containing microparticles that are coated with immobilized oligonucleotide probes to capture amplified product. After washing, a molecular conjugate of avidin and horseradish peroxidase is added, which binds to captured, biotinylated product. A peroxidase substrate (tetramethylbenzidine) is added, and color development is read in a standard microplate EIA reader. Overall assay times for the current generation of Roche assays range from two to four hours. The COBAS system is currently being employed in the U.S. and worldwide in NAT screening of pooled samples from blood donors.
Roche has further devised a probe-independent PCR detection strategy, kinetic PCR, based on the intercalation of ethidium bromide (included in each PCR reaction tube) into the double-stranded DNA that accumulates during each PCR cycle [55]. This results in exponentially increasing fluorescence in the tube that parallels PCR product formation. The fluorescence can be seen directly through the caps of the PCR reaction tubes by monitoring the reaction plate in real time with the use of a fluorescence video camera built into the same instrument system as the thermal cycler. Roche is developing a multiplex assay system that uses this technology that could be applied to screening donors for HIV, HBV and HCV. Once a donor sample is shown to be reactive with the kinetic PCR screening system, the viral specificity could be confirmed by testing the amplified product with type-specific virus probe assays.

Roche has also developed a method for the rapid, specific detection of amplified products that employs homogenous reaction chemistry that enables amplification and probe-based detection in a single tube. Termed TaqMan™, the method is based on the principle of nick-translation PCR with fluorogenic probes [56]. Two fluorescent dyes (a rhodamine and a fluorescein) are affixed to each oligonucleotide probe. While the two dyes are attached to the probe, the rhodamine (designated the quencher) absorbs the excited state energy of the fluorescein (the reporter). However, if the probe anneals to a target sequence (either the primary target or an amplicon), it is subsequently cleaved by the 5' to 3' exonuclease activity of Taq DNA polymerase during PCR. In the process, the dyes separate, which allows an increase in reporter dye emission in proportion to the starting copy number. Since multiple fluorescein dyes are available as reporters, differently labeled probes can be used to detect different sequences in a single multiplex reaction. Roche and Perkin-Elmer Corporation (Norwalk CT) have each developed luminescence reader instruments that detect these multiple fluorescent label emissions simultaneously. These systems will likely see use in next generation NAT screening for blood donors.

Transcription-Mediated Amplification / Nucleic Acid Sequence-Based Amplification

Transcription-mediated amplification (TMA) and nucleic acid sequence-based amplification (NASBA) are similar techniques involving a one-step, isothermal amplification reaction that utilizes a continuous cycle of reverse transcription and RNA transcription to replicate the RNA target [57-60]. In addition to the complementary sequence to the target pathogen, primers contain the promoter sequence for an RNA polymerase. The initial reverse transcription of complementary DNA (cDNA) from RNA is followed by RNAse treatment, which degrades the initial RNA target and permits cDNA synthesis to proceed, ultimately generating duplex cDNA. RNA polymerase then binds to the promoter sequence and transcribes antisense RNA, which serves as a target for primer hybridization, leading to a second cycle of cDNA synthesis, RNA degradation and second-strand cDNA synthesis.

Organon Teknika (The Netherlands) has patent rights to NASBA and has developed an automated sample processing, amplification and detection system. The patented sample preparation procedure, which uses silica and guanidium thiocyanate, enables rapid and efficient extraction of RNA and DNA from virtually any specimen type. The detection of amplified product is achieved by hybridization to probes tagged with an electro-chemiluminescent moiety. NASBA assay kits for detection and quantitation of HIV and HCV RNA are available in Europe and have completed clinical trials in the USA [57-60].

Gen-Probe (San Diego, California, USA) has developed a similar homogenous, one-tube amplification and detection system. Greater than a billion-fold amplification is carried out under isothermal conditions using the company's patented transcription-mediated TMA system. In this system, only two enzymes, reverse transcriptase and RNA polymerase, are added to the reaction mixture. The target sequences are copied by reverse transcriptase to produce a transcription complex. The transcription complex is used to make numerous RNA transcripts of the sequence, and the process repeats autocatalytically. The detection of amplified product is achieved by the hybridization protection assay (HPA) technology, whereby probes labeled with acridinium ester are added to the amplification reaction, a selection reagent is added to hydrolyze differentially the label associated with the unhybridized probe and chemiluminescence of the hybridized probe is detected following the addition of peroxide and alkali [61,62]. No wash steps or the transfer of amplicon to separate reaction vessels is required, minimizing carry-over contamination. In addition, Gen-Probe has developed a rapid specimen-processing method that uses a probe to capture target sequences and eliminates the need for the extraction of plasma with organic solvents or centrifugation steps prior to amplification. Gen-Probe has a semi-automated system (Magenta) that is currently being employed in large-scale clinical trials for analysis of pooled plasma samples. A fully automated system (Tigris) for testing single donations is slated to enter clinical trials in 2000.

Ligase Chain Reaction

The ligase chain reaction (LCR), which is being developed and commercialized by Abbott Laboratories (Abbott Park, Illinois, USA), uses a thermostable enzyme that ligates (covalently joins) the phosphate backbone of two adjoining nucleotide probes bound to a target DNA or RNA sequence [63]. Probes that are not hybridized, or are hybridized incorrectly at the junction between them, are not ligated. In the presence of a second set of adjoining probes that are complementary to the first set of probes, the first set, when ligated, acts as a template for ligation of the second set. The ligated products thus multiply in a chain reaction through a series of denaturing and annealing temperature cycles. Abbott has developed an LCR assay system that combines pre-aliquoted unit dose reagent packs with a sample preparation instrument, a thermal cycler and an LCR analyzer. The LCR analyzer initiates detection by having its probe pierce through the lid of each reaction tube to remove a portion of the contents. Detection is based on the binding of fluorescently labeled probes and fluorometry.

Similar to PCR, LCR is capable of multiplex detection of multiple target sequences in a single reaction tube. Carry-over control is achieved by eliminating pipetting steps and by the use of reagents that inactivate previously ligated probes following detection. Abbott has focused initial commercialization of LCR on the detection of non-blood-borne pathogens, including Neisseria gonorrhea, Mycobacterium tuberculosis, and human papilloma viruses. These assays have performed with sensitivities and specificities exceeding 99% in FDA clinical trials. Preliminary data on LCR detection of HIV and HCV is encouraging. Abbott Laboratories also has license rights to PCR, which can be run on LCR instruments.

Branched DNA Signal Amplification Assay

The branched DNA (bDNA) assay developed by Chiron Corporation (Emeryville, California, USA) is based on the principle of an amplifying signal from annealed DNA probes through the formation of controlled networks of synthetic oligonucleotide probes, rather than amplification of the target nucleic acid [64]. The method involves a series of hybridizations of target probes specific to the sequence of interest (e.g., the HIV assay uses 39 binding sites to gag or pol sequences). A subset of the probes mediates capture of the target molecules to a capture oligomer-modified microplate, while the second set binds to target as well as to a bDNA amplifier molecule. The bDNA amplifier is a chemically synthesized structure containing 45 copies of a sequence that can bind a labeled probe, but just one copy of a sequence that binds to target probes. Multiple copies of bDNA amplifiers are bound to the probe-target complex on the microplate to incorporate up to 1,800 labeled (alkaline phosphatase) probes on each target sequence. A chemiluminescent substrate, dioxetane, is used to generate a visible light signal that is quantitated with the use of a luminometer.

Since no amplicon is produced, bDNA assays are much less prone to false positivity than target amplification assays and, consequently, physical containment and product destruction methods developed for target amplification procedures are not required. On the other hand, the sensitivity of bDNA assays is lower than can be achieved by target amplification methods [65]. Consequently, ultracentrifugation of relatively large volumes of plasma (1-2 mL) is required for the sensitive detection of viruses such as HIV. At present, commercialized bDNA assays are available for research use for the detection and quantitation of HIV, HBV and HCV virions in plasma, serum, blood cells or tissue, and an assay for cytomegalovirus is in development [40,46,66-68]. The sensitivity of bDNA assays is generally one to two logs less than other NAT methods, given comparable input sample volumes [40,46]. Consequently, development of bDNA systems for donor screening is not as promising as the other methods discussed.

Current Pooled DNA NAT Screening Strategies

Although the previous section documents rapid advances in NAT technologies with regard to ease of use, reproducibility and sensitivity, these methods are not yet ready to be implemented on an individual donation level. Limitations yet to be overcome include complete automation and scale-up of sample processing, and validation that amplification and detection methodologies retain high sensitivity and specificity when applied to thousands of individual donor samples on a daily basis. As reviewed above, these limitations are being addressed by several companies which are developing high-throughput, fully automated, multiplex viral RNA/DNA detection systems to be submitted for FDA review in a two to five year time-frame. However, political, regulatory, and competitive pressures have driven the implementation of NAT screening on a more rapid timetable. The plasma derivatives manufacturing industry, driven by pressures from European regulatory organizations, has introduced NAT in order to detect high-titer viremic donations prior to pooling of plasma for further manufacture (Cohn fractionation). Because of the complexity, low-throughput, and high cost of NAT technology as it presently exists (including both test kits and the construction and maintenance of specialized facilities for testing), strategies involving testing of pooled samples are now in widespread use [69].

Pooling Algorithms and Options

The first consideration regarding pooling algorithms and options for NAT is the advantage that pooling has over single-donation testing, in that it allows the performance of fewer tests per day. Pooling is required for NAT because single-donation testing is not yet feasible, as a result of limitations in automation available for all current NAT technologies. The most significant consideration in regard to pool size is assay sensitivity, because sensitivity decreases as pool size increases. In addition, test specificity and contamination issues during pooling and testing are major influences on component availability, and, as such, the use of large pools will affect a greater amount of component. The relationship between pool size and component availability is apparent by calculations involving different pool sizes. The greatest number of tests is performed by using a small pool size (e.g. 16-24 donations/pool), but the number decreases with the use of a larger pool (e.g., 512 donations/pool). However, consideration must be given to the amount of component on hold if a reactive result occurs (or if there are multiple reactive results/day), and, if the management of components is based on reactive pool results, the impact of using large pools is much more significant.

Sensitivity issues related to pool size vary, depending on which agent is being considered. For example, data generated by the American Red Cross (ARC) with two different NAT formats and assays have demonstrated that the impact of pool size is more significant in connection with HIV than HCV. If a pool size of 500 is evaluated by using the PCR assay services of National Genetics Institute (NGI), the RNA-positive window period for HCV is reduced by 50 to 90% and the window period for HIV is reduced by 30 to 50%; for HBV, a reduction in the window period does not occur [23]. In contrast, with a pool size of 16, the transcription-mediated amplification assay of GenProbe projects viremic window period reductions approximating 100% for HCV and greater than 90% for HIV [23]. HBV has not been studied as of yet with the GenProbe assay, but it is predicted that, even with small pools, there will be relatively little benefit over HBsAg testing, especially as newer, more sensitive HBsAg assays are introduced.

Whatever assay is selected for NAT, its performance must be validated. Sensitivity decreases as pool size increases, and qualification of sensitivity of the selected assay must be performed. Positive samples diluted to end-point determine the assay's analytic sensitivity, and the pool size chosen determines window period reduction (clinical sensitivity). Testing of pools must also consider the effect of interfering substances, including possible matrix effects and amplification inhibitors. However, current NAT technologies all include the use of an internal amplification control to verify that the amplification process has in fact occurred as expected.

One obvious consideration in pooling is the pooling method. The simplest but most impractical method is the use of manually created pools. Manual methods involve human error and are labor intensive. In contrast, automated pooling methods involve less risk of human error, but require the development and execution of complex validation protocols. Pooling can occur in one, two, or three dimensions; the creation of pools by complicated pooling algorithms increases the dependence on and complexity of automated pooling methods.

Another consideration in creating pools is whether samples that are serologically reactive are included or whether the initially or repeatedly reactive samples are removed before pooling. Removal of seroreactive samples before pooling increases turnaround times, because pooling cannot be initiated until all of the serologic results are known. Increases in turnaround time due to waiting for serologic results have an impact on the release of cellular components, but removal of serologically reactive samples will decrease the potential for contamination. Contamination resulting from the inclusion of serologically reactive samples has already been observed in US NAT screening programs. False-positive results due to contamination at the individual donation level never occurred in the Phase I NAT program of the ARC in which seroreactive donations were removed, but they are occurring at a frequency of 1 in 28,500 in a Phase II NAT program in which seroreactive donations are included. The impact of contamination will be addressed later in this section.

Resolution algorithms must be developed for testing of pooled donations. One method of resolution involves the use of a crossover matrix design. That is, if pools are created in two dimensions, the implicated donation exists at the crossover of an intersecting row and column. If pools are created in three dimensions, the implicated donation exists at the crossover of an intersecting row, column and layer. When a pool is created in rows, columns, and layers, resolution of the reactive pool occurs by testing all of the constituent subpools of the rows, columns, and layers (for a pool of 512 donations, this totals 24 tests). The advantages of this resolution algorithm include the ability to locate the specific donation at the reactive intersection, and testing of the subpools allows a more concentrated sample to be tested (as compared to the dilution factor of the final pool). Therefore, testing that occurs during resolution serves to confirm the initial reactivity observed and also eliminates the need to retest the final pool.
Straight-line resolution of pools is operationally the simplest to create and to resolve. For example, pools created ARC's Phase I NAT contained a total of 128 donations each; when a 128-member pool reacted, the eight subpools, each containing 16 donations, were evaluated, and the 16 individual donations in the reactive subpool were tested.

Another resolution method involves leaving the seroreactive donations in the pools subject to primary screening, but also to create for resolution testing a second pool from which the seroreactive donations have been removed [11]. Therefore, if a pool reacts, the second pool, from which the seroreactive donations have been removed, can be tested. In almost all cases, the NAT reactivity of the pool results from a single seroreactive donation, so that, if the re-testing of the pool that contains only seronegative donations is negative, those donations may be cleared for release.

Pooling and NAT strategies must also consider confirmatory algorithms. Because it will not be possible to enroll in follow-up studies every donor identified as NAT-reactive and seronegative, another method is needed for confirmation that can be performed at the time that the NAT reactivity is observed. A second or supplemental NAT, of a different type, should be used to test all NAT-reactive donations. The concept is similar to that of the dual enzyme immunoassay strategy in use for testing for human T-lymphotropic virus antibody. In this way, NAT reactivity can be confirmed by a second test using different primers and probes and a different technology. One further recommendation is that the second NAT be performed with a sample from an independent source to exclude the possibility of source-tube contamination. The supplemental NAT in use by the ARC is PCR (versus the screening that occurs using transcription-mediated amplification); the PCR test used (which is performed by NGI) has sensitivity for HIV and HCV RNA comparable to that of TMA [23].

Sources of laboratory contamination must also be considered, because contamination affects not only the number of pools that require resolution, but also the amount of component on hold awaiting the resolution of test results. Pooling should occur in a separate, well-contained facility, using techniques that minimize aerosolization and droplet formation. Separate areas with controlled air flow are required for amplification and detection. Training and proficiency of staff in the areas of amplification technology are also requirements for valid pooling and testing to minimize potential contamination. However, even with all the process controls required by amplification technology, contamination in some cases cannot be avoided, especially if seroreactive samples are combined and tested simultaneously with seronegative samples.

Yield of NAT during First Year of Screening in the USA.

NAT screening was initiated in the United States of America in the first half of 1999. Three major testing programs were established, two of which are using the GenProbe technology and one of which uses the Roche assay system. Table 3 compares the key characteristics of the three NAT testing programs now under way in the USA and presents the yield over the first six to nine months of screening. The yield of HCV RNA-positive, antibody negative is consistent across the three testing programs, with detection rates in the 1 per 300,000 donation range. This yield is similar to the predictions of yield based on the incidence-window period model. The yield of HIV has been more variable. At the American Red Cross (ARC) only a single HIV RNA+/antibody negative donation was detected; however, the ARC was screening pools of 128 donations for the majority of this period, so the sensitivity for HIV detection was somewhat compromised. Since conversion to pools of 16 in late 1999, the yield of HIV at the American Red Cross NAT lab has been in the range of 1 per million units.

Table 3. Yield of NAT screening of US blood donors, first year.

	ARC	ABC/AIBC	ABC
Assay system	GenProbe at 1:128 and 1:16	GenProbe at 1:24	Roche at 1:24
Number of screened donations	4.6 million	1.6 million	2.1 million
Rate and number of HCV NAT-positive donations	1 in 321,000 (n=15)	1 in 320,000 (n=5)	1 in 210,00 (n=10) Ab-negative
Rate and number of HIV NAT-positive, p24 Ag-negative, Ab-negative donations	1 in 4.6 million (n=1)	1 in 800,000 (n=2)	N/A

similar to that observed at non-ARC sites using the GenProbe test on pools of 24 donations per pool. Roche testing labs introduced HIV testing in early 2000, so it is too early to assess the yield at these sites.

Conclusions

In this chapter we have reviewed the current understanding of the magnitude and causes of residual risk for each of the major transfusion transmitted viral pathogens. We reviewed the dynamics of viral and immune parameters during the window periods for each agent since window period related risk is the major source of breakthrough infections from serologically screened transfusions. The molecular approaches for amplifying and detecting viral nucleic acids were then considered, and the status of automation of each of the major NAT systems discussed. The pros and cons of minipool versus single donations testing strategies were then examined. Finally, the rapid transition to implementation of minipool NAT screening into routine screening of blood donors reviewed. This rapid process has clearly been a remarkable accomplishment of which the blood banking community should be proud.

Acknowledgment

The material contained in this summary is continually being revised and updated. It has been presented in workshops at the 51st and 52nd Annual Meetings of the American Association of Blood Banks (1998 and 1999, respectively) and parts have previously appeared in and been adapted from Garratty, ed. [70]. The reader is also referred to the recently published Joint Report of the (USA) Interorganizational Task Force on NAT [71].

References

1. AuBuchon JP, Birkmeyer JD, Busch MP. Safety of the blood supply in the United States: opportunities and controversies. Ann Intern Med 1997;127:904-09.
2. Schreiber GB, Busch MP, Kleinman SH, Korelitz JJ for the NHLBI Retrovirus Epidemiology Blood Donor Study. The risk of transfusion-transmitted viral infections. N Engl J Med 1996;334:1685-90.
3. Pau C-P, Hu DJ, Spruill C, et al. Surveillance for human immunodeficiency virus type 1 group O infections in the United States. Transfusion 1996;36:398-400.
4. de Oliveira CF, Diaz RS, Machado DM, et al. Surveillance of HIV-1 genetic subtypes and diversity in the U.S. blood supply. Transfusion 2000 (submitted).
5. Imagawa DT, Lee MH, Wolinsky SM, et al. Human immunodeficiency virus type 1 infection in homosexual men who remain seronegative for prolonged periods. N Engl J Med 1989;320:1458-62.
6. Farzadegan H, Polis MA, Wolinsky SM, et al. Loss of human immunodeficiency virus type 1 (HIV-1) antibodies with evidence of viral infection in asymptomatic homosexual men: A report from the Multicenter AIDS Cohort Study. Ann Intern Med 1988;108:785-90.
7. Alter MJ, Margolis HS, Krawczynski K, et al., for The Sentinel Counties Chronic Non-A, Non-B Hepatitis Study Team. The natural history of community-acquired hepatitis C in the United States. N Engl J Med 1992;327:1899-1905.

8. Imagawa D, Detels R. HIV-1 in seronegative homosexual men. N Engl J Med 1991;325: 1250-51.
9. Roy MJ, Damato JJ, Burke DS. Absence of true seroreversion of HIV-1 antibody in seroreactive individuals. JAMA 1993;269:2876-79.
10. Lifson AR, Stanley M, Pane J, et al. Detection of human immunodeficiency virus DNA using the polymerase chain reaction in a well-characterized group of homosexual and bisexual men. J Infect Dis 1990;161:436-39.
11. Roth WK, Weber M, Seifried E. Feasibility and efficacy of routine PCR screening of blood donations for hepatitis C virus, hepatitis B virus, and HIV-1 in a blood-bank setting. Lancet 1999;353:359-63.
12. Busch MP, Watanabe KK, Smith JW, et al., for the Retrovirus Epidemiology Donor Study. False-negative testing errors in routine viral marker screening of blood donors. Transfusion 2000;40:585-89.
13. Kleinman SH, Busch MP, Hall L, et al., for the Retrovirus Epidemiology Donor Study. False positive HIV-1 test results in a low-risk screening setting of voluntary blood donation. JAMA 1998;280:1080-85.
14. Busch MP, Satten GA. Time course of viremia and antibody seroconversion following human immunodeficiency virus exposure. (Presented at CDC/NFID Workshop: "HIV Post-Exposure Management of Health Care Workers", Atlanta GA, Mar 1996, as "Kinetics of in vivo Virus Expansion and Host Immune Response Following Acute HIV Infection: Implications for Disease Progression and Therapeutic Management"). Am J Med 1997;102(Suppl 5B):117-24. (Discussion: pp 125-6)
15. Petersen LR, Satten GA, Dodd R, et al., and the HIV Seroconversion Study Group. Duration of time from onset of human immunodeficiency virus type 1 infectiousness to development of detectable antibody. Transfusion 1994;34:283-89.
16. Busch MP, Lee LLL, Satten GA, et al. Time course of detection of viral and serological markers preceding HIV-1 seroconversion; implications for blood and tissue donor screening. Transfusion 1995;35:91-97.
17. Niu MT, Stein DS, Schnittman SM. Primary human immunodeficiency virus type 1 infection: review of pathogenesis and early treatment intervention in humans and animal retrovirus infections. J Infect Dis 1993;168:1490-501.
18. Beach MJ, Meeks EL, Mimms LT, et al. Temporal relationships of hepatitis C virus RNA and antibody responses following experimental infection of chimpanzees. J Med Virol 1992;36:226-37.
19. Farci P, London WT, Wong DC, et al. The natural history of infection with hepatitis C virus in chimpanzees: comparison of serologic responses measured with first and second generation assays and relationship to viremia. J Infect Dis 1992;165:1006-11.
20. Murthy KK, Henrard DR, Eichberg JW, et al. Redefining the HIV-infectious window period in the chimpanzee model: evidence to suggest that viral nucleic acid testing can prevent blood-borne transmission. Transfusion 1999;39:688-93.
21. Busch MP, Satten GA, Herman SA, et al. Time course and kinetics of HIV viremia during primary infection. Abstract S-164. Transfusion 1996;36(9-Suppl):41s.
22. Busch MP, Satten GA, Sakahara NS, Garrett PE, Herman SA. Time course and dynamics of viremia during early stages of HIV seroconversion. 4th Conf on Retroviruses and Opportunistic Infections, Washington DC 1997. Abstractbook (oral) Session 78, Abstract 749, p 203.
23. Stramer SL, Porter RA, Brodsky JP, Waldman KJ, Dodd RY. Sensitivity of HIV and HCV RNA detection by pooled genome amplification testing (GAT). Abstract S-259. Transfusion 1998;38(10-Suppl):70s.
24. Busch MP, Rawal BD, Fiebig EW, et al. Dynamics of HCV viremia and seroconversion in transfusion-acquired HCV infections. Abstract S-265. Transfusion 1998;38 (10-Suppl):72s.
25. Mimms LT, Mosley JW, Hollinger FB, et al. Effect of concurrent acute infection with hepatitis C virus on acute hepatitis B virus infection. Brit Med J 1993;307: 1095-97.

26. Rawal BD, Kleinman SH, Kuhns MC, Busch MP. Infectious HBV window period and its projected reduction by genome amplification testing. Abstract S-343. Transfusion 1998;38(10-Suppl):91s.
27. Kuhns MC, McNamara AL, Peterson B, et al. Detection of hepatitis B seroconversion by highly sensitive assays for surface antigen and HBV DNA. Abstract S-342. Transfusion 1998;38(10-Suppl):91s.
28. Busch MP, Stramer SL, Kleinman SH. Evolving applications of nucleic acid amplification assays for prevention of viral transmission by blood components and derivatives. Chapter 6 in Garratty G, ed. Applications of Molecular Biology to Blood Transfusion Medicine (1997 Workshop book). American Association of Blood Banks. Bethesda, MD, USA. 1997:123-76.
29. Persing DR, Smith TF, Tenover FC, White TJ, eds. Diagnostic Molecular Microbiology: Principles and Applications. American Society for Microbiology, Washington DC; and Mayo Foundation, Rochester, MN, USA. 1993.
30. Wolcott M. Advances in nucleic acid-based detection methods. Clin Microbiol Rev 1992;5:370-86.
31. Mole L, Margolis D, Carroll R, et al. Stabilities of quantitative plasma culture for human immunodeficiency virus, RNA, and p24 antigen from samples collected in Vacutainer CPT and standard Vacutainer tubes. J Clin Microbiol 1994;32:2212-25.
32. Busch MP, Wilber JC, Johnson P, et al. Impact of specimen handling and storage on detection of hepatitis C virus RNA. Transfusion 1992;32:420-25.
33. Fong TL, Charboneau F, Valinluck B, Govindarajan S. The stability of serum hepatitis C viral RNA in various handling and storage conditions. Arch Pathol Lab Med 1993;117: 150-51.
34. Halfon P, Khiri H, Gerolami V, et al. Impact of various handling and storage conditions on quantitative detection of hepatitis C virus RNA. J Hepatol 1996;25:307-11.
35. Imagawa DT, Lee MH, Wolinsky SM, et al. Human immunodeficiency virus type 1 infection in homosexual men who remain seronegative for prolonged periods. N Engl J Med 1989;320:1458-62.
36. Kwok S, Higuchi R. Avoiding false-positives with PCR. Nature 1989;339:237-38.
37. Findlay JB, Atwood SM, Bergmeyer L, et al. Automated closed-vessel system for in vitro diagnostics based on polymerase chain reaction. Clin Chem 1993;39:1927-33.
38. Hu DJ, Dondero TJ, Rayfield MA, et al. The emerging genetic diversity of HIV. The importance of global surveillance for diagnostics, research, and prevention. JAMA 1996;275:210-16.
39. Soni P, Dusheiko GM, Dhillon AP, Harrison TJ. Genetic diversity of hepatitis C virus: implications for pathogenesis, treatment, and prevention. (Report - Grand Round) Lancet 1995;345:562-66.
40. Hawkins A, Davidson F, Simmonds P. Comparison of plasma virus loads among individuals infected with hepatitis C virus (HCV) genotypes 1, 2, and 3 by quantiplex HCV RNA assay versions 1 and 2, Roche Monitor Assay, and an in-house limiting dilution method. J Clin Microbiol 1997;35:187-92.
41. Hewlett IK, Ruta M, Cristiano K, et al. Co-amplification of multiple regions of the HIV-1 genome by the polymerase chain reaction: potential use in multiple diagnosis. Oncogene 1989;4:1149-51.
42. Sunzeri FJ, Lee TH, Brownlee RG, Busch MP. Rapid simultaneous detection of multiple retroviral DNA sequences using the polymerase chain reaction and capillary DNA chromatography. Blood 1991;77:879-86.
43. Heredia A, Soriano V, Weiss SH, et al. Development of a multiplex PCR assay for the simultaneous detection and discrimination of HIV-1, HIV-2, HTLV-I and HTLV-II. Clin Diag Virol 1996;7:85-92.
44. Piatak M Jr., Saag MS, Yang LC, et al. High levels of HIV-1 in plasma during all stages of infection determined by competitive PCR [see comments]. Science 1993;259:1749-54.

45. Mulder J, McKinney N, Christopherson C, et al. Rapid and simple PCR assay for quantitation of human immunodeficiency virus type 1 RNA in plasma: application to acute retroviral infection. J Clin Microbiol 1994;32:292-300.
46. Lin HJ, Myers LE, Yen-Lieberman B, et al. Multicenter evaluation of quantification methods for plasma human immunodeficiency virus type 1 RNA. J Infect Dis 1994;170:553-62.
47. Schochetman G, Sninsky JJ. Direct detection of HIV infection using nucleic amplification techniques. In: Schochetman G, George JR, eds. AIDS testing: A Comprehensive Guide to Technical, Medical, Social, Legal, and Management Issues, 2nd ed. Springer Verlag, New York, NY. 1994:141-69.
48. Blackbourn DJ, Ambroziak J, Lennette E, et al. Infectious human herpesvirus 8 in a healthy North American blood donor. Lancet 1997;349:609-11.
49. Operskalski EA, Busch MP, Mosley JW, Kedes DH. Blood donations and viruses. (Letter) Lancet 1997;349:1327-28.
50. Fischer SH, Masur H. Laboratory monitoring of cytomegalovirus disease - is polymerase chain reaction the answer? (Editorial) Clin Infect Dis 1997;24:841-42.
51. Brecher ME, Boothe G, Kerr A. The use of a chemiluminescence-linked universal bacterial ribosomal RNA gene probe and blood gas analysis for the rapid detection of bacterial contamination in white cell-reduced and nonreduced platelets. Transfusion 1993;33:450-57.
52. Dodd RY. Transmission of parasites by blood transfusion. [State-of-the-Art Paper] Vox Sang 1998;74(suppl 2):161-63.
53. Mullis KB, Faloona FA. Specific synthesis of DNA in vitro via a polymerase-catalyzed chain reaction. Methods Enzymol 1987;155:335-50.
54. Ou CY, Kwok S, Mitchell SW, et al. DNA amplification for direct detection of HIV-1 in DNA of peripheral blood mononuclear cells. Science 1988;239:295-7. [published erratum: Science 1988;240:240].
55. Higuchi R, Fockler C, Dollinger G, et al. Kinetic PCR analysis: real-time monitoring of DNA amplification reactions. Biotechnology 1993;11:1026-30.
56. Lee LG, Connell CR, Bloch W. Allelic discrimination by nick-translation PCR with fluorogenic probes. Nucleic Acids Res 1993;21:3761-6.
57. Kievits T, van Gemen B, van Strijp D, et al. NASBATM isothermal enzymatic in vitro nucleic acid amplification optimized for the diagnosis of HIV-1 infection. J Virol Meth 1991;35:273-86.
58. van Gemen B, Kievits T, Schukkink R, et al. Quantification of HIV-1 RNA in plasma using NASBA during HIV-1 primary infection. J Virol Meth. 1993;43:177-87.
59. van Gemen B, Kievits T, Nara P, et al. Qualitative and quantitative detection of HIV-1 RNA by nucleic acid sequence-based amplification. AIDS 1993;7(Suppl 2):s107-10.
60. Bruisten S, van Gemen B, Koppelman M, et al. Detection of HIV-1 distribution in different blood fractions by two nucleic acid amplification assays. AIDS Res Hum Retrovir 1993;9:259-65.
61. Arnold LJ, Hammond PW, Wiese WA, Nelson NC. Assay formats involving acridinium-ester-labeled DNA probes. Clin Chem 1989;35:1588-94.
62. Shinnick TM, Jonas V. Molecular approaches to the diagnosis of tuberculosis. In: Bloom B, ed. Tuberculosis: Pathogenesis, Protection and Control. Washington, DC. American Society of Microbiology Press 1994:517-30.
63. Barany F. Genetic disease detection and DNA amplification using cloned thermostable ligase. Proc Natl Acad Sci USA 1991;88:189-93.
64. Urdea MS. Synthesis and characterization of branched DNA (bDNA) for the direct and quantitative detection of CMV, HBV, HCV, and HIV. Clin Chem 1993;39: 725-26.
65. Busch MP, Operskalski EA, Mosley JW, et al., for the Transfusion Safety Study Group. Factors influencing HIV-1 transmission by blood transfusion. J Infect Dis 1996;174: 26-33.

66. Mellors JW, Kingsley LA, Rinaldo CR, et al. Quantitation of HIV-1 RNA in plasma predicts outcome after seroconversion. Ann Intern Med 1995;122:573-79.
67. Davis GL, Lau JY, Urdea MS, et al. Quantitative detection of hepatitis C virus RNA with a solid-phase signal amplification method: definition of optimal conditions for specimen collection and clinical application in interferon-treated patients. Hepatology 1994;19: 1337-41.
68. Dewar RL, Highbarger HC, Sarmiento MD, et al. Application of branched DNA (bDNA) signal amplification to monitor HIV burden in human plasma. J Infect Dis 1994;70: 1172-79.
69. Roth WK, Weber M, Seifried E. Feasibility and efficacy of routine PCR screening of blood donations for hepatitis C virus, hepatitis B virus, and HIV-1 in a blood-bank setting. Lancet 1999;353:359-63.
70. Garratty G, ed. Applications of Molecular Biology to Blood Transfusion Medicine (1997 Workshop book). American Association of Blood Banks. Bethesda, MD, USA. 1997:123-76.
71. Nucleic acid amplification testing of blood donors for transfusion-transmitted infectious disease. Report of the Interorganizational Task Force on Nucleic Acid Amplification Testing of Blood Donors. Transfusion 2000;40:143-59.

ROUTINE EXPERIENCE AND FUTURE DEVELOPMENTS OF VIRUS NAT APPLICATION

J. Coste[1], C.Defer[2], C. Saura[3]

Introduction

In February 1999, the French Health Authorities required progressive introduction of Nucleic Acid Testing (NAT) in blood screening, so as to become general in all Blood Transfusion Centres (BTC) in the year 2000. Because greatest concern revolves around HCV, priority is given to HCV-NAT, with extension to other viruses according to advances in the technology. This decision was taken after the results of different reports realised by the experts of the Agence Française du Sang and by the Ministry of Health, including those of a feasibility study concerning routine screening of blood donations for HCV RNA, which was conducted between February and August 1998 in two BTC (Lille and Montpellier). The residual risk due to window period donations has been estimated by A.M. Courouce, J. Pillonel and the members of the working groups of the French Society of Blood Transfusion [1]. For the overall blood supply the residual risk is estimated at 2.6 HCV infectious donations per 1 million, at 5.8 per 1 million donations for HBV and at 0.6 per 1 million donations for HIV. On the basis of 2.7 million blood donations collected each year, the estimated number of undetected window period donations is 7 HCV, 16 HBV and 2 HIV, which makes a total of 25 [2]. Considering that NAT could reduce at by 59 days the HCV window period, by 25 days the HBV and by 11 the HIV window period (Table 1), NAT implementation would reduce the residual risk by 89 %, 45 % and 50 % respectively for HCV, HBV and HIV. In that way, NAT is projected to detect an additional 14 infected blood donations in France. When taking into account the natural history of each virus and the fact that in France, one blood donation is involved in the production of 1.2 labile blood components, the benefit of NAT in terms of avoided chronic infections and severe diseases in recipients has been estimated to be very low: NAT would prevent each year 7 HCV infections, among which 6 chronic hepatitis and 1 severe disease; 8 HBV infections from which less than 1 will become chronic, and 1 severe HIV disease (Table 2). In conclusion, the experts estimated that the

1. Ets de Transfusion Sanguine de Languedoc-Roussillon, F.
2. Ets de Transfusion Sanguine de Nord Pas-de-Calais, F.
3. Agence Française du Sang, F.

Table 1. Estimated residual risks associated with window period donations by seroconverting donors* during the 1995-1997 period (Courouce AM, Pillonel J, SFTS).

	Window period (days)	Residual risk per million donations	Nb of window period donation per year**
HCV	**66** (38 – 94)	**2.6** (1.0 – 4.9)	**7** (3 – 13)
HBV	**56** (25 – 109)	**5.8** (2.1 – 13.7)	**16** (6 – 37)
HIV	**22** (6 – 38)	**0.6** (0.1 – 1.5)	**2** (0 – 4)
Total			**25**

* Adapted from Schreiber et al.
** Estimation on 2.7 millions BD per year.

Table 2. Projected benefit of adding NAT to blood donor screening (Agence Française du Sang data source).

	Estimated reduction of window period (days)	Seronegative PCR positive detected by NAT (nb per year)	Infections severe diseases avoided (per year)
HCV	**-59** (-89%)	**6**	**7 infections** – 6 chronic – 1 severe disease
HBV	**-25** (-45%)	**7**	**8 infections** – < 1 chronic
HIV	**-11** (-50%)	**1**	**1 severe disease**
Total		**14**	**7 infections** – 6 chronic – 1 severe disease

benefit of NAT implementation in blood screening is low, in terms of avoided chronic and severe diseases.

Feasibility Study of Real-Time NAT Screening of Blood Donations

In this context, it was important to check that routine NAT screening for HCV RNA was feasible. The aim of the study was first to assess the feasibility of a methodology for real-time screening of blood donations in parallel with the serology testing, second to provide results before release of all blood components including platelets, and third to display the technical and logistical impediments encountered, from the blood donation, to the release of the blood components. The various gene amplification technologies have three steps in common: sample preparation or nucleic acid extraction, amplification of the target and detection of the amplified product. At present, no front-end automation is yet available able to realise these three steps. However, at the end of 1997,when the study was developed, the following equipment was available in France: the Qiagen BioRobot 9604 for the sample preparation and the Roche Cobas HCV Amplicor (v.2.0) for HCV RNA amplification and detection. The Cobas automate is designed to process 24 samples in 4 hours and the BioRobot can prepare 96 samples in less than two hours. Because of the low throughput of this system, single unit testing could not be achieved and a strategy involving testing of pooled samples had to be proposed. This was performed using the Tecan Genesis. A software called "Pool Management Software" (PMS) from Organon Teknika, was linked to the Genesis handling the complete tracking of samples in pools.

Study Design

We choose a pool-size of 48 samples, because it was a good compromise between feasibility and sensitivity.

The automate was set up to take 850 microlitres from the primary tube, to dispense 800 microlitres into a well of an archive plate, and the remaining 50µl in the pool tube. This plate was stored in the refrigerator and was used only when a pool was found positive and launched secondary cross-pool testing in order to identify the individual positive sample.

The workflow was the following: the first day, an additional vacutainer (5 ml without anticoagulant) was collected from each donor. The tubes were kept at 2-8°C within 6 hours after puncture, and then travelled in isothermal containers at the same temperature to the NAT laboratory. The latest arrival time of the tubes was day 2 at 6 am. At day 2, 7 am, the tubes were centrifuged, pools were made with help of the Tecan automate. Then, HCV RNA extraction was performed with the QIAamp Viral RNA 96 test reagent on the BioRobot. At last, amplification and detection was realised with the Cobas HCV Amplicor 2.0 version (Table 3).

The results obtained from the Cobas were imported into the Pool Management Software. Two situations could occur: first case, when the pool was negative, PMS would assign a negative result to the 48 individual members of the pool; in the second case, when the pool was positive, PMS would handle the setting up of minipools and also identify the individual donation after the testing of the minipools and the import of the results into the PMS software. The individual positive or negative results were exported to a laboratory software designed by Ortho Clinical Diagnostics, which takes into account the results of the serology tests. At last, final results were transmitted to the central computer and the donor records were filled-in: in case of a positive result, the blood components were discarded, and when the donor was HCV RNA negative, the products were released. The detection limit of the BioRobot combined with Cobas HCV Amplicor 2.0 was estimated with help of reference panels. For single testing the detection limit was 80 copies/ml with a hit rate of 96.6 %, and 4000 copies/ml (96.4 % hit rate) when a positive sample was diluted in a pool of 48 negative samples.

Table 3. Study design. Workflow.

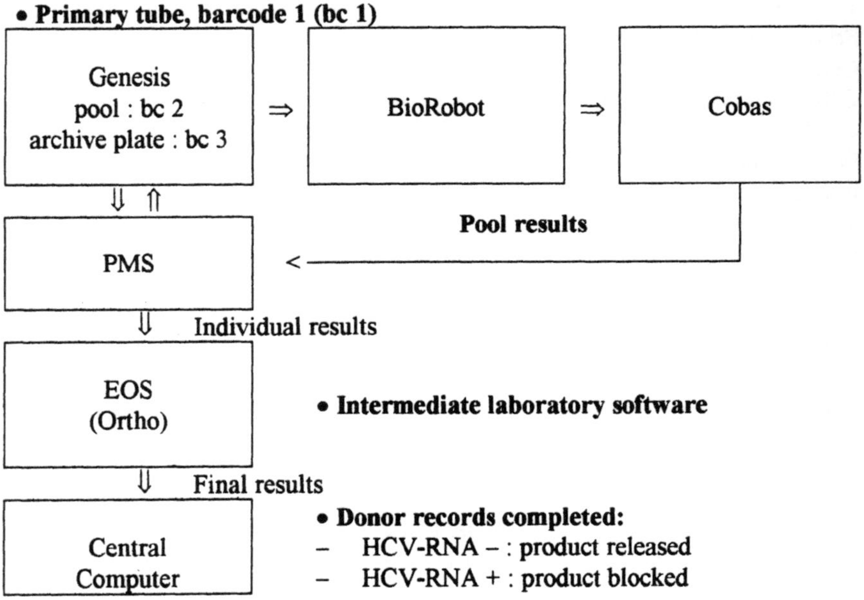

The real-time study was conducted in 2 BTC of different size and epidemiology: the Languedoc-Roussillon region is 2 times greater in surface than is the Nord Pas-de-Calais region (27 000 versus 12 000 km^2), but the number of inhabitants per km^2 is ca 4 times less (80 per km^2 versus 300 per km^2 in Nord Pas-de-Calais). In the same way, the number of blood donations collected per year are 2 times less in this region than in Nord Pas-de-Calais (ca 100 000 versus 200 000).But to the opposite, the anti-HCV prevalence is 6 times higher in the Languedoc-Roussillon region, 0.12 % versus 0.02 % for the Nord Pas de Calais region.

Results

NAT testing was performed in parallel with the serology screening without sorting out the anti-HCV positive samples prior to testing.

A total of 60 265 blood donations in 1643 pools of 48 were analysed in both BTC. Day 2, at 3 PM, 96 % of the results were released whereas 4 % were blocked. The released results corresponded to the negative pools. Further analysis of the blocked pools (which were released day 3 at 8 AM), showed that 28 (1.7 %) were PCR positive and 36 (2.2 %) were PCR negative, but the results was invalidated because of a negative internal control. Cross-pool testing of the 28 PCR positive pools, identified 14 (0.85 %) HCV RNA positive individual blood donations, whereas the 14 (0.85 %) others were false positive reactions. The 14 HCV RNA positive blood donations issued from first-time anti-HCV antibody positive blood donors. When the 36 invalidated pools were tested again, all were again PCR negative and the results were this time validated because the IC was positive. Further investigations of these results demonstrated that in 26 (72 %) of the cases, these results were due to a failure of the BioRobot (low extraction yield), and in 10 (28 %) to other causes such as cross contamination due to the manual steps or to a failure of the Cobas. At last, 6 HCV RNA negative pools contained an anti-HCV antibody positive sample. When the 6 anti-HCV positive samples were tested by PCR, 5 (0.3 %) were HCV RNA negative, but 1 (0.06 %) was a weak positive (< 1000 copies/ml) genotype 3a blood donation undetectable in pools of 48 samples.

The technicians received a one month theoretical and technical training course prior to NAT real-time screening. The technical training involved running the automates and the reagents, to become aware of the lack of total traceability due to many manual steps and to get sensitised to cross-contamination risks. The opening hours of both facilities were from 7 AM to 7 PM, 7 days a week for Lille BTC and 6 days a week for Montpellier BTC.

Daily pool screening of 500 to 1000 BD needed 4 technicians in Lille and 2.5 in Montpellier for 300-500 BD. Indeed, in case of a positive pool, cross-pool testing was carried on in the afternoon. Both facilities comprised 3 sectors: sector I for the sorting out and centrifugation of the tubes; sector II for the pre-amplification tasks which contained one area for the pooling, a second for the extraction and a third for the setting up of the A-rings. Amplification and detection were performed in sector III.

Conclusions

Real-time HCV RNA screening of blood donations in pools of 48 appears to be feasible: blood components including platelets could be released without delay.

However, the experts concluded that this first generation technology was not robust enough to be generalised to all the BTC and needed improvements. In particular, cross-contamination and low extraction yield were displayed for the BioRobot, and detection failures were suspected with the Cobas. At last, to many manual steps prevent complete tracking of the samples.

Future Developments in Viral NAT

Improved technology must be evaluated: other equipment designed for the extraction such as the Nuclisens Extractor of Organon as well as the upgraded version of the QIAgen BioRobot are available. At present these two automates can be linked to the Cobas. New tools for blood screening must also be evaluated such as the Chiron HIV/HCV TMA multiplex assay. This Chiron assay was developed for Blood Transfusion Centres and is able to detect the 2 analytes with a single test and with no liquid transphase. New feasibility studies with these improved technologies will be carried out by the viral NAT expert group set up in June 1999 by the Agence Française du Sang. The task will be to select the most suitable methods for the French Blood Transfusion Network, which will be progressively introduced in screening of blood donations current the year 2000.

References

1. Pillonel J, Saura C, Courouce AM. Screening of viral markers for HIV, HBV and HIV infections in blood donors in France and residual risk of viral transmission by blood transfusion. Eurosurveillance 1998;3:76-79.
2. Saura C, Courouce AM, Pillonel J, et al. Risk Benefit Analysis of Nucleic Acid testing: the French experience. Biologicals 1999;27:343-48.

DEVELOPMENT OF A NEW DIAGNOSTIC METHOD FOR THE DETECTION OF HCV ANTIBODIES: THE RAPID LATERAL FLOW TEST

H. A. Fields[1], Y.E. Khudyakov[1], E. Lopareva[1], T. Ulanova[2], G. Talekar[3], Y. Devash[3]

Introduction

After the introduction of serologic assays for the detection of hepatitis A and hepatitis B markers of infection, it became apparent that many hepatitis patients remained seronegative for both infections. Infections of this type were called non-A, non-B hepatitis (NANB) and occurred most frequently following a blood transfusion [1]. The agent was identified by Choo and coworkers in 1989 [2] following a series of sophisticated molecular biological experiments that led to the first isolation and cloning of an RNA virus from acute-phase sera obtained from a chimpanzee experimentally infected with NANB containing material. This new RNA virus was called the hepatitis C virus (HCV).

The HCV genome is a positive sense RNA molecule of approximately 9400 nucleotides containing a 3' poly-A tail, a 5' untranslated region of 341 bases, and a long open reading frame encoding for a polyprotein of more than 3000 amino acids followed by a 3'-untranslated region of greater than 120 bases [2,3]. This genomic organization is most similar to that of other members of the *flaviviridae* family with three N-terminal structural proteins (C, E1, E2/NS1) and four C-terminal nonstructural protein (NS2, NS3, NS4, NS5) essential for viral replication.

Many HCV isolates from different parts of the world have now been sequenced and compared [4–8], which has led to the identification of distinct types that may differ from each other by as much as 33% over the entire genome [9]. A uniform classification was agreed upon in August 1994 at the Second International Conference of HCV and Related Viruses [10]. In this new classification, HCV genotypes are placed into at least 6 major genotypes with each genotype containing multiple subtypes based upon sequence homology. Sequence heterogeneity among various HCV genotypes have been associated with variant antigenic and biological properties [11]. This is especially significant as it relates to various serologic and virologic assays for HCV detection and their interpretation [12].

---

1. Hepatitis Branch, Div. of Viral and Rickettsial Diseases, Centers for Disease Control and Prevention, Atlanta, GA, USA
2. Diagnostic System, Nizhny Novgorod, Russia
3. Devaron Inc., Dayton, NJ, USA

Cloning of the HCV genome led to the identification of the first diagnostically relevant protein, designated 5-1-1 from the NS4 gene. A recombinant protein called c100-3 containing the 5-1-1 region was expressed in yeast and used to develop the first generation screening assay licensed by the US Food and Drug Administration (FDA) in 1990. This first generation enzyme immunoassay (EIA) was reactive in 80 - 90% of serum specimens obtained from patients with chronic post transfusion NANB hepatitis, but not in those from patients with acute or early infections [13]. Furthermore, screening in a low-risk population resulted in a high rate of false-positive results, which led to the introduction of a strip immunoassay (SIA), also called recombinant immunoblot assay (RIBA; Chiron Corporation, Emeryville, CA). Because of their intrinsic high specificity compared to EIA, SIA or RIBA are used as confirmatory tests for those sera yielding EIA positive results.

A second generation EIA (EIA-2) and SIA (SIA-2 or RIBA 2.0 were approved by the US FDA in 1992 and 1993, respectively. These assays incorporated the c22-3 antigen derived from the core gene and the c33c antigen derived form the NS3 gene. Although these assays are highly sensitive for the detection of HCV antibodies (anti-HCV) and have almost eliminated post transfusion hepatitis, caution must be exercised in interpreting negative results in certain groups of patients including those with recent infection since seroconversion may take as long as 12 weeks following HCV exposure [14]. SIA-2 is widely used for the validation of EIA-2 in the serologic diagnosis of HCV infection. The current criteria used for validation by SIA-2 is immunoreactivity to at least two different HCV antigens. A common problem is indeterminate results, which is defined as reactivity to only a single antigen or reactivity to two antigens derived from the same gene such as 5-1-1 and c100-3. Five to 10% of samples that were repeatedly reactive by EIA-2 yielded indeterminate results by SIA-2 and some of these patients with indeterminate results had detectable HCV RNA by the polymerase chain reaction (PCR), a common method to confirm HCV viremia [15].

Third generation assays (EIA-3 and SIA-3 or RIBA 3.0) became commercially available in 1996. In these assays, a recombinant NS5 full length antigen was added, while c33c and c100-3 was combined into a new antigen designated c200. These assays demonstrated higher sensitivities and specificities than second-generation assays [16]; however, these assays still have similar shortcomings to those of EIA-2 including significant rates of indeterminate results [17].

All of the commercially available screening assays are enzyme immunoassays and require the use of a photometer to quantify the results over a statistically established cutoff optical density value. Although ideally suited for the industrialized nations of the world and for larger centers in developing countries, these assays are less than ideal for resource limited regions of the world. The characteristics of an appropriate format for such regions include instrument independence, rapidity with test results completed within 20 minutes, technical forgiveness, stability where there may not be an adequate cold chain, and cost-effectiveness. One such format is the lateral flow technology (LF), which relies on the use of lyophilized immuno-gold labeled protein A as the detector in front of the sample application region. The sample flows laterally across the membrane toward a wicking pad located at the opposite end of the test. The detector is resuspended by the sample, which then binds to IgG in the sample. This complex continues to flow laterally where it comes into contact with the antigenic target(s). The red pattern, which is visually observed, determines the interpretation of the test. The present study demonstrates the utility of this format and introduces several new antigens as antigenic targets resulting in a rapid test that combines a screening test and a confirmation test into an affordable stable package ideally suited for resource limited regions of the world.

Materials and Methods

Synthetic oligodeoxyribonucleotides

Oligodeoxyribonucleotides were synthesized with an automatic synthesizer (model 380A; Applied Biosystems, Foster City, CA) and purified by either high-performance liquid chromatography or electrophoresis in a 10% polyacrylamide gel containing 7 M urea in TBE buffer (0.045 M Tris-borate, 0.001 M EDTA [pH8.3]). Oligodeoxynucleotides were recovered from the gel by electroelution using a model 230A HPEC system (Applied Biosystems) according to the manufacturer's protocol.

Recombinant antigens

Five recombinant antigens were prepared as antigenic targets for the lateral flow rapid test. The core antigen was derived from a PCR fragment of approximately 270 nucleotides from the 5'-end of the gene. The technique of PCR walking was used for both the NS3 and NS5a genes. A total of 23 overlapping fragments of approximately 100 nucleotides each were prepared from the NS3 gene and 17 overlapping fragments of the same size were prepared from the NS5a gene (details of the polymerase chain reaction are described elsewhere). The core, NS3 and NS5a fragments were derived from genotype 1b sequences. For the NS4 region, an artificial gene encoding a mosaic NS4 antigen was assembled incorporating 11 divergent sequences from the 5-1-1 region and 6 divergent sequences from the C-terminus of NS4 (designated as region 59). Sequences were chosen based upon each fragment having significant sequence divergence from each other and representing all 6 known genotypes [18]. Finally, a combined chimeric gene, designated as T-clone, was assembled using the most diagnostically relevant fragments from each gene (Figure 1).

Protein expression and purification

Each PCR fragment was cloned with the plasmid pGEX-2T (Pharmacia Biotech Inc., Piscataway, NJ) and expressed as a fusion protein with glutathione S-transferase (GST). *E. coli* JM109 competent cells (Invitrogen Co., San Diego, CA) were transformed with plasmids containing various HCV-specific inserts and the cells were grown in LB medium containing 100 ug of ampicillin per ml until the optical density at 600 nm was equal to 0.6, after which the promoter controlling the expression of the fusion proteins was activated by the addition of isopropyl-\$-D-thiogalactopyranoside at a final concentration of 1 mM. After 4 to 6 h of growth at 37°C, the cells were harvested and a lysate was prepared [19].

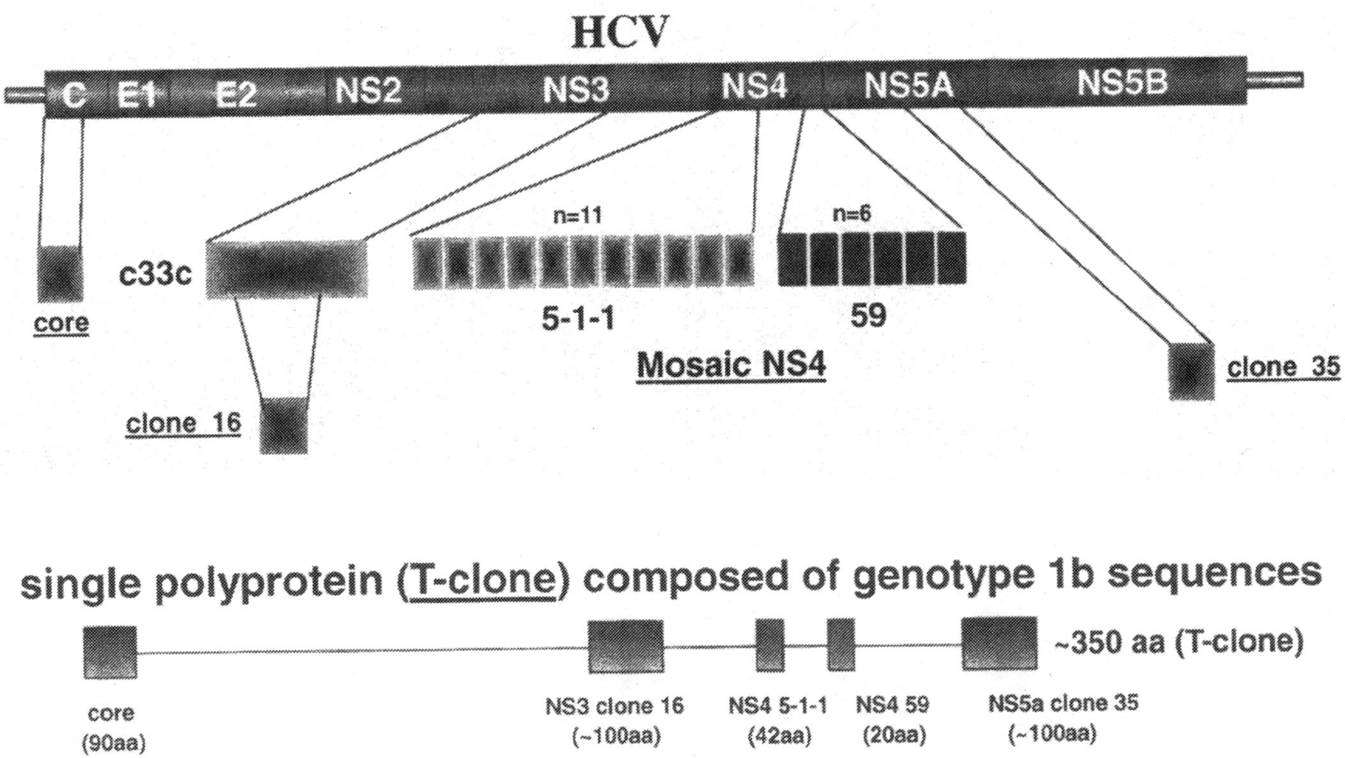

Figure 1. New HCV-specific antigens used in the lateral flow rapid test. The underlined words indicate new recombinant antigens.

The GST-fusion proteins were purified by ligand-affinity chromatography [20] using glutathione-Sepharose 4B columns (Pharmacia Biotech).

Western Blot

Aliquots of each lysate were analyzed by Western blotting [21]. Nitrocellulose membranes containing immobilized-proteins were incubated at 20°C for 2 hr with different anti-HCV positive sera diluted 100 times in 50 mM Tris-HCl (pH 7.5) containing 0.5% Triton X-100, 1% gelatin, and 1% BSA (NET). The membranes were washed with NET three times and then incubated for 1 h with affinity-purified anti-human immunoglobulin G (IgG) coupled to horseradish peroxidase (Boehringer Mannheim, Indianapolis, IN) diluted 1:5,000 in NET. After the membranes were washed, diaminobenzidine (Sigma, St. Louis, MO) and hydrogen peroxide were used to develop the reaction.

Enzyme immunoassay (EIA) for anti-HCV

Ligand-affinity purified antigens (110 ul) diluted to the optimal concentration in 0.1 M phosphate-buffered saline (pH 7.5) was adsorbed to microtiter wells (Immunolon II; Dynatech Laboratories, Inc.) at room temperature for 12 h. Human anti-HCV positive and negative sera, previously determined by EIA-3 (Ortho, Raritan, NJ), were diluted 1:100 in 0.1 M phosphate-buffered saline (pH 7.5) containing 0.1% Tween 20 and 10% normal goat sera. Binding of antibodies to the recombinant proteins adsorbed to the surface of the wells of microtiter plates were identified with optimally diluted affinity-purified anti-human IgG coupled to horseradish peroxidase (Boehringer Mannheim). A statistically derived cutoff was determined and shown to be equal to an OD value greater than 0.10 and a P/N value greater than 5.0. This value was equivalent to an optical density value greater than the mean of negative controls plus 10 standard deviations of the mean.

Human sera

Sera positive, indeterminate, and negative for anti-HCV were obtained from those reposited at the Centers for Disease Control and Prevention, Atlanta, GA. All specimens were tested and confirmed by EIA-3 and RIBA 3.0. Indeterminate specimens were tested for the presence of HCV RNA by PCR (HCV Amplicor, Roche Molecular Systems, Alameda, CA).

Lateral flow rapid test for anti-HCV

A lateral flow (LF) rapid test for anti-HCV (Devaron Inc., Dayton, NJ) was assembled using the above antigens (Figure 2). Gold labeled protein A was used as the detector, which was lyophilized onto a membrane in front of the sample application region. The sample (50 ul) flows laterally across the membrane toward a wicking pad located at the opposite end of the device. The detector is re-suspended by the sample, which then binds to IgG in the sample. This complex continues to flow laterally where it comes into contact with the various antigenic targets. The red pattern, which is visually observed after 20 minutes, determines the interpretation of the test. Reactivity to more than one antigen was

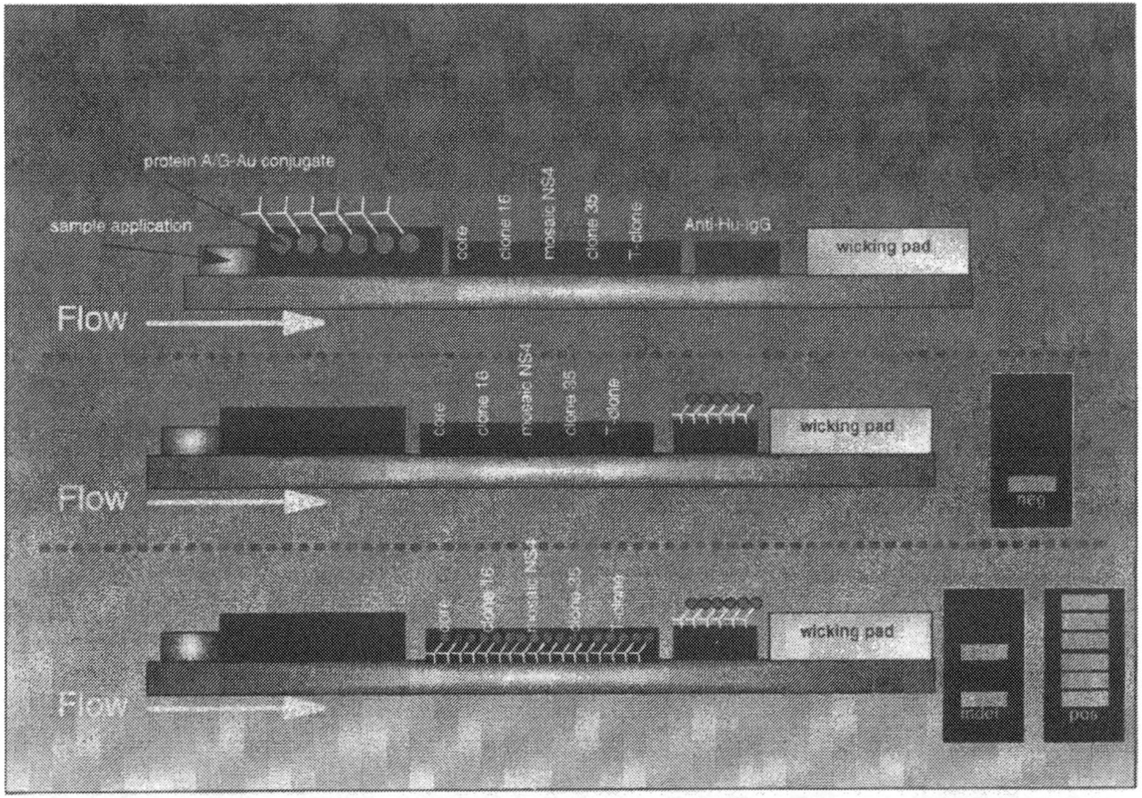

Figure 2. Lateral flow rapid test for the detection and confirmation of anti-HCV activity in human sera.

considered positive, while reactivity to a single antigen was considered indeterminate.

Results

Selection of diagnostically relevant clones

Each PCR fragment obtained by the PCR walking technique covering the NS3 and NS5a genes were cloned, expressed as GST-fusion proteins, and the breadth of immunoreactivity determined by Western blot analysis and EIA. Clone 16 from the NS3 gene and clone 35 from the NS5a gene demonstrated broad immunoreactivity to anti-HCV positive sera of greater than 90% and 80%, respectively. These clones were chosen for inclusion into the preparation of T-clone, all of which were used in the development of the LF rapid test (Figures 1 and 2).

Immunoreactivity of T-clone

To assess whether each antigenic region designed into T-clone remained accessible to bind to specific antibodies, T-clone was used as the antigenic target in an EIA format and tested against a panel of indeterminate sera. Table 1 demonstrates that each of the antigenic regions reacted with most, but not all, RIBA 3.0 indeterminate sera. All of the indeterminate sera positive to the RIBA 3.0 core antigen also reacted with T-clone, while only 67% of sera positive to either the 5-1-1 antigen of the NS5 antigen reacted with T-clone. Initially, only 43% of indeterminate sera to the c33c protein reacted with T-clone. However, when retested against a panel of indeterminate c33c positive sera derived from the acute-phase from whom follow-up specimens revealed complete seroconversion, 82% immunoreacted.

Table 1. Antigenic reactivity of T-clone with RIBA 3.0 indeterminate specimens.

RIBA antigen	Percent indeterminate
c22P	100%
c33c	43% (82%*)
5-1-1	67%
NS5	67%

* Acute-phase sera from whom follow-up sera confirm HCV infection.

Performance of LF rapid test

The LF test was tested against a panel of human sera positive, indeterminate, and negative for anti-HCV activity. Figures 3a and 3b compare the results of RIBA 3.0 with the LF test using a panel of HCV positive sera (n=40), while figures 4a and 4b compare the results of EIA-3 with the LF test using a panel of HCV negative sera (n=40). Among the 40 HCV positive sera, 26 tested positive

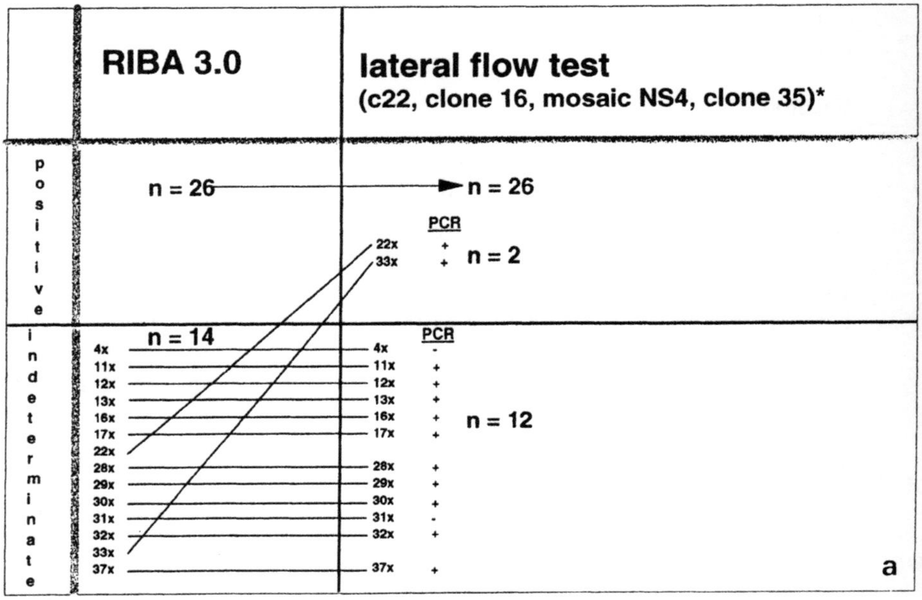

* analysis without regard to reactivity of T-clone

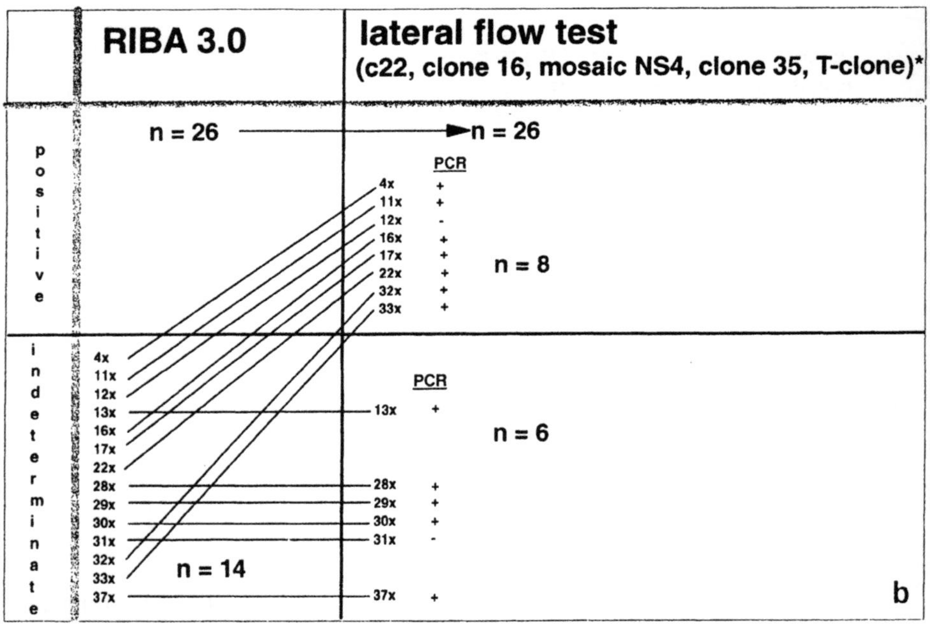

* analysis including reactivity of T-clone

Figures 3a and 3b. A comparison of RIBA 3.0 results with the lateral flow test with and without regard to the immunoreactivity of T-clone using a panel of 40 anti-HCV positive sera.

and 14 tested indeterminate by RIBA 3.0. All 26 positive sera also tested positive by the LF test. Among the 14 indeterminate sera, 12 tested indeterminate and 2 tested as positive by the LF test. This analysis was made without regard to the immunoreactivity of T-clone (Figure 3a). The two sera that tested positive by the LF test were confirmed as positive by PCR. When T-clone was included in the analysis (Figure 3b), a total of 8 indeterminate sera among the 14 RIBA 3.0 positive tested positive by the LF test, all of which were confirmed as positive by PCR except for one specimen. Figures 4a and 4b compare EIA-3 to the LF test using a panel of HCV negative sera with and without regard to T-clone reactivity, respectively. Among the 40 EIA-3 negative sera, 36 tested negative by the LF test. However, the LF test identified two specimens as indeterminate and two specimens as positive, all of which were only 1+ reactive against different antigens in the LF test. When T-clone was included in the analysis (Figure 4b), an additional specimen was identified as a weak indeterminate.

Table 2. Comparison of antigens used in various tests for anti-HCV activity.

Gene	EIA-3	RIBA 3.0	Lateral flow
NC	c22 (119aa)	c22P (44aa)	core (1b;92aa)
NS3	c200 (740aa)	c33c (266aa)	clone 16 (1b;100aa)
NS4	c200 (740aa)	5-1-1 (42aa)	mosaic NS4 (~350aa)
NS5	NS5a (942aa)	NS5a (942aa)	clone 35 (1b:102aa)
Other			T-clone

Discussion

A rapid lateral flow test for the detection of anti-HCV activity in sera has been developed using several different antigens than those used in EIA-3 and RIBA 3.0 as shown in Table 2. While all of the antigens used in EIA-3 and RIBA 3.0 are derived from genotype 1a, most of the antigens used in the LF test are derived from genotype 1b. The mosaic NS4 antigen, however, is very different from any other antigen. This antigen is composed of 11 5-1-1 and 6 region 59 divergent sequences representing all six genotypes. This antigen has been shown to detect anti-NS4 activity from patients infected with HCV genotypes 1 through 5, even in sera that tested negative for anti-NS4 activity by an FDA licensed assay [18]. Other investigators have shown that sera from individuals infected with non-type 1 genotypes react infrequently with recombinant proteins derived from 5-1-1 and c100-3 clones [22, 23]. In another report, investigators demonstrated that differences in serological reactivity between genotypes do exist and that a large proportion of epitopes of the type 1a or 1b recombinant proteins used in current assays are genotype-specific. This variation in sensitivity of screening assays for different genotypes is of potential concern when used in countries where non-type 1 genotypes predominate in the blood donor or patient population [24, 25]. The mosaic NS4 antigen represents a unique recom-

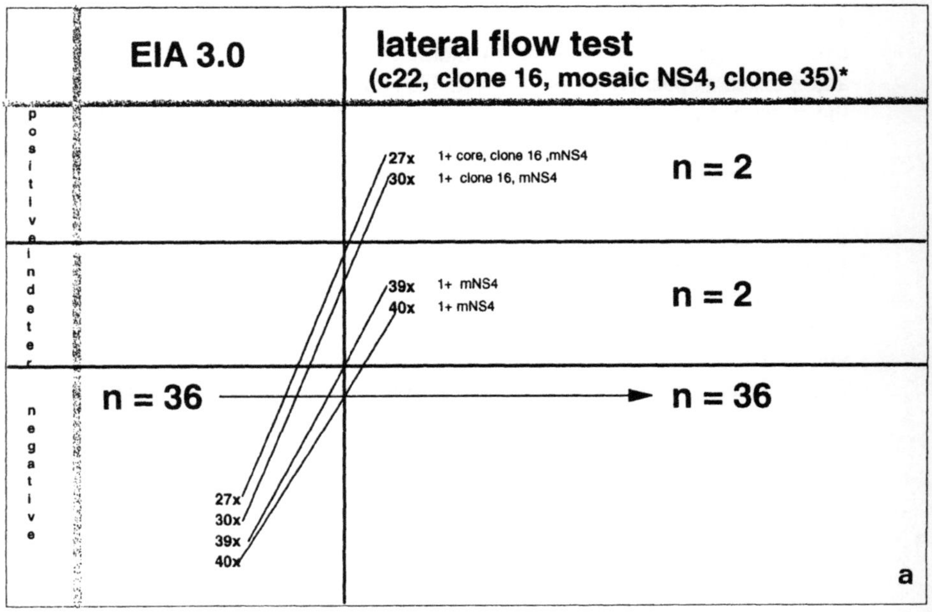

* analysis without regard to reactivity of T-clone

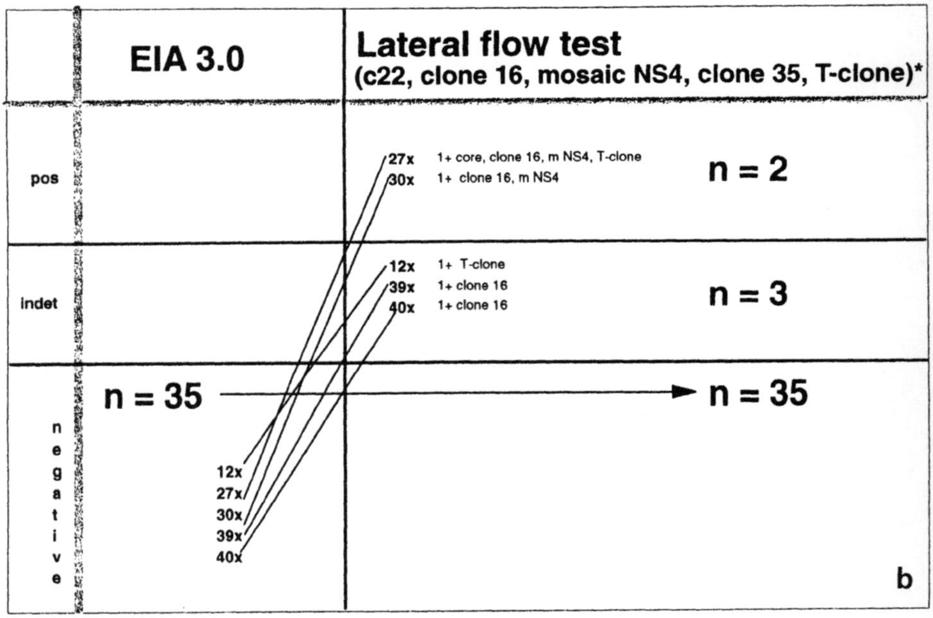

* analysis including reactivity of T-clone

Figures 4a and 4b. A comparison of EIA-3 results with the lateral flow test with and without regard to the immunoreactivity of T-clone using a panel of 40 anti-HCV negative sera.

binant protein designed to successful overcome these limitations imposed by utilizing antigens from only a single genotype. This increased immunoreactivity to NS4 epitopes, in addition to genotype 1b sequences representing core, clone 16 from NS3, and clone 35 from NS5a, may have collectively contributed to the increased sensitivity of the LF test compared to RIBA 3.0. Even though many of these same epitopes are also represented in T-clone, these epitopes may be displayed in a somewhat different conformation because of its chimeric nature allowing for antibodies of slightly different specificities to react with it. Nonetheless, there is little doubt that the LF test utilizing the antigens shown in Figure 1 and Table 2 demonstrate a significant increase in sensitivity as evidenced by converting many indeterminate results to positive results, all of which except for a single specimen were confirmed as HCV positive by PCR. This increased sensitivity of the LF test, however, appears to have come at the expense of specificity (Figures 4a and 4b) as evidenced by detecting up to 5 EIA-3 negative specimens as either indeterminate or positive. All five specimens tested 1+ positive and it is likely that these sera may retest negative after adjusting the amount of antigen on the membrane or the concentration of protein A labeled gold.

By testing T-clone against a panel of RIBA 3.0 indeterminate sera by EIA, all of the epitopes designed into this chimeric protein remain accessible to interact with antibody. However, not all RIBA 3.0 indeterminate specimens reacted with T-clone. There may be several reasons for this observation. First, some RIBA 3.0 antigens are larger than the corresponding antigens in T-clone and, therefore, may have more antigenic sites. Such sera may be true positive specimens obtained from individuals either early in the infection during the process of seroconversion, or late in the infection during the process of seroconversion which may occur in up to 10% - 20% of patients. Second, RIBA 3.0 antigens and T-clone are derived from different HCV genotypes and subtypes and many of the antigens have significantly different sequences. Sera that react with RIBA 3.0 antigen and not with T-clone may also be true positive sera, but their immunoreactivity may be genotype- or subtype-dependent. Finally, some RIBA 3.0 antigens, because of their larger size, may react with some sera nonspecifically and may actually be true negative sera for HCV. Although PCR would help resolve indeterminate specimens collected in the acute-phase during the process of seroreversion since such specimens should always be PCR positive, PCR would not, however, be very useful in discriminating between specimens collected during the process of seroreversion and specimens that are nonspecifically indeterminate. In both cases, PCR should remain be negative.

The LF test presented in this paper represents a highly sensitive and rapid screening and confirmation test combined into a single cost-effective format that is extremely simple to use. This test, as well as other rapid test formats, is being considered by a newly formed organization called the International Consortium for Blood Safety (ICBS), as the appropriate technology in resource limited regions of the world. As part of its first demonstration project, ICBS will evaluate various test formats and purchase several hundred thousand tests through a bid and tender process and distribute the tests in a developing country. At the same time, ICBS will be involved in technology transfer for local manufacturer so that test will remain available at a reasonable cost into the foreseeable future.

Finally, ICBS plans to assist various National Control Authorities by providing a fully characterized panel of HCV positive and negative specimens to be used to license domestically produced tests. ICBS will continue to develop additional strategies to render the worldwide blood supply free of blood borne pathogenic viruses.

References

1. Feinstone SM, Kapikian AZ, Purcell RH, Alter HJ, Holland PV. Transfusion-associated hepatitis not due to viral hepatitis type A or B. N Engl J Med 1975;292: 767-70.
2. Choo QL, Kuo G, Weiner AJ, Overby LR, Bradley DW, Houghton M. Isolation of a cDNA clone derived from a blood-borne non-A, non-B viral hepatitis genome. Science 1989;244: 359-62.
3. Houghton M, Weiner AJ, Han J, Kuo G, Choo QL. Molecular biology of the hepatitis C viruses; implications for diagnosis, development and control of viral disease. Hepatology 1991;14:381-88.
4. Delisse AM, Descurieux M, Rutgers T, et al. Sequence analysis of the putative structural genes of hepatitis C virus from Japanese and European origin. J Hepatol 1991; 13(Suppl 4);S20-S23.
5. Li JS, Tong SP, Vitvitski L, Sepot D, Trepo C. Evidence of two major genotypes of hepatitis C virus in France and close relatedness of the predominant one with the prototype virus. J Hepatol 1991;13(Suppl 4):S33-S37.
6. Chen PJ, Lin MH, Tai KF, Liu PC, Lin CJ, Chen DS. The taiwanese hepatitis C virus genome: sequence determination and mapping the 5' termini of viral genomic and antigenomic RNA. Virology 1992;188:102-13.
7. Enomoto N, Takada A, Nakao T, Date T. There are two major types of hepatitis C virus in Japan. Biochem Biophys Res Commun 1990;170:1021-25.
8. Kato N, Ootsuyama Y, Ohkoshi S, et al. Distribution of plural HCV types in Japan. Biochem Biophys Res Commun 1991;181:279-85.
9. Okamoto H, Kurai K, Okada S, et al. Full-length sequence of a hepatitis C virus genome having poor homology to reported isolates: comparative study of four distinct genotypes. Virology 1992;188:331-41.
10. Simmonds P, Alberti A, Alter HJ, et al. A proposed system for the nomenclature of hepatitis C viral genotypes (letter). Hepatology 1994;19:1321-24.
11. Zein NN, Persing DH. Hepatitis C genotypes: current trends and future implications. Mayo Clin Proc 1996;71:458-62.
12. Zein NN, Rakela J, Persing DH. Genotype-dependent serologic reactivities in patients infected with hepatitis C virus in the United States. Mayo Clin Proc 1995; 70:449-52.
13. Alter HJ, Purcell RH, Shih JW, et al. Detection of antibody to hepatitis C virus in prospectively followed transfusion recipients with acute and chronic non-A, non-B hepatitis. N Engl J Med 1989;321:1494-500.
14. de Medina M, Schiff ER, Hepatitis C: diagnostic assays. Semin Liver Dis 1995;15: 33-40.
15. Gretch D, Lee W, Corey L. Use of aminotransferase, hepatitis C antibody, and hepatitis C polymerase chain reaction RNA assays to establish the diagnosis of hepatitis C virus infection in a diagnostic virology laboratory. J Clin Microbiol 1992;30:2145-49.
16. Kao JH, Lai MY, Hwang YT, et al. Chronic hepatitis C without anti-hepatitis C antibodies by second-generation assay: a clinicopathologic study and demonstration of the usefullness of a third-generation assay. Dig Dis Sci 1996;41:161-65.

17. Pawlotsky JM, Bastie A, Pellet C, et al. Significance of indeterminate third-generation hepatitis C virus recombinant immunublot assay. J Clin Microbiol 1996; 34:80-83.
18. Chang JC, Ruedinger B, Cong M-E, et al. Artificial NS4 mosaic antigen of hepatitis C virus. J Med Virol 1999;59:437-50.
19. Sambrook J, Fritsch F, Maniatis T, eds. Molecular cloning: a laboratory manual. Cold Spring Harbor Laboratory, Cold Spring Harbor, NY. 1989, vol. 3:18.40-18.41.
20. Smith DB, Johnson KS. Single-step purification of polypeptides expressed in Escherichia coli as fusions with glutathione S-transferase. Gene 1988;67:37-40.
21. Harlow E, Lane D, eds. Antibodies. A laboratory manual. Cold Spring Harbor Laboratory, Cold Spring Harbor, NY. 1988:471-510.
22. McOmish F, Chan SW, Dow BC, et al. Detection of three types of hepatitis C virus in blood donors: investigation of type-specific differences in serologic reactivity and rate of alanine aminotransferase abnormalities. Transfusion. 1993;33:7-13.
23. Chan SW, Simmonds P, McOmish F, et al. Serological responses to infection with three different types of hepatitis C virus [letter]. Lancet 1991;338:1391.
24. Dhaliwal SK, Prescott LE, Dow BC, et al. Influence of viraemia and genotype upon serological reactivity in screening assays for antibody to hepatitis C virus. J Med Virol 1996;48:184-90.
25. Zein NN, Germer JJ, Wendct NK, et al. Indeterminate results of the second-generation hepatitis C virus (HCV) recombinant immunoblot assay: significance of high-level c22-3 reactivity and influence of HCV genotypes. J Clin Microbiol 1997; 35:311-12.

THE RELEVANCE OF PLATELET CD62P (P-SELECTIN) TESTING

A.D. Michelson[1]

Introduction

P-selectin (also known as CD62P and formerly known as GMP-140 and PADGEM protein) is a component of the α granule membrane of resting platelets that is only expressed on the platelet surface membrane during and after platelet degranulation and secretion [1,2]. P-selectin has therefore been regarded as a useful marker for platelet activation during processing, storage, and manipulation (filtration, washing) [3-6].

Because P-selectin mediates the adherence of degranulated platelets to leukocytes *in vitro* [7,8] and *in vivo* [9], it has been postulated that surface P-selectin-positive (degranulated) platelets may be rapidly cleared from the circulation by leukocytes [2,4,5,7,8]. In apparent contradiction to this postulate, other investigators have reported that degranulated platelets are no more rapidly cleared from the circulation than control platelets [10,11]. Methods have not previously been available to directly address this issue. However, in a recent study, we used novel three color whole blood flow cytometric methods for tracking of circulating degranulated platelets and measurement of their function *in vivo* [12]. We thereby demonstrated that circulating degranulated platelets rapidly lose surface P-selectin to the plasma pool, but continue to circulate and function *in vivo*. Thus, our study [12] reconciles the results of previous studies [4,5,10,11] by demonstrating that platelet surface P-selectin molecules, rather than degranulated platelets, are rapidly cleared.

Methods

Studies were performed in baboons (adult male *Papio anubis*), because these primates are hemostatically similar to man [13]. PKH2 (Zynaxis, Malvern, PA), a non-radioactive lipophilic fluorescent dye that binds to the lipid bilayer of the plasma membrane [14], was used as a platelet label. Three color whole blood flow cytometric analysis of peripheral blood samples was used to simultaneously 1) identify platelets with a monoclonal antibody directed against GPIIb-IIIa, 2) distinguish infused platelets by their PKH2 fluorescence, and 3) analyze

1. Center for Platelet Function Studies, Univ. Massachusetts Medical School, Worcester, MA, USA.

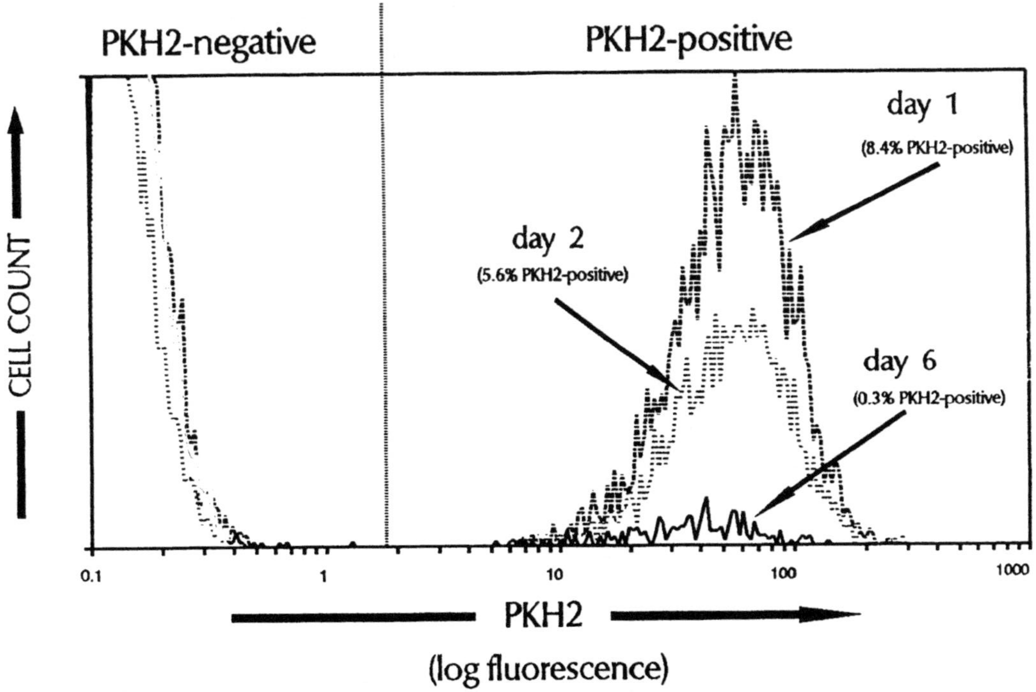

Figure 1. Ability of whole blood flow cytometry to distinguish infused, PKH2-labeled platelets *in vivo*. Autologous washed thrombin-activated PKH2-labeled platelets were transfused into a baboon. Peripheral blood was drawn from the baboon on days 1, 2, and 6. Platelets were identified in whole blood by light scatter and binding of the GPIIb-IIIa-specific monoclonal antibody 7E3. The proportion of all circulating platelets that were PKH2-positive is indicated. The experiment is representative of 10 so performed. (Reproduced with permission from Michelson et al. Proc Natl Acad Sci USA 1996;93:11877-82).

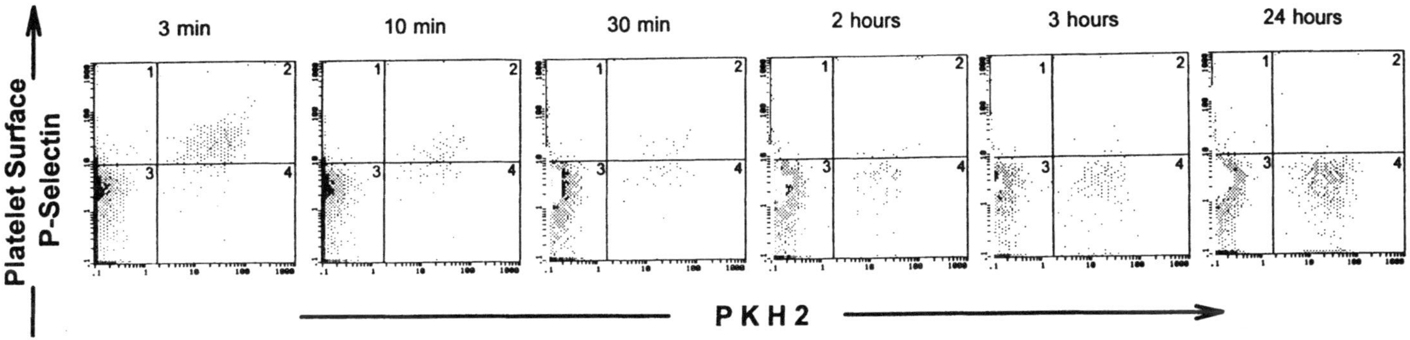

Figure 2. Platelet surface expression of P-selectin on degranulated platelets infused *in vivo*, as determined by three-color whole blood flow cytometry. Autologous washed thrombin-activated PKH2-labeled platelets were transfused into a baboon. Peripheral blood was drawn from the baboon at the indicated time points and immediately fixed with formaldehyde 1% (final concentration). Platelets were identified in whole blood by light scatter and binding of the phycoerythrin-conjugated GPIIb-IIIa-specific monoclonal antibody 7E3. The platelet surface expression of P-selectin was determined by the binding of biotinylated monoclonal antibody S12 followed by streptavidin-RED613 (vertical axis). On each individual platelet, PKH2 fluorescence was measured simultaneously (horizontal axis). The experiment is representative of 6 so performed. (Reproduced with permission from Michelson et al. Proc Natl Acad Sci USA 1996;93:11877-82).

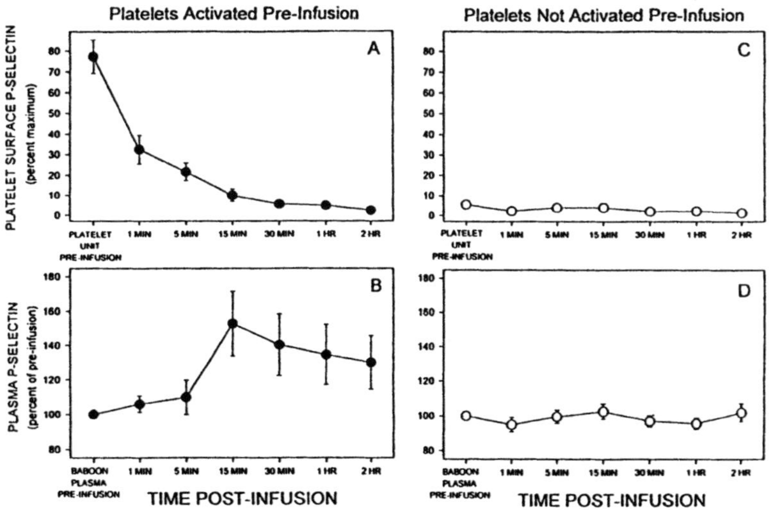

Figure 3. Platelet surface P-selectin (panels A and C) and plasma P-selectin (panels B and D) in peripheral blood samples from experiments in which platelets were thrombin-activated pre-infusion (panels A and B) or not thrombin-activated pre-infusion (panels C and D). Autologous washed PKH2-labeled platelets were transfused into baboons. Peripheral blood, drawn from the baboons at the time points indicated on the horizontal axis, was immediately either formaldehyde fixed for flow cytometric analysis or centrifuged to separate the plasma. Panels A and C (three color whole blood flow cytometry): Platelets were identified in whole blood by light scatter and binding of the phycoerythrin-conjugated GPIIb-IIIa-specific monoclonal antibody 7E3. The infused platelets were identified by their PKH2 fluorescence. The platelet surface expression of P-selectin (vertical axis) was determined by the binding of the biotinylated monoclonal antibody S12, followed by streptavidin-RED613. Maximum platelet surface P-selectin was defined as the binding of S12 to platelets in pre-infusion peripheral blood samples after stimulation at 22°C for 15 minutes with thrombin 10 U/mL. Panels B and D: plasma P-selectin was determined by ELISA. Data are mean ± S.E.M., n = 5. (Reproduced with permission from Michelson et al. Proc Natl Acad Sci USA 1996;93:11877-82).

platelet function with various monoclonal antibodies (as described in detail in reference [12]).

PKH2-labeled autologous infused platelets were easily discriminated from non-infused platelets by whole blood flow cytometry, because PKH2-labeled platelets were approximately 300-fold more fluorescent than unlabeled platelets (Figure 1). A subpopulation of PKH2-labeled platelets as small as 0.3% of all platelets could be identified *in vivo* (Figure 1). PKH2 binding to platelets was stable *in vivo*, as shown by the fact that PKH2-labeled platelets still circulating on day 6 had 85.0 ± 6.1% (mean ± S.E.M., n = 8) of the fluorescence of the PKH2-labeled platelets circulating on day 1 (Figure 1).Results

Circulating Degranulated Platelets Rapidly Lose Surface P-selectin

As demonstrated in Figure 2, infused thrombin-activated PKH2-labeled platelets initially circulated in a P-selectin-positive form (quadrant 2 of the 3 minute time point) but, after a period of sequestration (as previously described [15]) (lower density of dots in quadrants 2 and 4 of the 10 minute and 30 minute time points), gradually reappeared as P-selectin-negative platelets (quadrant 4 of the 2 hour, 3 hour, and 24 hour time points). Two hours after the infusion of the autologous thrombin-activated platelets (P-selectin-positive, PKH2-labeled), 95 ± 1% (n = 5) of the circulating PKH2-labeled platelets had become P-selectin-negative (Figure 2) and there was a 97.0 ± 0.6% (n = 5) average decrease in the platelet surface expression of P-selectin on the circulating PKH2-labeled platelets (Figure 3A).

Etiology of the Decreased Platelet Surface Expression of P-selectin on Circulating Degranulated Platelets

The decreased platelet surface expression of P-selectin on circulating degranulated platelets did not appear to be the result of a conformational change in P-selectin, because parallel results were obtained with two different P-selectin-specific monoclonal antibodies known to be directed against different epitopes [8]: S12 (Figures 2 and 3A) and G1 (data not shown). The loss of platelet surface P-selectin was associated with a 52.7 ± 18.8% (67.1 ± 16.7 ng/mL) maximal increase in the plasma concentration of soluble P-selectin (Figure 3B), which accounted for 93.5 ± 23.8% of the calculated loss of platelet surface P-selectin [12]. At all measured time points, plasma P-selectin was not reduced by centrifugation of the plasma at 100,000 g for 2 hours, thereby demonstrating that the assay was not measuring P-selectin in platelet-derived microparticles [16]. In control experiments in which platelets were not thrombin-activated pre-infusion, the circulating PKH2-labeled platelets were P-selectin-negative at all time points (Figure 3C) (demonstrating that PKH2 labeling of platelets did not result in significant degranulation) and there was no increase in plasma P-selectin (Figure 3D).

Despite Their Loss of Surface P-selectin, Degranulated Platelets Are Not Rapidly Removed from the Circulation

Platelets that were thrombin-activated pre-infusion sequestrated [15] earlier and longer than platelets not thrombin-activated pre-infusion (Figure 4). How-

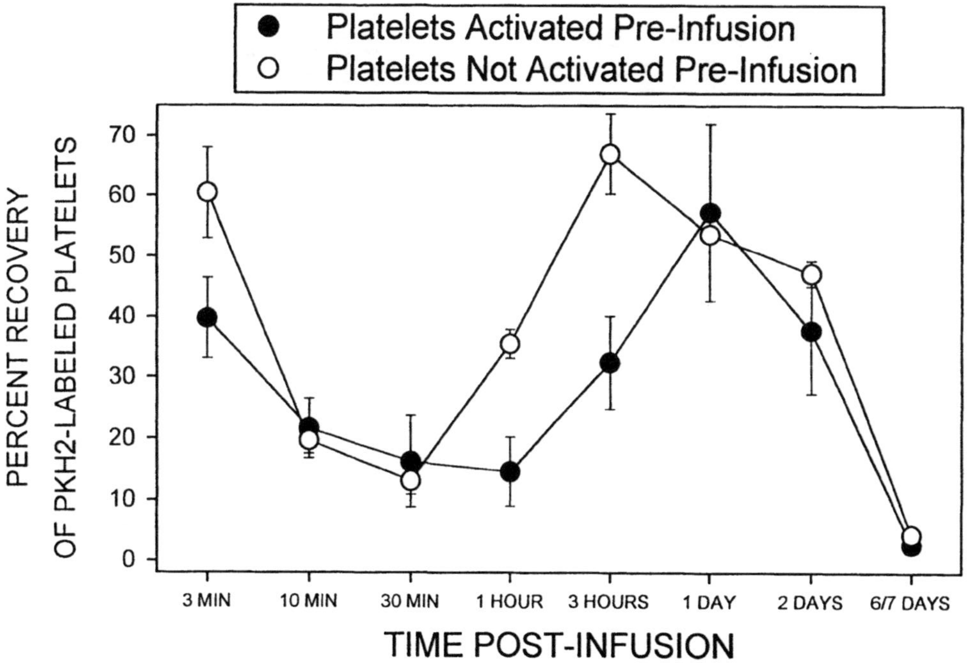

Figure 4. Recovery of platelets thrombin-activated pre-infusion compared with platelets not thrombin-activated pre-infusion, as determined by whole blood flow cytometry. Autologous washed PKH2-labeled platelets were transfused into baboons. Peripheral blood was drawn from the baboons at the time points indicated on the horizontal axis and immediately fixed with formaldehyde. Platelets were identified in whole blood by light scatter and binding of the phycoerythrin-conjugated GPIIb-IIIa-specific monoclonal antibody 7E3. The infused platelets were identified by their PKH2 fluorescence. Percent recovery of PKH2-labeled platelets (vertical axis) was determined as described in ref. [12]. Data are mean ± S.E.M., n = 5. (Reproduced with permission from Michelson et al. Proc Natl Acad Sci USA 1996;93:11877-82).

ever, 24 hours after infusion, the recovery of platelets thrombin-activated pre-infusion was similar to the recovery of platelets not thrombin-activated pre-infusion, and was only slightly less 48 hours after infusion (Figure 4).

Despite Their Loss of Surface P-selectin, Circulating Degranulated Platelets Continue to Function

Despite their degranulation and subsequent loss of surface P-selectin, circulating PKH2-labeled platelets that had been thrombin-activated pre-infusion were still able to function 24 hours after infusion, as determined by four independent methods. First, platelets thrombin-activated pre-infusion were demonstrated to function *in vivo* 24 hours after their infusion, as determined by their presence (dots in Figure 5B) in platelet aggregates in the shed blood emerging from a standardized bleeding time wound. As determined by percent positive analysis on the log forward light scatter signal (a measure of percent aggregate formation [17]), PKH2-positive platelets formed aggregates in shed blood to at least the same extent as PKH2-negative platelets in the same sample, irrespective of whether the platelets were or were not thrombin-activated pre-infusion (n = 6). Secondly, as determined by platelet surface exposure of the fibrinogen binding site on the GPIIb-IIIa complex (reported by monoclonal antibody PAC1), platelets thrombin-activated pre-infusion were able to function normally in 24 hour post-infusion peripheral blood stimulated *ex vivo* by ADP and epinephrine, although poorly in response to a second challenge with thrombin [12]. Thirdly, platelets thrombin-activated pre-infusion were able to function as determined by their generation of procoagulant platelet-derived microparticles [18] in response to *ex vivo* stimulation of peripheral blood drawn 24 hours post-infusion [12]. Fourthly, platelets thrombin-activated pre-infusion were able to function normally as determined by their adherence to Dacron in an arteriovenous shunt. Three color flow cytometry was used to calculate the ratio of PKH2-positive to PKH2-negative platelets a) adherent to the Dacron and b) in peripheral blood. By these means, we determined that platelets thrombin-activated pre-infusion (PKH2-positive) adhered to the Dacron 106.0 ± 1.9% (n = 3) as well as the non-infused (PKH2-negative) platelets.

PKH2 Does Not Interfere With Platelet Function

PKH2 did not interfere with platelet function, as demonstrated by the following six independent methods in which infused autologous PKH2-labeled platelets not preincubated with thrombin functioned as well as circulating non-PKH2-labeled platelets simultaneously analyzed in the same sample [12]: i) participation in platelet aggregates emerging from a bleeding time wound (Figure 5D); ii) exposure of fibrinogen binding sites on the GPIIb-IIIa complex in response to *ex vivo* stimulation of peripheral blood; iii) platelet surface expression of P-selectin in response to *ex vivo* stimulation of peripheral blood with either thrombin, phorbol myristate acetate, or ADP/epinephrine; iv) generation of procoagulant platelet-derived microparticles in response to *ex vivo* stimulation of peripheral blood; v) infused (PKH2-positive) platelets adhered to Dacron in the arteriovenous shunt 112.3 ± 5.1% (n = 3) as well as the non-infused

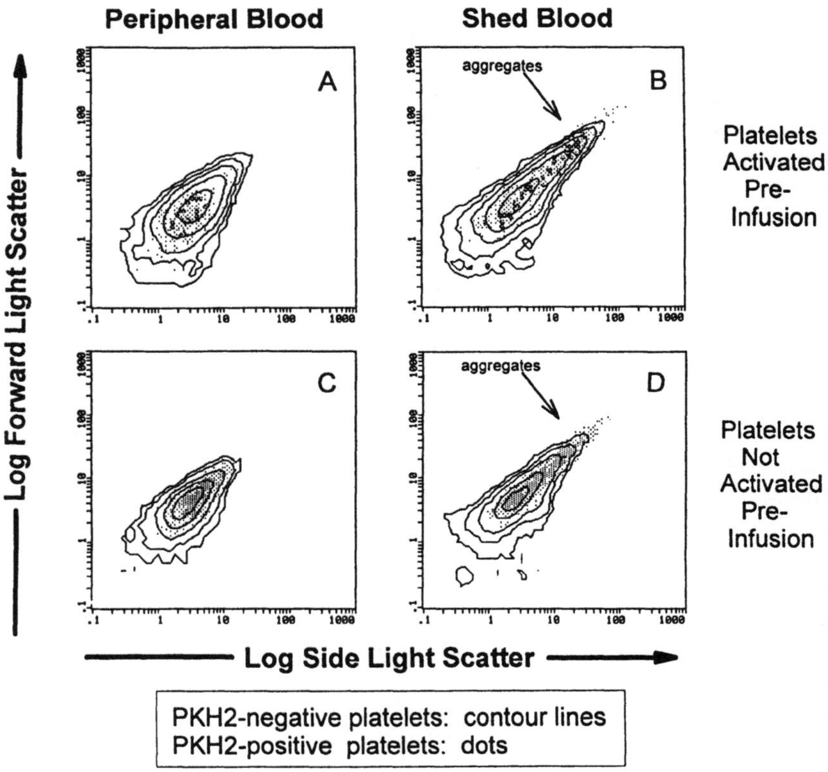

Figure 5. *In vivo* function of platelets thrombin-activated pre-infusion (panels A and B) or not thrombin-activated pre-infusion (panels C and D), as determined by their presence in platelet aggregates emerging from a standardized bleeding time wound. Twenty-four hours after infusion of autologous PKH2-labeled platelets, peripheral blood (panels A and C) and the blood emerging from the wound 2 minutes after the bleeding time incision (shed blood, panels B and D) was drawn, immediately fixed with formaldehyde, and analyzed for platelet aggregation by whole blood flow cytometry. Platelets were identified in whole blood by light scatter and binding of the phycoerythrin-conjugated GPIIb-IIIa-specific monoclonal antibody 7E3. The infused platelets (dots) were identified by their PKH2 fluorescence. The non-infused platelets in the same blood sample (contour lines) were identified by their lack of PKH2 fluorescence. (The different density of dots in panels A and B compared with panels C and D resulted from infusion of different numbers of PKH2-positive platelets in these different experiments.) The experiment is representative of 6 so performed. (Reproduced with permission from Michelson et al. Proc Natl Acad Sci USA 1996;93:11877-82).

(PKH2-negative) platelets; and, vi) expression of surface P-selectin in response to adherence to Dacron in the *ex vivo* shunt.

Discussion

In summary, we have demonstrated that circulating degranulated platelets rapidly lose surface P-selectin to the plasma pool, but continue to circulate and function *in vivo*. Thus, our study [12] reconciles the results of previous studies (2,4,7,8,10,11] by demonstrating that platelet surface P-selectin molecules, rather than degranulated platelets, are rapidly cleared.

Our results [12] have recently been independently confirmed by Berger et al. [19], who found that the platelets of both wild-type and P-selectin knockout mice had identical life spans. When platelets were isolated, activated with thrombin, and re-injected into mice, the rate of platelet clearance was unchanged. The infused thrombin-activated platelets rapidly lost their surface P-selectin in circulation, and this loss was accompanied by the simultaneous appearance of a 100-kD P-selectin fragment in the plasma [19]. Storage of platelets at 4°C caused a significant reduction in their life span *in vivo*, but again no significant differences were observed between the two genotypes. Thus, the results of Berger et al. [19] confirm that P-selectin does not mediate platelet clearance.

These studies [12,19] strongly suggest that the measurement of platelet surface P-selectin in platelet concentrates stored in the blood bank should not be used as a predictor of platelet survival or function *in vivo* [12,19]. However, platelet surface P-selectin could be a useful measure of quality control during processing, storage, and manipulation (filtration, washing) [3–6]. This is because, in contrast to the situation *in vivo* [12,19], the activation-dependent increase in platelet surface P-selectin is not reversible over time under standard blood banking conditions [20].

Despite recent progress in understanding platelet function *in vitro*, there remains a paucity of methods to study platelet function *in vivo*. In this study, we used novel three color whole blood flow cytometric methods for *in vivo* tracking of infused platelets and measurement of their function. These methods should provide the means to answer many previously difficult to address questions about *in vivo* platelet function in transfusion medicine.

References

1. McEver RP. Properties of GMP-140, an inducible granule membrane protein of platelets and endothelium. Blood Cells 1990;16:73-80.
2. Celi A, Furie B, Furie BC. PADGEM: an adhesion receptor for leukocytes on stimulated platelets and endothelial cells. Proc Soc Exp Biol Med 1991; 198:703-09.
3. Fijnheer R, Modderman PW, Veldman H, et al. Detection of platelet activation with monoclonal antibodies and flow cytometry. Changes during platelet storage. Transfusion 1990;30:20-25.
4. Rinder HM, Murphy M, Mitchell JG, Stocks J, Ault KA, Hillman RS. Progressive platelet activation with storage: Evidence for shortened survival of activated platelets after transfusion. Transfusion 1991;31:409-14.

5. Triulzi DJ, Kickler TS, Braine HG. Detection and significance of alpha granule membrane protein 140 expression on platelets collected by apheresis. Transfusion 1992;32:529-33.
6. Divers SG, Kannan K, Stewart RM, et al. Quantitation of CD62, soluble CD62, and lysosome-associated membrane proteins 1 and 2 for evaluation of the quality of stored platelet concentrates. Transfusion 1995;35:292-97.
7. Larsen E, Celi A, Gilbert GE, et al. PADGEM protein: a receptor that mediates the interaction of activated platelets with neutrophils and monocytes. Cell 1989;59:305-12.
8. Hamburger SA, McEver RP. GMP-140 mediates adhesion of stimulated platelets to neutrophils. Blood 1990;75:550-54.
9. Palabrica T, Lobb R, Furie BC, et al. Leukocyte accumulation promoting fibrin deposition is mediated in vivo by p-selectin on adherent platelets. Nature 1992;359:848-51.
10. Reimers HJ, Packham MA, Kinlough-Rathbone RL, Mustard JF. Effect of repeated treatment of rabbit platelets with low concentrations of thrombin on their function, metabolism and survival. Br J Haematol 1973;25:675-89.
11. Reimers HJ, Kinlough-Rathbone RL, Cazenave JP, et al. In vitro and in vivo functions of thrombin-treated platelets. Thromb Haemost 1976;35:151-66.
12. Michelson AD, Barnard MR, Hechtman HB, et al. In vivo tracking of platelets: circulating degranulated platelets rapidly lose surface p-selectin but continue to circulate and function. Proc Natl Acad Sci USA 1996;93: 11877-82.
13. Michelson AD, Benoit SE, Barnard MR, MacGregor H, Valeri CR. A panel of platelet mAb for the study of haemostasis and thrombosis in baboons. In Schlossman SF, Boumsell L, Gilks W, et al., eds. Leucocyte Typing V. Oxford University Press, Oxford. 1995:1230-31.
14. Horan PK, Slezak SE. Stable cell membrane labelling. Nature 1989;340:167-68.
15. Franco RS, Lee KN, Barker-Gear R, Gates R, Menitove JE. Use of bi-level biotinylation for concurrent measurement of in vivo recovery and survival in two rabbit platelet populations. Transfusion 1994;34:784-89.
16. George JN, Thoi LL, McManus LM, Reimann TA. Isolation of human platelet membrane microparticles from plasma and serum. Blood 1982;60:834-40.
17. Michelson AD, MacGregor H, Barnard MR, Kestin AS, Rohrer MJ, Valeri CR. Reversible inhibition of human platelet activation by hypothermia in vivo and in vitro. Thromb Haemost 1994;71:633-40.
18. Sims PJ, Faioni EM, Wiedmer T, Shattil SJ. Complement proteins C5b-9 cause release of membrane vesicles from the platelet surface that are enriched in the membrane receptor for coagulation factor Va and express prothrombinase activity. J Biol Chem 1988;263:18205-12.
19. Berger G, Hartwell DW, Wagner DD. P-selectin and platelet clearance. Blood 1998; 92:4446-52.
20. Holme,S, Sweeney JD, Sawyer S, Elfath MD. The expression of P-selectin during collection, processing, and storage of platelet concentrates: relationship to loss of in vivo viability. Transfusion 1997;37:12-17.

RECIPIENT MECHANISMS OF ALLOIMMUNITY: ANTIGEN PROCESSING MEDIATES IgG IMMUNITY AGAINST DONOR PLATELETS[1]

J.W. Semple[2], K.W.A. Bang[3], E. Speck[2], V.S. Blanchette[2], J. Freedman[2]

Introduction

Advances in transfusion and transplantation medicine have provided much of the impetus for recognition and subsequent study of alloimmune responses and it now appears that the immune mechanisms responsible for alloimmunization are dependent on the recipient's T cell responses [1-3]. Two recipient T helper (Th) cell recognition mechanisms have been postulated to initiate the alloimmune response [4-7] and Figure 1 summarizes these pathways. The first is termed the direct pathway and occurs when the recipient's Th cells directly interact with MHC class II molecules on donor antigen presenting cells (APC) [4]. This abnormally strong immune mechanism is the result of crossreactivity of recipient T cell receptors with polymorphic residues of allogeneic MHC class I and class II molecules [4]. In transfusion medicine, strategies such leukoreduction are aimed at eliminating this pathway and have been shown to be effective in reducing the incidence of alloimmunization in multiply transfused patients [8]. The second T cell recognition mechanism leading to alloimmunization is termed the indirect pathway and is analogous to the normal immune response [5-7]. This mechanism occurs when allogeneic cells are administered to the recipient and it involves the uptake, processing and presentation of donor antigens (e.g. MHC class I molecules) by recipient APC to the recipient's Th cells [5-7]. A major question still to be answered regarding the indirect pathway is whether it is active in human transfusion recipients who often are in an immunocompromised state. Since a proportion of leukemic patients receiving only leukoreduced platelets still become alloimmunized [8] it may be that indirect allorecognition is competent in these patients. In addition, it is unknown at present how indirect allorecognition might be involved in the alloimmune response induced in those transfusion recipients who have been previously alloimmunized (e.g. by previous pregnancy or transfusion). Knowing the answers to these questions may allow for the development of more specific immunotherapeutic approaches to further eliminate the incidence of alloimmunization.

---

1. Supported by a grant from the Canadian Blood Services R&D Fund.
2. Dept Lab Med Pathobiol, St. Michael's Hosp, Div Hematol/Oncol, The Hosp for Sick Children, Depts Pharmacol, Med and Pediatr, Univ of Toronto and the Toronto Platelet Immunobiol Group, Toronto, Ontario, C.
3. KWAB is the recipient of a Research Studentship from the Canadian Blood Services.

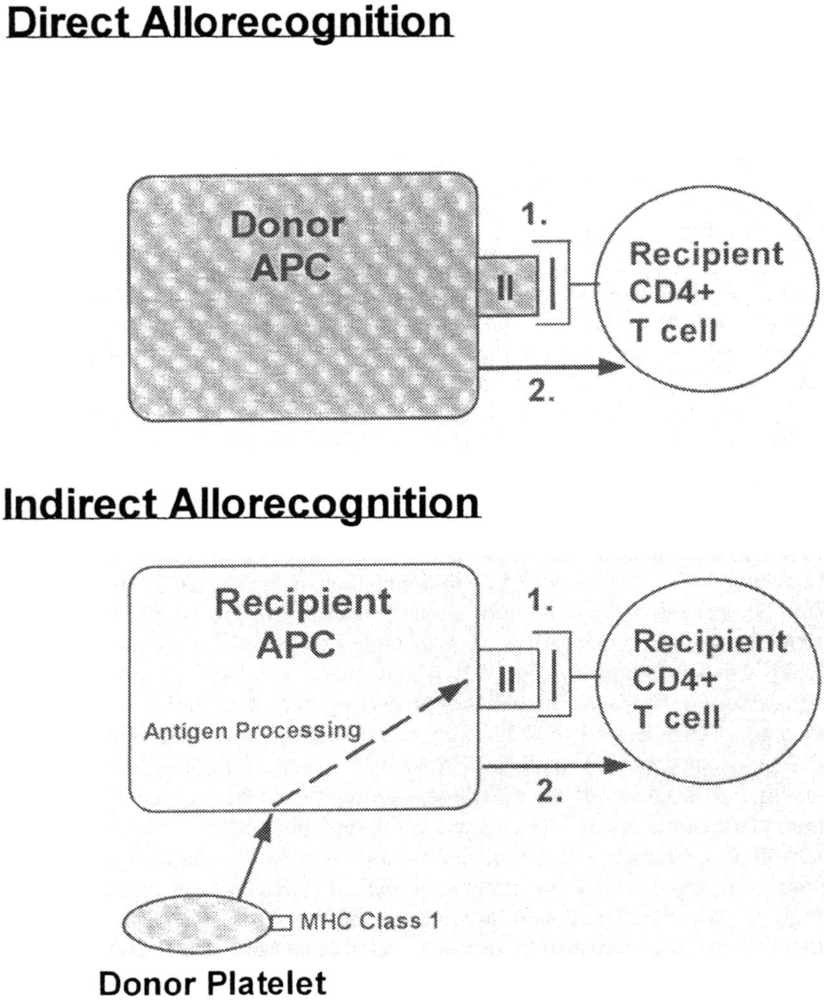

Figure 1. Direct and indirect allorecognition. (A) Direct allorecognition occurs when the TcR of recipient Th cells directly interacts with the MHC class II molecules on donor APC. The activated T cells subsequently stimulate MHC class I antigen primed B cells to differentiate into plasma cells and secrete IgG alloantibodies. (B) Indirect allorecognition takes place when donor platelet-derived MHC class I alloantigens are processed, reexpressed and presented by MHC class II molecules on recipient APC. These activated T cells subsequently"help" IgG alloantibody production.

Recently, studies in animal models of platelet immunity have determined that leukoreduced allogeneic platelets can stimulate the production of IgG anti-donor antibodies via indirect allorecognition [9-12]. It was found that indirect alloimmunity against platelets was dependent on the activation of the enzyme inducible nitric oxide synthase (iNOS) within recipient macrophages [11]. These results suggested that recipient macrophages may mediate donor platelet immunity via their known roles as a phagocyte and as APC. However, the mechanisms by which donor platelets are processed and presented to the recipient's immune system remain unknown.

T cells recognize protein antigens that are degraded or processed and combined with molecules encoded by the major histocompatibility complex (MHC) [13,14]. Antigen processing is critical for generating antigenic determinants to be loaded into the antigen binding grooves of either MHC class I or II molecules [15]. Exogenous proteins (e.g. bacterial glycoproteins) are generally processed by antigen presenting cells (APC) via endosomal compartments and are shunted to intracellular compartments rich in MHC class II molecules [16]. This pathway is necessary for the activation of CD4+ Th cells and generation of "help" for eventual IgG antibody production [16]. Endogenous antigens (e.g. virally derived proteins), on the other hand, are processed by large molecular weight proteasomes within the APC's cytosol and are subsequently transported to the luminal surface of the endoplasmic reticulum for loading onto MHC class I molecules [16,17]. This pathway is responsible for the stimulation of CD8+ cytotoxic T cells (CTL) [18]. Experimentally, the major distinction between these pathways has been that exogenous antigen processing is generally susceptible to pH-raising lysomotrophic agents such as chloroquine whereas the non-endosomal pathway is not [17,19-21]. Understanding antigen processing pathways of clinically relevant protein antigens such as platelet alloantigens may be fundamental to developing efficacious antigen-specific therapies for alloimmunization. To study platelet antigen processing pathways in recipient APC, we used an in vitro system in which adherent APC from BALB/c recipient mice were pulsed with allogeneic platelets from C57BL/6 mice and then transfused weekly into naive BALB/c mice (Figure 2).

The Immunogenicity of Platelet-Pulsed Adherent APC

When platelet-pulsed APC were transfused into naive recipient mice, they stimulated IgG antibodies in 100% of the recipients by the sixth transfusion (Figure 3) and the antibodies were found to be composed of high titred IgG1 and IgG2a anti-donor antibodies (Table 1). These results initially confirmed that allogeneic platelet-pulsed APC could stimulate recipient IgG anti-donor antibodies and suggested that the pulsed APC could be used to validly assess antigen processing pathways.

The Role of iNOS and NO in Platelet Antigen Processing.

Previously we observed that when AMG, a selective inhibitor of iNOS [22,23], was administered to recipient mice, it completely prevented formation of IgG alloantibodies against transfusions of intact platelets [11]. Using the pulsed APC cultures, we tested the effect of AMG and found that compared with untreated

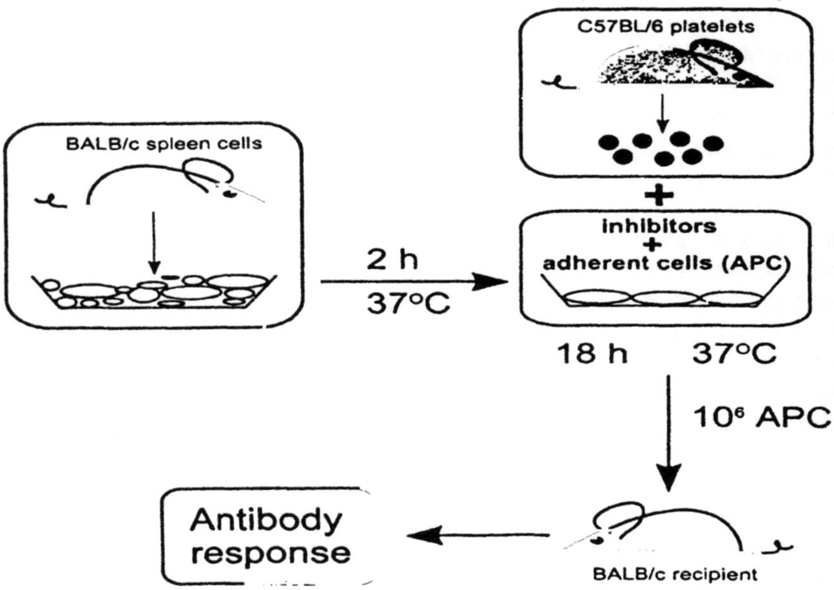

Figure 2. Summary of the platelet-pulsed APC protocol. BALB/c 2 h adherent APC were incubated with C57BL/6 platelets for 18 hr at 37°C, washed and 10^6 APC were transfused into naive BALB/c mice weekly. Recipient sera were tested for IgG anti-donor antibodies by flow cytometry.

pulsed APC immunity, the ability of AMG-treated pulsed APC to stimulate IgG alloantibodies was significantly inhibited (Figure 3). These data confirmed our previous results and suggest that iNOS activation is an essential platelet-processing step within APC responsible for IgG alloantibody production. However, how iNOS and its product nitric oxide (NO) actually mediate platelet antigen processing is unknown. Perhaps phagolysosome-associated NO [24,25] directly causes platelet membrane glycoprotein damage (e.g. due to nitration, unfolding or cleavage) to generate MHC class II binding motifs. As endosomes containing MHC class II molecules normally fuse with phagolysosomes [26], this may allow for the NO-processed platelet antigens to be loaded into the antigen binding grooves of the MHC molecules for subsequent transport to the surface. Our current results support the conclusion that manipulating NO levels

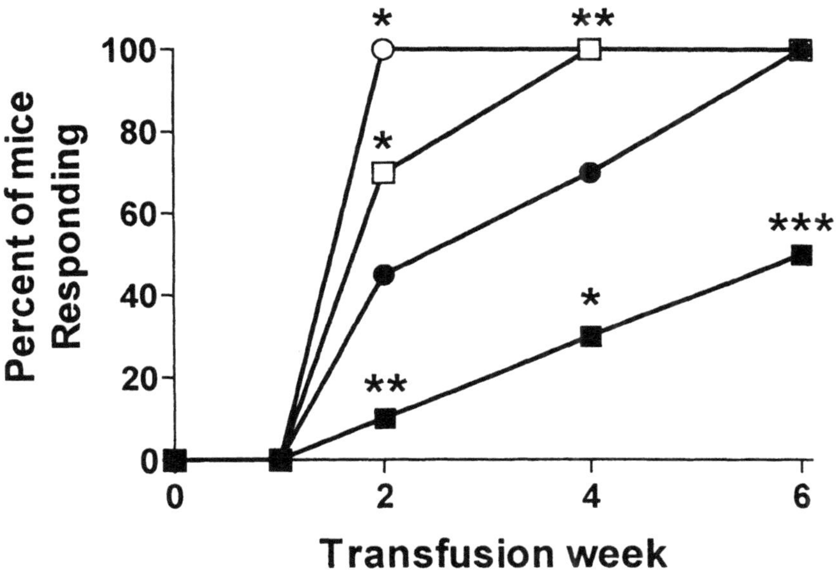

Figure 3. Percentage of mice with detectible antidonor allantibody. APC were pulsed with donor platelets for 18 h in the absence (control,●) or presence of 1 mM AMG (■), 0.1 mM chloroquine (□) or 5μM LLnL (○), washed twice and then transfused weekly as previously described (27). Data is expressed as the percentage of recipients with detectible antidonor antibody by flow cytometry (rounded to nearest 5%). The $\_^2$ test of independent proportions was used to compare the agent-treated pulsed APC groups against the non-treated pulsed APC group (* = $P < 0.05$, ** = $P < 0.01$, *** = $P < 0.0005$).

within recipient APC may be an effective and selective immunotherapy for the total reduction of alloimmunization against leukoreduced platelets.

The Role of Endosomal and Non-Endosomal Compartments in Platelet Processing [27]

To further identify platelet processing pathways, we distinguished between endosomal (pH dependent) and non-endosomal (pH independent) pathways by incubating the platelet/APC cultures with chloroquine. The chloroquine treated APC induced an accelerated total IgG anti-donor response (Figure 3) which paradoxically inhibited IgG1 anti-donor production while stimulating IgG2a alloantibodies (Table 1). These results suggested that endosomal or chloroquine-sensitive (pH dependent) processing pathways of platelet antigens appear to

Table 1. Effect of chloroquine on the APC's ability to modulatie IgC alloantibody titres.[1]

Antibody test	APC treatment (N)	Trranfusion week 2	4	6
Total IgG	None (10)	95±34[2]	2,366±596	11,880±1,849
	Chloroquine (10)	161±62	3,445±730	9,940±1,200
		n/s[3]	n/s	n/s
IgG	None (10)	50±39	1,292±564	4,940±1,213
	Chloroquine (10)	N/D[4]	314±138	1,074±641
		n/s	P<0.01	P<0.01
IgG2a	None (10)	23±12	491±189	1,933±629
	Chloroquine (10)	323±134	762±190	1,546±331
		P = 0.05	n/s	n/s

1. APC/platelet cultures were treated with chloroquine and then used in the APC pulsing assay, washed twice and 106 APC were transfused weekly.
2. Data is expressed as the mean titre ± SEM determined by flow cytometry.
3. The unpaired t test with Welch's correction for variance was used to compare the treated pulsed APC groups against the non-treated pulsed APC group for each isotype (n/s = not significant).
4. N/D = not detected.

Table 2. Effect of chloroquine pretreatment on antidonor alloantibody production (data is expressed as the percentage of recipients with detectable antidonor antibody by flow cytometry (rounded to nearest 5%)[1]

Untreated cell	Pretreated cell	Chloroquine concentration	N	Transfusion week Pre	2	4
APC/platelets[2]	–	0	33	0	45	70
APC	platelets	400 mM	8	0	0^{++3}	0^{+++}
APC	platelets	0.1 mM	14	0	$55^{n/s}$	$85^{n/s}$
platelets	APC	0.1 mM	8	0	100^{+}	$100^{n/s}$

1. APC or donor platelets were pretreated with the indicated concentrations of chloroquine and then used in the APC pulsing assay, washed twice and 106 APC were transfused weekly.
2. Data from Figure 1 for untreated APC/platelet cultures (N = 33) is included for statistical comparison.
3. The _2 test of independent proportions wa used to compare the agent-treated pulsed APC groups against the non-treated pulsed APC group (+ = P<0.01, ++ = P<0.005, +++ = P<0.0001, $^{n/s}$ = not significant).

favour production of IgG1 whereas chloroquine-insensitive (pH independent) pathways exist and may be associated with the IgG2a anti-donor responses induced by the pulsed APC. To ensure that chloroquine was not directly affecting the donor platelet MHC molecules and primarily acting on the APC in the cultures, titrations of chloroquine were pre-incubated with the donor platelets before they were added to the APC cultures. Preincubation of donor platelets with high doses (400 mM) of chloroquine which completely inhibited the reactivity of a conformation-dependent anti-MHC class I monoclonal (mAb) antibody (indicating loss of β_2M or MHC, Figure 4) prevented the platelets from positively converting the pulsed APC's immunity (Table 2). However, the lower dose (0.1 mM) of chloroquine used in our culture conditions did not affect the mAb reactivity (Figure 4) nor did it affect the ability of platelet pulsed APC to induce IgG antibodies (Table 2). These experiments verified that donor platelet MHC antigens in the cultures were the primary stimulus for the pulsed APC immunity and that the chloroquine concentrations in the APC cultures were primarily affecting recipient APC processing.

Classically, for MHC class II presentation and subsequent stimulation of antibody production, exogenous protein antigens need to be shunted to endosomal compartments within the APC for processing by pH-dependent proteases (e.g. sensitive to the lysomotrophic agent chloroquine) such as the cathepsins [13, 15]. The experiments using chloroquine showed that it actually enhanced the APC's ability to stimulate total IgG alloantibodies (Figure 3), however. In addition, we found that chloroquine significantly reduced the pulsed APC's ability to stimulate non-complement-fixing IgG1 but not complement-fixing IgG2a antibodies (Table 1). Thus, both endosomal and non-endosomal processing pathways can act on platelet antigens; endosomal (chloroquine-sensitive) processing may be responsible for stimulation of IgG1 alloantibody responses whereas non-endosomal (chloroquine-insensitive) pathways regulate IgG2a production. This data represents one of the few examples where an exogenous protein antigen can be handled by an APC for subsequent MHC class II-dependent antibody production without the requirement for endosomal (chloroquine-independent) processing; to our knowledge, this has only been shown to occur for influenza matrix protein [19,20] and measles virus protein [21]. The significance of the non-endosomal processing of platelet antigens is not known, however, since IgG1 and IgG2a isotypes are closely associated with Th2 and Th1 activation respectively, it may suggest that the different processing pathways (endosomal vs non-endosomal) affect the isotype patterns via differential T cell activation. Taken together, the data suggest the intriguing possibility that different processing pathways may shunt platelet antigens to different intracellular compartments or MHC class II molecules, which can induce differential stimulation of IgG isotypes.

The Cytosolic Processing Pathway may Regulate Platelet-Induced Humoral Immunity

Exogenous antigens generally do not enter the cytosolic (proteasome) pathway of antigen processing which is primarily responsible for the generation of MHC class I-restricted cytotoxic T lymphocytes (CTL). We previously reported, however, that allogeneic platelet transfusions are associated with the generation of CD8+ CTL and that anti-platelet immune non-responsiveness in C57BL/6 recipients is reversed when their CD8+ T cells are deleted (e.g. CD8 knockout mice) [28]. To examine whether donor platelet antigens could be processed in the cytosolic compartment, APC were pulsed with donor platelets in the pres-

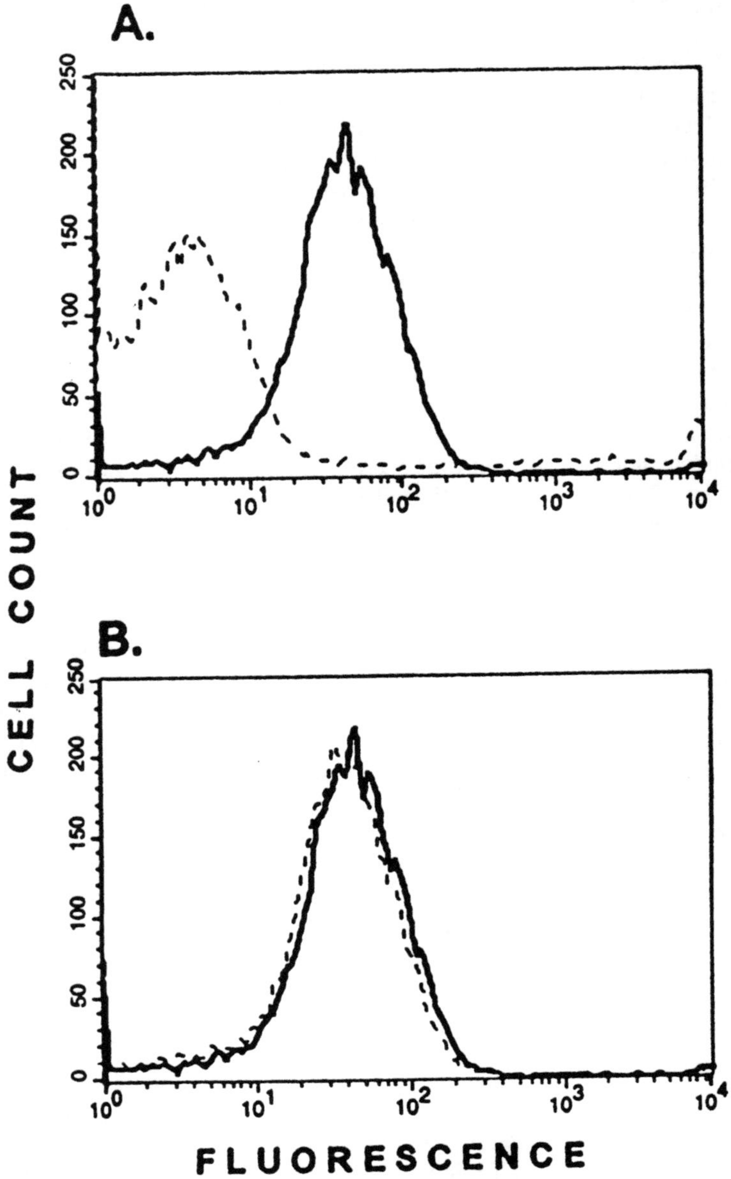

Figure 4. β_2-Microglobulin-associated MHC class I (H-2 K^d) expression on murine platelets after chloroquine pretreatments. Each panel represents fluorescence histograms of BALB/c platelets labeled with PE-conjugated anti-H-2 K^d . A shift of the fluorescence to the left represents reduced reactivity to anti-K^d monoclonal antibody. Untreated control platelets (-----) (A) Pretreatment with 400 mM chloroquine (- - - -). (B) Pretreatment with 0.1 mM chloroquine (- - - -).

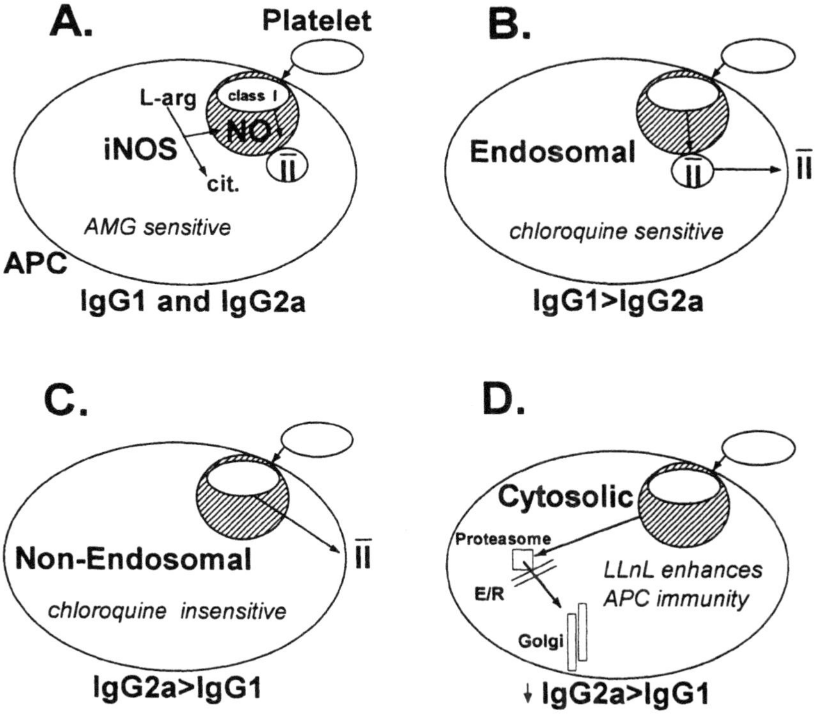

Figure 5. Proposed model (based on the inhibitor studies) of four potential platelet antigen trafficking pathways (-->) within recipient APC and how these affect IgG immunity. (A) Donor platelets (shaded oval) are engulfed within phagolysosomes (hatched areas) and donor MHC class I molecules are exposed to **NO** produced by conversion of L-arginine (L-arg) to citrulline (cit) by iNOS. This pathway is essential for the production of both IgG1 and IgG2a anti-donor antibodies; inhibition by AMG significantly inhibits total IgG production. (B) Platelet class I antigens could subsequently be transported to **endosomal** compartments, exposed to pH-dependent enzymes and loaded onto the APC's MHC class II molecules (II) for expression on the surface; chloroquine sensitivity suggests that this pathway favors production of IgG1 over IgG2a. (C) Platelet class I antigens can also be acted on independently of the pH-dependent proteases in a **non-endosomal** (chloroquine-insensitive) compartment which favours the production of IgG2a antibodies over IgG1. (D) Some antigens may be transported through the **cytosolic** compartment and interact with proteasomes; inhibition of this pathway by LLnL, however, enhances IgG anti-donor antibody production suggesting that this pathway may naturally reduce donor platelet immunity, particularly IgG2a production.

ence of the proteasome inhibitory peptide LLnL. We found that LLnL significantly enhanced the pulsed APC's ability to stimulate total IgG alloantibodies so that 100% of recipients were antibody positive after the second transfusion (Figure 3). This data may suggest that donor platelet antigens enter the cytosolic pathway which somehow reduces the APC's ability to stimulate humoral immunity e.g. the enhancement of the pulsed APC immunity by LLnL may be due to sparing more of the platelet MHC antigens for the B cell stimulatory MHC class II pathways. Alternatively, the cytosolic pathway may negatively regulate humoral immunity by stimulating the generation of CD8+ T cells e.g. the activated CTL may lyse APC presenting platelet antigens to Th cells. We are currently examining these possibilities.

Conclusions

For donor platelets to induce an IgG anti-donor antibody response, their MHC class I molecules need to be exposed to iNOS activation and can either utilize an endosomal or non-endosomal processing pathway in order to generate immunogenic motifs which control the stimulation of IgG alloantibody formation. These pathways are summarized in Figure 5. Overall, the current data suggests manipulating platelet processing pathways, particularly related to iNOS, may be a strategy for the development of specific immunotherapies to prevent alloimmunization in against platelets. These results may have application for those clinical situations where platelet recipients become alloimmunized despite receiving leukoreduced platelets, or potential application in those already immunized recipients (e.g. due to prior pregnancy or transfusion) who receive leukoreduced platelet products.

References

1. Rosenberg AS, Mizuochi T, Singer A. Evidence for involvement of dual-function T cells in rejection of MHC class I disparate skin grafts. Assessment of MHC class I alloantigens as in vivo helper determinants. J Exp Med 1988;168:33-45.
2. Rosenberg AS, Singer A. Cellular basis of skin allograft rejection: an in vivo model of immune-mediated tissue destruction. Annu Rev Immunol 1992;10:333-58.
3. Auchincloss H Jr, Sykes M, Sachs DH. Transplantation Immunology. In: Paul WE (ed). Fundamental Immunology. New York, NY: Lippincott-Raven, 1999:1175-235.
4. Gould DS, Auchincloss H Jr. Direct and indirect recognition: the role of MHC antigens in graft rejection. Immunol Today 1999;20:77-82.
5. Shoskes DA, Wood KJ. Indirect presentation of MHC antigens in transplantation. Immunology Today 1994;15:32-38.
6. Reed EF, Hong B, Ho E, Harris PE, Weinberger J, Suciu-Foca N. Monitoring of soluble HLA antigens and anti-HLA antibodies identifies heart allograft recipients at risk of transplant-associated coronary artery disease. Transplantation 1996; 61:566-74.
7. Benham AM, Sawyer GJ, Fabre JW. Indirect T cell allorecognition of donor antigens contributes to the rejection of vascularized kidney allografts. Transplantation 1995; 59:1028-32.
8. The Trial to Reduce Alloimmunization to Platelets Study Group. Leukocyte reduction and ultraviolet B irradiation of platelets to prevent alloimmunization and refractoriness to platelet transfusions N Engl J Med 1997;337:1861-915.

9. Semple JW, Speck ER, Milev YP, Blanchette V, Freedman J. Indirect allorecognition of platelets by T helper cells during platelet transfusions correlate with anti-MHC antibody and cytotoxic T lymphocyte formation. Blood 1995;86:805-12.
10. Oh JH, Taysavang, Whelchel JD. Conferring immunogenicity to platelets by preincubation with recipients' adherent cells. In: Kappler JW ed. Proc IX Intern Congress of Immunology, 1995:69.
11. Bang A, Speck ER, Blanchette VS, Freedman J, Semple JW. Recipient humoral immunity against leukoreduced allogeneic platelets is suppressed by aminoguanidine, selective inhibitor of inducible nitric oxide synthase. Blood 1996;88:2959-66.
12. Semple JW, Speck ER, Cosgrave D, Lazarus AH, Blanchette VS, Freedman J. Extreme leukoreduction of major histocompatibility complex class II positive B cells enhances allogeneic platelet immunity. Blood 1999;93:713-20.
13. Ziegler HK, Unanue, ER. Decrease in macrophage antigen catabolism caused by ammonia and chloroquine is associated with inhibition of antigen presentation to T cells. Proc Natl Acad Sci USA 1982;79:175-82.
14. Shimonkevitz R, Kappler J, Marrack P, Grey H. Antigen recognition by H-2 restricted T cells. I. Cell free antigen processing. J Exp Med 1983;158:303-16.
15. Allen PM, Unanue ER. Differential requirements for antigen processing by macrophages for lysozyme-specific T cell hybridomas. J Immunol 1984;132:1077-79.
16. Paul WE. Antigen processing and presentation. In: Gallen JL, Goldstein I, Snyderman R (eds). Inflammation. New York: Raven, 1992: 776-89.
17. Bersofsky JA and Berkower IJ. Immunogenicity and antigen structure. In: Paul WE (ed). Fundamental Immunology. Philadelphia, Lippincott-Raven, 1999:651-700.
18. Henkert PA. Cytotoxic T lymphocytes. In: Paul WE ed. Fundamental Immunology, Philadelphia, Lippincott-Raven, 1999:1021-49.
19. Morrison LA, Llukacher AE, Braciale VL, Fan D, Braciale TJ. Differences in antigen presentation to MHC class I and class II-restricted influenza virus specific cytolytic T lymphocyte clones. J Exp Med 1986;163:903-21.
20. Nuchtern JG, Biddison WE, Klausner RD. Class II MHC molecules can use the endogenous pathway of antigen presentation. Nature 1990;343:74-76.
21. Jacobson S, Sekaly RP, Jacobson CL McFarland HF, Long EO. HLA class II-restricted presentation of cytoplasmic measles virus antigen to cytotoxic T cells. J Virol 1989;63:1756-62.
22. Griffiths MJD, Messent M, MacAllister RJ, Evans TW.: Aminoguanidine selectively inhibits inducible nitric oxide synthase. Br J Pharmacol 1993;110:963-70.
23. Wolff DJ, Lubeskie A. Aminoguanidine is an isoform-selective, mechanism-based inactivator of nitric oxide synthase. Arch Biochem Biophys 1995;316:290-96.
24. Fang FC. Mechanisms of nitric oxide-related antimicrobial activity. J Clin Invest 1997;99:2818-23.
25. Rotrosen D. The respiratory burst oxidase. In: Gallen JL, Goldstein I, Snyderman R (eds). Inflammation, basic principles and clinical correlates. New York, Raven, 1992: 589-601.
26. Kaufmann SHE. Immunity to intracellular bacteria. In: Paul WE (ed). Fundamental Immunology. Philadelphia, Lippincott-Raven, 1999:1335-72.
27. Bang KWAB, Speck ER, Blanchette VS, Freedman J, Semple JW. Unique processing pathways within recipient antigen presenting cells determine IgG immune responsiveness against donor platelet MHC antigens. Blood 2000;95:1735-42.
28. Semple JW, Speck ER, Blanchette V, Freedman J. Immune non-responsiveness to allogeneic platelets in murine strains which lack MHC class II I-E molecules is due to the presence of CD8+ T cells. Blood 1996;88(suppl 1):162a.

CD34+ CELL ANALYSIS: MULTICENTER COOPERATIVE TRIALS

Sunny Dzik[1]

During the last decade there has been tremendous expansion in the application of hematopoietic progenitor cell therapy in hematology/oncology. Two fundamental advances have supported this therapy: the development of growth factors to mobilize progenitor cells and the discovery of cell-surface markers that identify progenitors. The CD34 antigen identifies a subpopulation of cells with properties characteristic of hematopoietic progenitors. Progenitor cells are able to self-renew and are able to differentiate into mature myeloid and erythroid lineages. Antibodies to CD34 can be used to quantify the concentration of CD34+ cells in patient samples. This information has become an essential element in the decision to begin harvesting progenitors from the donor, in determining the threshold dose of cells needed for engraftment, in assessing the impact of laboratory cell manipulations designed to purify progenitor cells, and in the prediction of time-to-engraftment of the patient. While numerous assays for measuring CD34+ cells exist, there have been difficulties in the standardization of measurement of this important analyte. This paper presents information on a multicenter study focused on standardization of CD34 cell measurement that was conducted by the Biomedical Excellence for Safer Transfusion (BEST) working party of the International Society of Blood Transfusion (ISBT) [1].

Background

CD34 antigen is found on a 110 kDa highly glycosylated cell-surface protein. The CD34 glycoprotein appears to be involved in the adhesion of hematopoietic progenitor cells to the marrow stromal environment. Several monospecific monoclonal antibodies reacting with the CD34 antigen have been developed. These have been divided into 3 classes based on their reactivity with enzyme-treated CD34+ cells. Antibodies of the Class III have proved to be the most useful for determination of CD34 concentrations in clinical samples.

Assays for CD34+ Cells

Flow cytometry

Flow cytometry is the most widely used method for determination of CD34+ cell counts [2]. A variety of different assays exist. The most commonly used assays include the Milan protocol, the ISHAGE protocol, the ProCOUNT assay

---

1. Co-Director, Blood Transfusion Service, Massachusetts General Hospital, Boston, MA, USA.

Table 1. Prior multi-site trials of CD34+ cell enumeration. Studies shown included at least 5 shared samples for testing.

References	**Number of sites**	**Number of samples**	**Sample sources**	**Observed CD34+ cell (%)**	**Inter-site CV**	**Sources of variability**
Johnsen [17]	11	5	Apheresis Blood	0–3.5	28–198	Lysing reagents, monoclonals, number of events analyzed, gating, calculating methods
Brecher [7]	10	21	Apheresis Marrow	0.17–2.23	62–1055	Sample detorioration during transport, lysing and staining reagents, gating calculatin methods
Johnsen & Knudsen [18]	24	6	Apheresis Blood	0.17–3.52	16–295	Sample preparation, number of events analyzed, calculation methods
Lodwell & Brainbridge [8]	15	28	Blood Apheresis Marrow	0.08–19.31	63–100	Sample deterioration during transport, staining methods, number of events analyzed, instruments used
Brando [19]	8	34	Blood Apheresis	0.05–2.94	NA	Instruments used for WBC couts

(Beckton-Dickinson), the Stem-Kit (Coulter Immunotech) and the Ortho fluorosphere assay. Flow methods either directly determine the CD34+ cell count by indexing the fluorescent events to an internal control (such as fluorescent beads) or determine the CD34+ cell count as a percentage of leukocytes and then multiply this percentage by the number of leukocytes determined on another instrument. The ISHAGE and Milan protocol employ the latter strategy in which the flow result is indexed to a separately determined WBC count. In contrast, the ProCOUNT method includes internal beads that allow the user to directly determine the concentration of CD34+ cells [3]. This latter feature is generally regarded as an advantage.

A second major distinction in flow methods is whether or not they employ a cell lysing and/or wash step during sample preparation. Both the nature of the lysing agent and the need to do cell washing may interfere with the precision and reproducibility of CD34+ cell assays. The ProCOUNT assay was developed as a lyse no wash method. The Milan and ISHAGE methods include both a lyse step and wash step although a recent modification of the ISHAGE protocol has been developed which incorporates not only beads in the reaction tube for direct measurement of the CD34+ cell concentration but also a lyse no-wash assay strategy [4].

Microvolume fluorimetry

The STELLer assay is a CD34+ cell assay designed for use with the IMAGN 2000 device [5, 6]. This device is a microfluorimeter. In contrast to a flow cytometer, the microfluorimeter device analyses a fixed volume of sample which remains motionless in a disposable cuvette. The laser moves across the cuvette and scans a two-dimensional array. Any cell labeled with a fluorescent antibody to CD34 emits light when struck by the laser. The emitted signal is amplified by a photomultiplier tube and analyzed by a computer. The device directly reports the concentration of CD34+ cells/µL.

Prior Multicenter Studies

Prior to the BEST Working Party study, there had been 11 multicenter studies of CD34 standardization. These are listed in Table 1. All studies used flow cytometry and no study used volumetric microfluorimetry. Of the 1 prior studies few analyzed more than 10 samples. Brecher et al studied 21 samples consisting of leukapheresis harvests and marrow at 10 different sites [7]. Inter-site CVs among the different samples ranged from 62% to over 1000%. Indeed, the lack of inter-site agreement was a major finding of this early work. The investigators noted that sample deterioration during shipment was a problem. In addition, there was wide variability in many of the details of the assays used including lysing and staining reagents, gating strategies, and methods used to determine the final CD34+ cell concentration. The Brecher study serves as an initial benchmark that highlighted the difficulties of CD34+ cell standardization across different medical centers.

Lowdell et al analyzed 28 samples consisting of peripheral blood samples, leukapheresis harvests, and bone marrow at 15 different sites [8]. Inter-site CVs ranged from 62% to 100%. Many of the technical variations in assay methodology that were seen in the Brecher study were also found in the Lowdell study. Lumley et al conducted a two-phase study in which 12 samples were analyzed at 8 different institutions and an additional 12 samples were analyzed at 12 institutions [9]. The samples were cryopreserved leukapheresis samples and so may have been even more subject to sample deterioration during shipment after thawing. Inter-site CVs were again quite high (23 – 235%) and participants did not standardize a number of key assay variables including the staining method used, flow cytometric gating strategies, and calculations used to determine the final CD34+ cell concentration. As a result of the extreme variability observed in the measurement of this important cell, several appeals were made for greater co-operation and standardization of CD34+ cell analysis [10].

BEST Working Party Multicenter Study

Because of the importance of standardization of CD34 measurement and because of the somewhat disappointing results of earlier trials to standardize CD34+ cell measurement, the Biomedical Excellence for Safer Transfusion (BEST) Working Party of the International Society of Blood Transfusion initiated a multicenter study of two commercially available methods for counting CD34+ cells. The primary aim of the study was to assess the variability of measurements across different institutions, so called inter-site precision. Secondary outcomes included an assessment of the intra-site precision, the accuracy, and the limit of detection of the assays. The BEST Working Party study focused only on the measurement of CD34+ cells in whole blood. Whole blood was chosen because the measurement of CD34+ cells in whole blood is key to the first decision in the care of the patient, namely to begin harvesting progenitor cells by apheresis. The study centered on samples with a CD34+ cell concentration in the range of 0 – 200 cells /μL as this is the typical range of mobilized patients at the time of apheresis harvest. The study did not address the standardization of CD34+ cell counts taken from apheresis collections, from bone marrow, or from cord blood. Participants in the study used either the ProCOUNT assay (flow cytometry) or the STELLer assay (microfluorimetry) or both.

Methods

The BEST Working Party study was conducted in 3 phases: In phase I, diskettes with list-mode data obtained from flow cytometric assays of a variety of samples were distributed to participants. The diskettes were blinded to the participants and used to develop familiarity with a common gating strategy to be used in the ProCOUNT assay.

In phase II, wet samples were sent to each of the participants. Samples were prepared in Duarte, California (City of Hope Medical Center) for shipment to test sites in the USA and were prepared in Duesseldorf, Germany (University Medical Center) for shipment to test sites in Europe. Test sites are listed in

Table 2. Participants in the BEST multicenter study of CD34+ cell enumeration.

Site location	**BEST Committee Investigator**	**Technical principal Investor**
Centre Regionale de Transfusion Sanguine Besancon, France	M Masse	E Racadot
BI-Deaconess Hospital Boston, MA	W Dzik	M Figoni
COBE BCT Laboratories Denver, CO	L Dumont	P Zumerchik A Hamstra
City of Hope National Med Center Duarte, CA	I Sniecinski	B Nowicki
Bone Marrow Donor Center Heinrick Heine University Duesseldorf	P Wernet	J Fischer
Red Cross Blood Bank Groningen	C Th Smit Sibinga	RL McShine MK Elias-Elias
Red Cross Blood Transfusion Service Helsinki	G Myllyla	R Kekomaki
National Blood Service London	M Contreras	S Armitage
Centro Trasfusionale e di Immunologia dei Trapianti, Ospedale Maggiore Policlinico Milan	P Rebulla	L Porretti
Biometric Imaging Inc Mt View, CA	B Manian	S Kunitake
Penn-Jersey American Red Cross Philadelphia, PA	S Murphy	F Garbrecht
Beckton-Dickinson Immunocytometry Systems San Jose, CA	J Lowder	K Manion

Table 2. Samples were coded so that the test sites did not know the expected concentration of CD34 cells in any sample. Samples were prepared by diluting an apheresis sample known to contain in large numbers of CD34+ cells in whole blood. Thus, the study assessed the performance of these assays when testing cells suspended in whole blood. Samples were sent by overnight courier and tested the day which they were received. The expected concentrations of CD34+ cells ranged from near zero (normal unmobilized whole blood) to 200/μL. The majority of samples (68%) were designed to have 5 – 50 CD34+ cells/μL. Samples included replicates and linear dilutions. Results were faxed to a central data center and analyzed at the completion of the study. No data were excluded.

Table 3. Inter-site CV (expressed as a %) results with the STELLer assay among four participating laboratories in North America.

	200 /µL (n = 3)	**50 /µL (n = 6)**	**25 /µL (n = 6)**	**10 /µL (n = 8)**	**5 / µL (n = 6)**
Sample #1	21 %	24 %	20 %	10 %	30 %
Sample #2	12 %	14 %	20 %	16 %	17 %
Sample #3	16 %	15 %	13 %	22 %	22 %
Sample #4	–	15 %	16 %	12 %	18 %
Sample #5	–	7 %	32 %	15 %	13 %
Sample #6	–	9 %	17 %	15 %	13 %
Sample # 7	–	–	–	16 %	–
Sample #8	–	–	–	8 %	–
Mean Intersite CV	16.6 %	13.9 %	20.1 %	14.4 %	23.8 %

In phase III, a commercially available CD34+ control cell (StemTROL, Coulter-Immunotech, Marseille, France) was distributed to participants. These cells are prepared from a KG1a cell line that has been modified to express the CD34 antigen with a brightness typical of that found in normal hematopoietic progenitors. The cells were locally diluted in normal blood to the following concentrations: 200/µL, 100/µL, 50/µL, and 25/µL. These samples were then coded and tested at sites using either the ProCOUNT assay of the STELLer assay.

Data analysis

Evaluation of the *inter-site precision* was determined as follows: for each expected cell concentration, the coefficient of variation was determined from the results obtained at the various sites. Because multiple samples were sent at a given cell concentration, the inter-site CVs for any specific concentration were then averaged and the mean inter-site CV at that concentration reported. *Intra-site precision* was determined as the average CV of the sites when testing replicate samples. The *lower limit of detection* was determined in two ways: first, the 95% confidence intervals of samples expected to contain 5 cells/µL and background cell concentrations (normal unstimulated blood) were determined from the results obtained at all sites. The assay was considered capable of distinguishing 5 cells/ µL from background if the 95% confidence intervals between these samples did not overlap. Second, for each sample the observed value of the background normal blood used as a diluent was subtracted from the value obtained for the sample expected to contain 5 cells/µL. The 95% confidence intervals of these results from all sites were then determined. The assay was considered able to distinguish 5 cells/µL from background if the 95% confidence interval did not overlap with zero. In the absence of any internationally recognized stable stan-

Table 4. Inter-site CV of the ProCOUNT assay among five laboratories in North America and six laboratories in Europe.

	200 /µL (n = 3)	50 / µL (n = 6)	25 / µL (n = 6)	10 / µl (n = 8)	5 / µL (n = 6)
European Sample #1	23 %	24 %	23 %	29 %	30 %
European Sample #2	27 %	16 %	19 %	29 %	28 %
European Sample #3	24 %	14 %	38 %	53 %	37 %
European Sample #4	–	15 %	25 %	20 %	63 %
European Sample #5	–	12 %	29 %	35 %	22 %
European Sample #6	–	11 %	44 %	18 %	25 %
European Sample #7	–	–	–	24 %	–
European Sample #8	–	–	–	22 %	–
American Sample #1	16 %	8 %	38 %	16 %	19 %
American Sample #2	9 %	19 %	35 %	12 %	19 %
American Sample #3	22 %	9 %	18 %	14 %	17 %
American Sample #4	–	37 %	35 %	34 %	30 %
American Sample #5	–	44 %	38 %	47 %	27 %
American Sample #6	–	35 %	32 %	30 %	37 %
American Sample #7	–	–	–	33 %	–
American Sample #8	–	–	–	54 %	–
Mean Intersite CV	20.4 %	20.2 %	31.2 %	29.4 %	29.6 %

dard, the *accuracy* of the assays is more difficult to determine. In the multicenter study, accuracy of the methods was assessed in three ways: first, we determined if the 95% confidence intervals of the observed results from all centers bracketed the expected value for each sample. Second, we determined if the linearity of results from samples prepared by serial dilution of the same original starting sample. Finally, we used a commercially-available stable antigen-positive cell line to prepare samples containing a known concentration of CD34+ cells and determined the proportion of test sites whose results were within 20% of the expected value.

Results

The inter-site CV (inter-site precision) was a main outcome measure of the study. The results for "wet" samples mailed during phase II at each different concentration of CD34+ cells are shown in Tables 3 and 4. As expected from principles of counting infrequent events, the CVs were higher for low concentrations than for higher concentrations. Nevertheless, the overall mean CVs were under 25% at all concentrations tested for the STELLer assay and were below 30% for the ProCOUNT assay. The higher observed mean inter-site CVs with the ProCOUNT assay cannot be directly compared to the STELLer assay because their was a greater number of participants who used the ProCOUNT assay and thus there was greater opportunity for variability. Intra-site CVs were determined from a limited number of samples which were sent as blinded replicates. Despite the few number of determinations, the intra-site CVs were quite low as shown in Table 5. This demonstrates that any given laboratory is able to reproducibly replicate its own results.

Table 5. The mean intra-site CV of replicate samples.

Samples	**STELLer**	**ProCOUNT**
200 /μL (n = 2)	6.8 %	7.5 %
50 /μL (n = 2)	11.1 %	11.5 %
25 /μL (n = 2)	15.3 %	16.3 %
10 /μL (n = 2)	8.8 %	15 %
5 /μL (n = 2)	18.7 %	21.5 %

The lower limit of detection was an important secondary outcome of the study. Because the study focused on CD34+ cell concentrations in whole blood over a dynamic range that represents that found in patients who have been mobilized by growth factors, we were particularly interested in the ability to accurately measure 5 – 10 CD34+ cells/μL. Patients who have undergone prior chemotherapy or radiotherapy treatment may not mobilize well to growth factor protocols as evidenced by low CD34 cell concentrations. A decision to collect CD34 cells by apheresis on such patients depends on the ability to accurately measure low cell concentrations. The BEST send-around study found that either the ProCOUNT assay or the STELLer assay could reliably distinguish 5 CD34/μL from background. This was demonstrated by the fact that the 95% confidence intervals of the uncorrected results on samples prepared to contain 5 CD34/μL was 8.5 – 10.7/μL (STELLer) and was 6.1 – 7.3/μL (ProCOUNT) and these values did not overlap with the natural background concentration of CD34 cells present in the whole blood used as a diluent to prepare the samples (3.3 – 3.9 /μL, STELLer; and 1.2 – 1.9 /μL, ProCOUNT).

Three sets of coded samples were prepared by serial dilution from the same starting sample. Results of these samples were used to determine the linearity of results. For each test site, the correlation coefficient was determined. The results ranged from r^2 = 0.991 to r^2 = 0.993 (STELLer) and from r^2 = 0.895 to r^2 = 0.998 (ProCOUNT). Regression lines surrounding the results gave a slope of 1.0 for both assays.

Relative assay bias was investigated. In the absence of any bias, it would be expected that one assay would report a higher value than the other assay for approximately 50% of samples. In fact, over the range of 5 – 50 CD34/μL (n = 26 samples) the STELLer assay reported a higher value than the ProCOUNT on

Table 5. Results with commercially-available CD34+ control cells.

STELLer assay				
Expected value	**% of observed values within 20% of expected**	**Inter-site CV (n=4)**	**Mean observed CD34+ /µL**	**95% CI of observed**
200 /µL	90 %	12 %	181.0	167–194
100 /µL	54 %	20 %	93.9	82–105
50 /µL	90 %	14 %	51.2	47– 56
25 /µL	72 %	13 %	28.3	26– 30
10 /µL	27 %	39 %	16.1	12– 20
ProCOUNT assay				
Expected value	**% of observed values within 20% of expected**	**Inter-site CV (n=6)**	**Mean observed CD34+ /µL**	**95% CI of observed**
200 /µL	87 %	14 %	191.5	177–205
100 /µL	80 %	15 %	112.8	104–121
50 /µL	73 %	17 %	57.6	53– 63
25 /µL	80 %	18 %	29.7	27– 32
10 /µL	53 %	36 %	15.2	12– 18

more than 50% of occasions suggesting a slight relative negative bias of the flow cytometry method compared to the microfluorimetry method.

Finally, the study included a preliminary evaluation of a commercially-available preparation of fixed (stabilized) cells expressing the CD34 antigen (StemTROL cells). Both the STELLer assay and the ProCOUNT assay had to be modified to accommodate the different size and scatter characteristics of the stabilized cells. Cells were distributed to each site which locally prepared 5 different dilutions in normal blood. The results are shown in Table 6. Using these stabilized cells, the inter-laboratory CVs were lower (higher precision) than seen using fresh cells that had been distributed by overnight courier. The improved performance presumably is due the fact that the stabilized cells did not undergo any sample deterioration during shipment. Thus, the inter-laboratory variability observed using the stabilized cells may, in fact, more truly represent the inherent inter-site variability of the assays. Results with the stabilized cells also suggested that the assays were accurate as the 95% confidence intervals of the reported results by test sites bracketed the expected results.

Overall, the BEST Working Party study demonstrated that a high degree of accuracy, precision, and linearity could be achieved and that results obtained from different laboratories could be compared if the laboratories shared common commercially available assay methods. In addition, the performance of either the flow cytometric assay or the microfluorimetric assay appeared comparable with no striking disadvantage to either method.

Subsequent Studies of CD34+ Cell Standardization

In addition to the BEST Working Party study, three additional groups have performed multicenter investigations of standardization of CD34+ cell analysis. Johnsen et al published results of the third Nordic workshop on CD34 analysis [11]. Using a modification of the Milan/Mulhouse flow method, they reported inter-site CVs below 19% in 7 of 8 samples that were analyzed. However, the 8 "samples" consisted of diskettes with listmode data (n=2), prestained fixed samples (n=4), and fresh samples (n=2). Thus, the CVs observed may not reflect the true variability that would be seen in actual practice.

Gratama et al published results of the European Working Group on clinical cell analysis studies of CD34+ cell counting [12]. Their comparison of several different methods highlighted areas for improvement that would lead to better agreement among different laboratories. These included bright fluorochrome conjugates, use of a nucleic acid dye, staining with anti-CD45 to be included in the analysis, Boolean gating, and inclusion of both $CD34+^{bright}$ and $CD34+^{dim}$ cells.

Barnett et al published results from the experience of the United Kingdom national external quality assessment scheme (UK NEQAS) [13]. This is an extensive multi-national exercise in which fixed samples are provided for quality control of flow cytometry. The UK NEQAS program distributes stabilized whole blood samples bi-monthly to 91 laboratories in 20 countries. Participants determine the CD34+ cell count according to their own in-house methods. CVs ranged from as low as 24% for CD34+ cells > 50/µL to CVs > 70% for CD34+ cell counts < 15/µL. It was noted that the majority of laboratories did not directly determine the absolute CD34+ cell count but rather indexed the result to a measurement of total WBC count. Among the various participating laboratories, twenty different methods were used to determine WBC counts. The CVs for laboratories that directly determined the absolute CD34+ cell count (by the use of beads or exact volumes) was approximately half that of laboratories that required a separate WBC cell determination. A survey conducted among the participants documented wide variation in the technical details used to determine the CD34+ cell concentration. Given that the cells in the UK NEQAS system are stabilized, the observed variability can be attributed almost exclusively to differences in assay technique. The authors concluded that an international consensus on methods of cell staining and analysis was urgently needed in order to standardize CD34 counting.

Concluding Remarks

Enumeration of CD34+ cells remains the single most important analyte in the field of hematopoietic transplantation. Accurate and precise methods with a lower limit of detection or the order of 5 – 10 cells/µL are valuable assays for determining the optimum time to collect stem cells, for determining the results of processing progenitor harvests in the laboratory, and for predicting the time from transplant to engraftment. Despite continuing advances in the methodology to detect CD34+ cells, there is still wide variation in the assay methods used among laboratories worldwide. This variability has contributed to a striking lack of precision and agreement among most multi-center studies conducted to date. The study by the BEST Working Party documented that cooperating laboratories could provide results that were in good agreement provided that reagents and assay methods were standardized. As noted by Olivero et al. [14] and by the BEST Working Party study, the use of commercial assay systems provides an important means to achieve uniformity in the measurement of CD34+ cells. The availability of stabilized cells expressing the CD34+ antigen represents an additional valuable tool for assay standardization among different laboratories.

Recent research has suggested that identification of a more primitive hematopoietic stem cell may be forthcoming. Indeed a recent editorial has called into question the importance of the CD34+ cell antigen as a marker for hematopoietic progenitors [15]. Among the new candidate markers is the KDR+ antigen, also known as vascular endothelial growth factor receptor. KDR+ cells appear to be capable of complete hematopoietic reconstitution [16]. The identification of a new stem cell phenotype will give laboratories around the world a second chance to agree *in advance* on a consensus methodology for counting this vitally important cell.

References

1. Dzik WH, Sniecinski I, Fischer J and the Biomedical Excellence for Safer Transfusion Working Party. Toward standardization of CD34+ cell enumeration: an international study. Transfusion 1999;39:856-63.
2. Law P, Traylor L, Recktenwald DJ. Cell analysis for hematopoietic stem/progenitor cell transplantation. Cytometry 1999;38:47-52.
3. McNiece I, Kern B, Zilm K, et al. Minimization of CD34+ cell enumeration variability using the ProCOUNT standardized methodology. J Hematother 1998;7: 499-504.
4. Keeney M, Chin-Yee I, Weir K, Popma J, Nayar R, Sutherland DR. Single platform flow cytometric absolute CD34+ cell counts based on the ISHAGE guidelines. International Society of Hematotherapy and Graft Engineering. Cytometry 1998;34:61-70.
5. Sims LC, Brecher ME, Gertis K, et al. Enumeration of CD34-positive stem cells: evaluation and comparison of three methods. J Hematother 1997;6:213-26.
6. Read EJ, Kunitake ST, Carter CS, Chau Q, Yu MY, Klein HG. Enumeration of CD34+ hematopoietic progenitor cells in peripheral blood and leukapheresis products by microvolume florimetry: a comparison with flow cytometry. J Hematother 1997; 6:291-301.
7. Brecher ME, Sims L, Schmitz L, et al. North American multicenter study on flow cytometric enumeration of CD34+ hematopoietic stem cells. J Hematother 1996;5: 227-36.
8. Lowdell MW, Bainbridge DR. External quality assurance for CD34+ cell enumeration—results of a preliminary national trial. Royal microscopical society clinical flow cytometry group QA schemes. Bone Marrow Transplant 1996;17:849-53.
9. Lumley MA, McDonald DF, Czarnecka HM, et al. Quality assurance of CD34+ cell estimation in leucapheresis products. Bone Marrow Transplant 1996;18:791-96.
10. MartiG, Johnsen H, Sutherland R, et al. A convergence of methods for a worldwide standard for CD34+ cell enumeration (letter). J Hematother 1998;7:105-09.
11. Johnsen HE, Baech J, Nikolajsen K. Validation of the Nordic flow cytometry standard for CD34+ cell enumeration in blood and autografts: report from the third workshop. J Hematother 1999;8:15-28.

12. Gratama JW, Orfao A, Barnett D, et al. Flow cytometric enumeration of CD34+ hematopoietic stem and progenitor cells. European Working Group on clinical cell analysis. Cytometry 1998;34:128-42.
13. Barnett D, Granger V, Storie I, et al. Quality assessment of CD34+ stem cell enumeration: experience of the United Kingdom National External Quality Assessment Scheme (UK NEQAS) using a unique stable whole blood preparation. Br J Haematol 1998;102:553-65.
14. Olivero S, AlarioT, Ladaique P, et al. CD34+ cell enumeration in peripheral blood and apheresis samples, using two laboratory diagnostic kits or an institutional protocol. Bone Marrow Transplant 1999;23:387-94.
15. Goodell MA. CD34+ or CD34- : Does it really matter? Blood 1999;94:2545-47.
16. Ziegler BL, Valtieri M, Porada GA, et al. KDR receptor: a key marker defining hematopoietic stem cells. Science 1999;285:1553-58.
17. Johnsen HE. Report from a Nordic workshop on CD34+ cell analysis: technical recommendations for progenitor cell enumeration in leukapheresis from multiple myeloma patients. J Hematother 1995;4:21-28.
18. Johnsen HE, Knudsen LM. Nordic flow cytometry-- standards for CD34+ cell enumeration in blood and leukapheresis products. Report from the second Nordic workshop. J Hematother 1996;5:237-45.
19. Brando B, Sommaruga E, Sienna S, et al. A multicenter quality evaluation of absolute CD34+ cell counts by two different flow cytometric methods (abstract) Cytometry 1996:23(Suppl 8):89.

DISCUSSION

Moderators: C.L. van der Poel and M.P. Busch

M.P. Busch (San Francisco, CA, USA): Dr. Coste, those who were positive, how did you double check them to rule out a false negative?

J. Coste (Montpellier, F): If a primary pool was positive, further testing of mini-pools was conducted. The Tecan Genesis pipettor would perform 14 cross-pools from the archive plate (6 vertical mini-pools of 8 sample donations and 8 horizontal mini-pools of 6). according to the design protocol, each individual sample is contained in two pools (a horizontal pool and a vertical pool), and may be identified by the PMS software at the intersection of these two pools. If no mini-pool was positive, the pool turned out to be a false negative and contamination was then suspected.

M.P. Busch: So you used exactly the same system?

J. Coste: Yes.

M.P. Busch: P-selectin in the plasma is responsible for inhibiting the clearance of these platelets. Is it competitively inhibiting the PSGL on macrophages for example.

A.D. Michelson (Worcester, MA, USA): I do not think plasma p-selectin is responsible for pathway? platelet clearance, because there is no difference in clearance whether or not there is release of soluble p-selectin.

M.P. Busch: My question was: does that increase in p-selectin competitively inhibit the platelets from binding for example to macrophages so that they do not get cleared.

A.D. Michelson: Again, there is no relationship between platelet clearance and plasma p-selectin (see Michelson et al)[1].

M.P. Busch: Dr. Coste, what I understand of this whole NAT business at is that we need not be so much concerned about the test; we were that in the beginning.

---

1. Michelson AD, Barnard MR, Hechtman AB et al. In vivo tracking of platelets: circulating degranulated platelets rapidly lose surface P-selectin but continue to circulate and function. Proc Natl Acad Sci USA. 1996;93(21):11877-82.

Is the main challenge of NAT that it generates too much manual work, too much logistic problems?

J. Coste: It is true that the technique is not a problem. What must be organised is the logistics. In France we have large regions, and these regions are becoming even larger because we are having a reorganisation of our transfusion network: the at present 43 blood transfusion centres will be reduced to 15 during the year 2000. So the transportation of the tubes to the NAT facility is really a major concern. The other part is of course the tracking down of the samples. It is true that during the feasibility study too many manual steps were performed and prevented complete tracking of the samples.

M.P. Busch: I would perhaps like to bring the discussion to the issue whether we should be doing this. It is truly an impressive contrast to see the data with respect of developing assays for developing countries, where still transmission is going on in an exorbitant rate of hepatitis C and HIV versus nucleic acid testing. The cost is impressive; in the US we are picking up as I indicated perhaps 1 in 300,000 screened donations and the cost from mini-pool testing that is being done now is in the range of US$ 5-8 per donation to actually accomplish the turn-around time. There is an enormous pressure to move towards single donation testing which will cost in the range of US$ 15 per donation. In the States we spend around 1.5 to 2 million dollars per infected donation detected, which is an outrageous expenditure for the very limited yield that we are observing. The data that is coming out of Europe is even lower in terms of yield and I think some recent data from Germany only had something in the range of 2 infections out of about 5 million tested over the last several years. Some of the pilot data I think from Great Britain show that they still have not detected an infected donation that was missed by standard testing. So, we are really at an asymptotic level of risk, spending an enormous amount of what is a limited resource for blood safety in our countries let alone in the world as a whole on detecting these rare residual infections. The problem is internationally we are leapfrogging each other, for the development in one country leads everyone to follow. So, it is really a problem.

H.G. Klein (Bethesda, MD, USA): I remember the days when we started looking at hepatitis B surface antigen with agar gel diffusion. Then, everyone wondered whether we would be able to introduce a sophisticated new technology called radio-immuno assay at hospitals in the United States. I guess we figured out how to do it. So, I think my question to you and to all the panellists this morning is this: With the advent of marvellous robotic technology, multiplex assays, and very sophisticated PCR's and rt PCR's, do you anticipate that the day may come when we all abandon all of these other assays as we did agar gel diffusion and look for the virus directly in all instances

M.P. Busch: It sounds appealing, but certainly historically the reality of getting rid of tests has been very difficult. I think as you look at specific assays the problem we have is that a small proportion of persons who are antibody positive do test negative by these nucleic acid amplification technologies. The question is would people be comfortable transfusing blood that was antibody positive for HIV or hepatitis C or antigen positive for HBV that was negative on nucleic acid tests, and particular recognising that we are only sampling the blood. The volume of specimen that is being analysed for viruses is a small fraction of the amount of blood in the components, which leads to the discussion around the relationship between the concentration of virus in terms of particles per ml against the infectivity. For some viruses like hepatitis B and probably hepatitis C it is really only a very small number of particles that results in transmission. As little as ten infectious units in a whole bag of blood is adequate to transmit and we cannot be fully confident that we can detect that infectivity in a small sample. So, I do not think that we will ever get rid of the major assays. There are some examples like p24, I think we can get rid of. But even something like hepatitis B surface antigen, only about 90-95% of surface antigen positive samples do test positive for nucleic acid, because of the excess production of antigen, it is part of biology. The answer is I think we will continue to do a battery of serologic tests. The question I am looking at is how many more of these more and more sophisticated nucleic acid type tests can we afford to do in a truly effective screening programme.

C.Th. Smit Sibinga (Groningen, NL): Dr. Coste, you mentioned that there were a couple of failures; some in the extraction where the yields were poor and others in the amplification detection. Were these basically caused by the methodology or was it a pure apparatus defined type of failure.

J. Coste: In our experience, this was really failure of the equipment. During the feasibility study we kept aliquots from each step of the assay: the pool tube, the archive plate, an aliquot from the purified RNA after extraction by the BioRobot and the A-rings from the Cobas. Upon testing again all these aliquots we could each time link the problem to a specific automatic failure: cross-contamination and low extraction yield were displayed for the BioRobot and detection failure was suspected with the Cobas.

C.Th. Smit Sibinga: Thank you very much.A question to dr. Fields. Fascinating technology; you spoke about the bad news side. Could you elaborate a little further about the consequences of that bad news side of your results? The second question: to what extent do you think the methodology might be further implemented and introduced to other markers, not only on the virology side but probably also on the blood grouping side.

H.A. Fields (Atlanta, CA, USA): The consequence of the bad news side of course is that you increase the number of false positives. In the blood bank arena I think you have to be most concerned about sensitivity and the false negative issue. The consequence of false positivity is that you have less donors to chose from. But as I tried to point out, the data I showed you was just our original prototype and by readjusting the concentrations of many of the components of the tests we now have excellent specificity as well. The technology is not a new technology actually. It is been around for many years. I think what drove the whole technology probably was the home pregnancy test which was the first of its genre. I certainly believe that this technology is adaptable and amenable for the detection of not only other markers of infectious diseases, but it certainly could be applied to blood grouping as well, which I think would have a very important market in the developing world. So, I think yes it can be extended much beyond infectious diseases.

A. Fletcher (Sidney, AUS): Dr. Busch, two questions. First, perhaps you would like to comment on the findings of the number of positives in the United States, using the NAT technology as compared with perhaps the calculated, theoretical findings and why the numbers are in fact somewhat less, particularly with HIV than expected. Secondly, a question which I am not sure whether you can answer for the whole of the United States, or just for part of the United States. What is your strategy for confirmatory testing, if you identify a NAT positive antibody negative sample?

M.P. Busch: In terms of actual yield versus projected yield the observed yield is as I summarised in the States we picked up a total of about 28 HCV RNA only cases and 1 HIV. The denominator I think is probably around 4 million or so now. So the yield of HCV is running around 1 in 300,000 and the yield of HIV 1 in 3-4 million, which based on our earlier projections are lower than we have projected. A couple of issues: one is the estimates of the window period that were used to estimate risk assumed that the entire window period might be infectious. We now are understanding, I think more precisely, that only a subset of the window period is viremic and infectious. It is only that portion which should be used to project the yield and probably the risk. The other is: the incidence data that we originally reported in the Schreiber paper[1] has been tracked over time. There is a new paper coming out in JAMA soon[2] that will show that the incidence rates for the viral markers in the US have actually declined significantly over the last 4 to 5 years. I think that probably also explains the discrepancy with respect to yield in Europe where the incidence rates are generally substantially lower than in the US. So, we are fine tuning our models to incorporate more accurate incidence rate estimates as well as more accurate understanding of the window period. We have also actually in the States picked up one test error case meaning a viremic donation that was by all the original test data serologically negative but when re-tested was actually determined to be antibody positive and strongly so and tracked back to an error performing the test. Most of the donors who we detected as HCV viremic have seroconverted on follow-up, but we have one case of a very atypical seroconverter developed low-level antibody reactivity, but then reverted down again. I understand from both

1. Schreiber GB, Busch MP, Kleinman SH, Korelitz JJ for the NHLBI Retrovirus Epidemiology Blood Donor Study. The risk of transfusion-transmitted viral infections. N Engl J Med 1996;334:1685-90.
2. Glynn SA, Kleinman SH, Schreiber GB, et al. for the Retrovirus Epidemiology Donor Study. Trends in incidence and prevalence of major transfusion-transmissible viral infections in United States blood donors, 1991 to 1996. JAMA 2000;(in press).

France, Australia and several other countries, that they picked up these occasionally immunosilent carriers with NAT testing. So, in terms of the overall model we factor in both window period and other sources of transmission risk. I think that the data is coming forward which pretty well supports those approaches for estimating risk and I think we are generally correct. The other question about confirmatory testing, we are concerned about this. In the US, where the FDA is such a difficult barrier to bring tests forward, we have been unable to get good confirmatory tests for our serologic assays, particularly as they have improved. The companies simply have not improved confirmatory testing and brought them forward because there is not money to be made. So, the problem with nucleic acid confirmation presented the same dilemma - how are we going to get companies to build totally separate tests that we can use for confirmation and try to get them approved. So, rather than trying to do that, we are developing a system that uses the two technologies to confirm one another. So we are using the Roche system to confirm the Genprobe and vice versa and are doing studies that are trying to demonstrate that they are comparable in terms of sensitivity and that the cross-over does detect false positives. It is looking quite promising and there are moderate amounts of false positives, say overall we probably have 10 false positives for every true positive, but most of them track back to low level contamination during either the pooling or the run stages. So, it is actually most useful to simply get an alternative sample source from reactive cases and determine quickly whether you can corroborate the positive result on a segment or tube and then of course follow the donor to verify infection.

R. Atamian (Beirut, Lebanon): Dr. Fields and whoever else who would like to answer and then a comment to dr. Michelson and a comment to dr. Fields. In vivo platelet activity, degranulated platelets and p-selectin, how we can look into these three aspects and to the following two cases. First, sometimes the harvest of platelets in apheresis shows microaggregates and second, in the case of AB0 mismatch platelet transfusion.

A.D. Michelson: If I got the question correct you are saying that platelets aggregate in the bag.

R. Atamian: No, sometimes when we are doing the apheresis and harvest platelets we see micro-aggregates in the tubing. So, it is a good practice that we have to put on the agitator for one hour before we transfuse. But when you have micro-aggregates what is the status of the in vivo activity, the p-selectin and the degranulated status of those platelets?

A.D. Michelson: Well, in short I do not know. I do not know anyone who studied that. P-selectin as I showed does not have a role in platelet-to-platelet aggregation although it certainly has a role in platelet-to-leukocyte aggregation.

R. Atamian: Concerning the ABO mismatch platelet transfusion.

A.D. Michelson: I would just say no-one has looked at humans in vivo. It is not easy to do in fact, that is why we looked at baboons; Berger and Wagner looked in mice[1], but no-one has looked at any more practical variations like AB0 and compatibility; it is just not known.

R. Atamian: Dr. Fields, concerning the lateral flow; your presentation is excellent and after I leave from here I would say I would go for the lateral flow. But since I have been dealing with the lateral flow rapid kits for the last 5 to 6 years I allow myself to comment a little. In the lateral flow kits we have two problems; first of all the scientific or the manufacturer part and then the market or the practical part. In the scientific part: by the time we have identification of new serotypes or new variants of the viruses; by the time they reach you in the CDC and you go into the studies and divide the variations, then it goes to the company who manufactures the ELISA kits, it takes 1 to 3 years. Now, if this pathway is applicable to the kits for the lateral flow automatically it will take 1-2 years more. For developing countries though the financial aspect is important, however, unfortunately there is going to be more business than science. I can tell you in some of the countries I have been to there is at least 7-8 types of the rapid kits and at least half of them are questionable, at least to me. If you look at the information presented by the agent it is wonderful, it is 99% sensitive; if you look into the scientific articles you will find sensitivity reaching as low as 85% and specificity sometimes reaching as low as to 65-70%. Now there is a problem; can we really trust these rapid kits. If yes which companies we have to elude to. I wish you could convey the message during your lectures. In order to implement the lateral flow kits in any country, there must be either a central laboratory or a quality assurance laboratory that will automatically test a sample from a major batch reaching the country, check for all kind of viruses and all kind of agents, before they are allowed to be marketed in that country.

H. Fields: You are absolutely right and I do appreciate your comments. I have spent my entire career in developing countries and I fully agree with what you are saying. So, it is part of the goals of the International Consortium for Blood Safety, which I really did not have a lot of time to go into this morning. One component of that is to assist national control authorities by way of giving them the technology, the tools, the reagents and the training so they can be in a better position to license not only domestically produced tests, but also those that are imported. We are developing strategies to assist national control authorities. We hope this will address the problem in an appropriate way. With regard to your first comment about finding different variants in different genotypes that may not behave as well with the current test format and the long lag time it takes to incorporate different antigens into a new test that they will be able to detect variants with equal sensitivity; that is a very important question; it is even more difficult in the hepatitis C field because of the patent rights assigned to Chiron corporation. In order to receive a license from Chiron actually you have to use

---

3. Berger G, Hartwell DW, Wagner DD. P-selectin and platelet clearance. Blood 1998: 92:4446-52.

their antigens or Chiron must approve new antigens. So, they really control the market and all these antigens, as I mentioned, are derived from genotype 1-A and there has been a number of publications clearly demonstrating that 1-A antigens may not always be optimal for the detection of antibody in individuals infected with other genotypes. We have tried to address this problem by looking at a variety of different sequences and you may recall that I did talk about the mosaic NS4 protein, which is composed of sequences that represent the most diverted sequences. We demonstrated that this particular antigen does have equal sensitivity to all the known six genotypes. We are looking at other genotype independent epitopes and we intend to include as many of those in our assays as we go forward in the future. So, we are very aware that this is a very important question and we are addressing it. I am not sure that Chiron is addressing it in the same way or at all and I worry about that.

M.S. Mincheff (Bethesda, MD, USA): Dr. Michelson, looking at your dot plots it seems to me that at 4, 6, 8 hours you had fewer dots of p-selectin positive platelets compared to 24 hours for example. Is it possible that these platelets are sequestered and then released?

A.D. Michelson: You are absolutely right. I think I mentioned that briefly. You are absolutely right that if the number of dots decreases during the mid time points, but they return and that was seen also in the graph of survival. So, you see sequestration both irrespective of whether or not you thrombin activated the platelets before, but we know that the platelets that return are the ones we infused, because they still have the PKH2 label on them. This has been shown also by other investigators that there is a sequestration phase.

M.S. Mincheff: What is p-selectin; I am not familiar with the molecule; is it a transmembrane protein, a single-pass or multi-pass protein; how is it released from the membrane.

A.D. Michelson: It is a single-pass protein.

M.S. Mincheff: Dr. Semple, I do not quite understand your experiment with pulsing dendritic cells with platelets. In order to have antibody formation the B-cells need to see the native antigen; where does the native antigen in these experiments come from.

J.W. Semple (Toronto, Ontario, C): Because there is still free platelets in the APC-preparation. We estimate that we gave between 1 to 5 million allo-platelets. We cannot get them all out; we are trying to do it, but it is very difficult. Theoretically, if we get rid of the free platelets we will not have an immune response. However, if you add back platelets to the pulse, you can in fact enhance the response.

M.K. Elias (Groningen, NL): Dr. Semple, thank you for your interesting presentation. We used to believe that only the donor antigen presenting cells are capable of presenting the allo-MHC molecules to the T-helper cells of the recipients for restricted recognition. As you said this is the rationale for leukocyte reduction. You have just the molecular changes of the indirect pathway and you suggest that manipulation of the recipient APC rather than the donor APC might be effective in reducing allo-immunisation for those patients will still become allo-immunised even after leukocyte reduction. But we do not know beforehand which patients are going to be allo-immunised. This is determined by the genetic constitution of the histocompatibility complex of an individual. We do not know either which pathway is going to be involved. So, manipulation of the patient's antigen presenting cells cannot be used as an alternative for leukocyte reduction; probably a dual approach?

J.W. Semple: I am not trying to say that we should not do leukocyte reduction. The fact is, in Canada for example, where we universally leukocyte reduce, I still get clinical colleagues who tell me about allo-immunisation that is taking place. So I am talking about those patients that are in fact still becoming allo-immunised.

M.K. Elias: Talking about prevention of allo-immunisation, not treatment of allo-immunisation probably a dual approach could be considered: elimination of the donor antigen presenting cells together with manipulation of the recipient antigen presenting cells.

J.W. Semple: Sure, it is a good point. Even with non-leukocyte reduced platelets you cannot predict who is going to respond because before the leukocyte reduction strategies came into effect in the mid eighties or so no-one could predict who was going to become immunised either. In those days only 50% of the patients became allo-immunised.

P.H.M.M. Kuijer (Utrecht, NL): Dr. Michelson, there is a sort of reasonable evidence that platelet concentrates which are transfused after a longer storage period have less increment on the platelet number in the patient. There was always the assumption that that was related to platelet activation because that usually increases during storage. How do you explain that phenomenon if the recovery is not related to activation; what is the explanation?

A.D. Michelson: I did not actually say it was not related to platelet activation. I said it was not related to platelet surface p-selectin and there is a big difference, because platelet activation is a complex process and what happens with different molecules is probably very different. Not only do they potentially express at different times; what happens after in vivo infusion is probably different also. I obviously focused on p-selectin, showing it was proteolysed. But to give another example: the activated confirmation of the GPIIb-IIIa complex which is very important as platelets aggregate. There is no doubt that during time of storage there is an increased number of activated IIb-IIIa on platelets. But the in vivo story appears to be different; this appears to be a reversible change; it is a confirmational change as opposed to proteolysis and that confirmation can apparently reverse back. I did not focus on that, but we did study that in in vivo studies the monoclonal antibody recognising its activated GPIIb-IIIa was actually normal at the first time point we measured in vivo at three minutes despite the fact it was elevated a 24 hours before we infused. So, activation covers a multitude of sins and there are many aspects to it.

P.H.M.M. Kuijer: Is the soluble p-selectin the one you measure in the plasma? If you try to do that in humans, you have the problem with dilution. If you transfuse the platelet concentrates in a small amount of plasma and you take peripheral blood samples after the transfusion you do not see it anymore, because it is so diluted in the whole plasma pool of the patient.

A.D. Michelson: In our studies in baboons we deliberately collected a large volume of platelets and infused autologously large volumes. That particularly enabled us to see a good signal of circulating platelets over the several days that we looked in vivo, but you are right the same applies to p-selectin and to my knowledge no-one has looked in humans in vivo what happens to soluble p-selectin in solution. I suspect that you are right that with atypical platelet transfusion you may not see a significant rise.

III. DEVELOPMENTS IN PROCESSING OF CELLS AND THERAPEUTIC PROTEINS

APPLICATION OF MOLECULAR HEMATOLOGY IN STEM CELL TRANSPLANTATION

C.E. van der Schoot[1], O.J.H.M. Verhagen[1], A.C. Lambregts[2], J.J.M. van Dongen[3], I.C.M. Slaper-Cortenbach[4]

Introduction

Stem cell transplantation is used to rescue patients after lethal irradiation or chemotherapy to overcome dose-limiting myelosuppression [1]. With the widespread availability of cytokines to mobilize haematopoietic progenitor cells from the marrow, peripheral blood has become the major source of stem cells. Other stem cell sources are bone marrow and cord blood. A potential complication of autologous haematopoietic progenitor cell infusion, regardless of its source, is the contamination of the stem cell graft with tumor cells. However, it should be noted that it is still debatable whether the presence of tumor cells in autografts has impact on clinical outcome. In this regard, it is remarkable that for example in Non Hodgkin Lymphoma (NHL) often relapses occur at the original site of disease after autologous stem cell transplantation with contaminated stem cell grafts. Moreover, the mobilization of haematopoietic cells may also lead to mobilization of tumor cells into the circulation. In the allogeneic setting the presence of high numbers of T-lymphocytes in peripheral blood stem cell (PBSC) grafts might increase the risk of graft versus host disease (GvHD). Therefore, several methods have been developed to minimize the number of contaminating cells. In principle, two approaches exist: stem cell enrichment or depletion/purging of contaminating cells. Theoretically, a higher log depletion can be reached by purging of contaminating tumor or T-lymphocytes than by enrichment of stem cells [2]. Even when stem cell enrichment leads to 99% purity of CD34+ cells, only a 2 log depletion is accomplished. However, in practice it is in most cases easier to enrich for stem cells, as for all purposes the same procedure can be applied. In addition, both approaches can be combined: mostly CD34+ cell isolation is followed by a purging procedure. Most isolation procedures make use of immunological methods, sometimes in combination with physical separation techniques. For acute myeloid leukemia also chemical purging procedures have been described.

---

1. Dept Experimental Immunohematology, Central Laboratory of the Blood Transfusion Service, Amsterdam, NL.
2. Dept Clinical Oncology, the Netherlands Cancer Institute, Amsterdam, NL.
3. Dept Immunology, Erasmus University, Rotterdam, NL.
4. Dept Hematology, University Medical Centre Utrecht, NL.

PCR TECHNOLOGY

The most sensitive method to detect the presence of contaminating tumor cells in stem cell grafts is the application of PCR technology. The major drawback of this approach has been the fact that in most cases only semi-quantitative results were obtained. However, the recently developed 'real-time' quantitative (RQ-PCR) PCR technology, based on Taqman chemistry, has greatly increased the abilities of the PCR approach [3]. This method is based on the 5'-3' nuclease activity of *Taq* DNA polymerase and an internal dubble-labeled fluorogenic probe with a 5'-reporter dye and a 3'-quencher dye. In intact probes, the quencher dye absorps the fluorescence signal of the reporter dye, resulting in no signal. During PCR, the 5'-3' nuclease activity of *Taq* DNA polymerase cleaves the hybridised probe, resulting in emission of a fluorescent signal that accumulates during the reaction. The fluorescence intensity is monitored during the PCR. This real time detection generates quantitative data based on the early cycles when PCR fidelity is the highest, resulting in a large dynamic range over five orders of magnitude. Using RQ-PCR it is possible to quantify the number of residual tumor cells with sensitivities of 10^{-5}, once a tumor specific target is present [4].

In the majority of hematological malignancies such a tumor specific target is present. In many lymphomas, in chronic myeloid leukemia and in about half of acute lymphoid and myeloid leukemia's specific chromosomal translocations are present. The genes involved in these translocations have been identified and it is therefore possible to use these genetic aberrations as RT- or DNA-PCR targets [5]. Although it should be noted that sometimes also in healthy individuals extremely low numbers of cells carrying these translocations are present. For example, in about 5% of blood donors we found positive results with a t(14;18) PCR, a translocation found in follicular NHL. Also the well-known bcr-abl transcript derived from the Philadelphia chromosome in CML has been detected in healthy, especially elderly individuals [6]. For all lymphoid malignancies it is possible to use antigen receptor rearrangements as patient specific tumor targets. Because recombination of the different gene segments occur at the genomic level these PCRs can be performed using genomic DNA. In a germ-line antigen receptor allele the distance between the two PCR-primers is too far and they will not lead to amplification of DNA. After rearrangement the in-between DNA sequences are spliced out and a PCR product is formed. In a population of polyclonal lymphocytes a smear of different PCR products is produced, whereas in a monoclonal proliferation a single PCR product is formed that can be sequenced. This DNA sequence can be used as tumor specific marker. In mature B-cell malignancies these sequences might also be used to produce tumor specific idiotype vaccines. In most non-hematological tumors no specific chromosomal translocations seem to be present. In that case, genes that are specifically expressed in e.g. epithelial cells or neuroblastoma cells, and therefore distinguishing these cells with certainty from the normal cells present in the stem cell graft, are used. However, the so-called 'illegitimate' expression at very low levels of epithelial cell specific genes in blood cells might be the cause of false positive

results [7]. Again, the quantitative results using RQ-PCR might resolve this problem.
Despite the availability of technical approaches for detection of contaminating leukemic or lymphoma cells in autologous stem cell grafts, only limited data are available on the clinical relevance of the eventual presence of these cells in the grafts. Much more evidence exists for the value of these techniques for monitoring response to therapy. Especially, in childhood ALL it has been shown that molecular response to induction therapy as determined by PCR is the most important independent prognostic factor [8,9]. Also in promyelocytic leukemia a clear distinction can be made, already in the early treatment phase, between patients that most likely will relapse later on and patients that will be cured from their disease. The most important study on the clinical significance of contaminating lymphoma cells in autologous stem cell grafts comes from the Dana Farber group in Boston. In a still ongoing study with very long follow up periods they demonstrate that PCR positivity of purged stem cell grafts in patients with follicular NHL is associated with a significantly increased risk of relapse [10]. Also our analysis of purged stem cell grafts of patients with childhood B-precursor ALL show that PCR-positivity of stem cell harvests is predictive for relapse [11]. However, from these studies it cannot be concluded that relapse is due to the re-infused tumor cells or due to concomitant higher tumor load in the patient itself. Gene-marking studies, in which the transplanted cells are retrovirally marked. suggest that the re-infused tumor cells at least contribute to relapse [12].

CONCLUSION

The development of new molecular techniques has improved the quality control of stem cell grafts. The same techniques can be used for following the patient after transplantation. In this regard, the recent results of donor lymphocyte infusions (DLI) are of special interest. When the molecular detection of impending relapse in CML patients or patients with Multiple Myeloma after allogeneic stem cell transplantation is followed by DLI, the occurrence of clinical relapse can in some cases be prevented.

REFERENCES

1. Thomas ED. A history of haemopoietic cell transplantation. Br J Haematol 1999;105:330-39.
2. Slaper-Cortenbach IC. Processing of hematopoietic stem cells: purging or selection? Vox Sang 1994;67:185-86.
3. Holland PM, Abramson RD, Watson R, Gelfand DH. Detection of specific polymerase chain reaction product by utilizing the 5'----3' exonuclease activity of Thermus aquaticus DNA polymerase. Proc Natl Acad Sci USA 1991;88:7276-80.
4. Verhagen OJHM, Willemse MJ, Breunis WB, et al. Application of germline IGH probes in real-time quantitative PCR for the detection of minimal residual disease in acute lympho-blastic leukemia. Leukemia 2000;14:(in press).
5. Rubnitz JE, Look AT. Molecular basis of leukemogenesis. Curr Opin Hematol 1998;5: 264-70

6. Biernaux C, Loos M, Sels A, Huez G, Stryckmans P. Detection of major bcr-abl gene expression at a very low level in blood cells of some healthy individuals: Blood 1995;86:3118-22.
7. Lambrechts AC, van 't Veer LJ, Rodenhuis S. The detection of minimal numbers of contaminating epithelial tumor cells in blood or bone marrow: use, limitations and future of RNA-based methods. Ann Oncol 1998;9:1269-76.
8. Steenbergen EJ, Verhagen OJ, van Leeuwen EF, et al. Prolonged persistence of PCR-detectable minimal residual disease after diagnosis or first relapse predicts poor outcome in childhood B-precursor acute lymphoblastic leukemia. Leukemia 1995;9: 1726-34.
9. van Dongen JJ, Seriu T, Panzer-Grumayer ER, et al. Prognostic value of minimal residual disease in acute lymphoblastic leukaemia in childhood. Lancet 1998;352: 731-38.
10. Freedman AS, Neuberg D, Mauch P, et al. Long-term follow-up of autologous bone marrow transplantation in patients with relapsed follicular lymphoma. Blood 1999; 94:3325-33.
11. Vervoordeldonk SF, Merle PA, Behrendt H, et al. PCR-positivity in harvested bone marrow predicts relapse after transplantation with autologous purged bone marrow in children in second remission of precursor B-cell acute leukaemia. Br J Haematol 1997;96:395-402.
12. Dilloo D, Rill DR, Grossmann ME, Leimig T, Brenner MK. Gene marking and gene therapy for transplantation medicine. J Hematother 1996;5:553-55.
13. Collins RH. Donor leukocyte infusions in 140 patients with relapsed malignancy after allogeneic bone marrow transplantation. J Clin Oncol 1997;15:433.

THE *IN VITRO* EFFECT OF CYTOKINES ON EXPANSION AND MATURATION OF MEGAKARYOCYTE PROGENITOR CELLS

A.L. Drayer[1], C.Th. Smit Sibinga[1], N.R. Blom[2], Joost Th.M. de Wolf[2], E. Vellenga[2]

Introduction

The availability of recombinant human haematopoietic growth factors has had a great impact on the field of stem cell transplantation. The use of cytokines to stimulate mobilization of bone marrow cells into the peripheral blood has facilitated the harvesting of stem and progenitor cells by apheresis procedures. In addition, the period of pancytopenia following high-dose chemotherapy is reduced when peripheral blood stem cells are transfused compared to bone marrow transplantation [1,2]. Despite this reduction there remains a cytopenic period of 1 to 3 weeks. The period of severe thrombocytopenia is one of the major problems following treatment with myeloablative chemotherapy, requiring frequent platelet transfusion support and increasing the risk of platelet refractoriness. The discovery of thrombopoietic growth factors in recent years has raised the hope that these might be used to reduce the need for platelet transfusions. Different approaches to accelerate platelet recovery are currently under investigation, involving *in vivo* treatment with growth factors [reviewed in Kuter 3] and supplementing stem cell transplants with autologous *ex vivo* expanded megakaryocyte progenitors [4,5]. Stimulation of platelet production by *in vivo* use of thrombopoietic growth factors does not appear to be effective in the myeloablative setting, possibly due to the low number of stem cells and high endogenous levels of platelet growth factors [3].

A variety of growth factors influence the proliferation of megakaryocyte precursor cells and the developments to mature megakaryocytes and platelets. Thrombopoietin (TPO), the ligand for the c-Mpl receptor, and its truncated form megakaryocyte growth and development factor (MGDF), are the most potent factors regulating megakaryopoiesis. *In vivo* and *in vitro* TPO and MGDF have been demonstrated to increase megakaryocyte and platelet production [6-9]. In addition, the TPO receptor (*c-mpl*) is expressed on $CD34^+$ cells and TPO stimulates proliferation of stem and progenitor cells of other lineages besides megakaryocytes [10-12]. TPO is not essential for megakaryopoiesis, as mice deficient in TPO production or the TPO receptor still produce megakaryocytes and platelets, although at a reduced level [13-16]. Combinations of TPO and early-acting

1. Blood Bank Noord Nederland, Groningen, NL.
2. Department of Haematology, University Hospital Groningen, NL.

cytokines such as interleukin (IL)-3, stem cell factor (SCF), granulocyte-macrophage colony-stimulating factor (GM-CSF), Flt3-ligand (FL), IL-l, IL-6 and IL-11 have been shown to induce optimal proliferation of megakaryocyte progenitors in suspension cultures [4,17-19].

A number of clinical studies have now indicated that progenitor cells expanded *in vitro* can be safely administered, although a more rapid haematopoietic recovery has not yet been unequivocally demonstrated [4,5,20,21]. Besides generating sufficient progenitors for clinical use on which most studies have concentrated, other factors may need to be taken into account such as the potential of expanded cells to migrate to the bone marrow, engraft, and mature under myelosuppressive conditions. Recent studies have demonstrated that the chemokine stromal cell-derived factor-l (SDF-l) and its receptor CXCR4 mediate migration of human CD34$^+$ cells [22,23] and engraftment in a NOD/SCID mouse model [24]. Interestingly, prior incubation of CD34$^+$ cells with SCF and IL-6 upregulated CXCR4 expression and expanded the number of SCID repopulating cells [24]. Inactivation of the SDF-l gene in mice resulted in a defect in B-cell lymphopoiesis and bone marrow myelopoiesis, indicating that SDF-l is one of the major chemoattractants of haematopoietic cells into the bone marrow [25]. In addition, the CXCR4 receptor was found to be expressed during megakaryopoiesis and SDF-l induced migration of megakaryocytes [26-29].

These data suggest that SDF-l and the CXCR4 receptor might play a role in homing of megakaryocyte progenitors in *ex vivo* expanded stem cell transplants. In this paper we compared the effect of different cytokine combinations on proliferation and differentiation of CD34$^+$ cells to megakaryocytic (CD61$^+$) cells and megakaryocyte progenitors, and on SDF-l induced migration of the expanded megakaryocytes.

Materials and methods

Suspension cultures

Peripheral blood cells were obtained from leukapheresis samples of patients undergoing autologous peripheral blood stem cell mobilization after informed consent. Purified CD34$^+$ cells were seeded in serum-free liquid medium (Cell-Gro, Boehringer Ingelheim, Heidelberg) plus growth factors as indicated. Final cytokine concentrations in cultures were 20 ng/ml for MGDF, 50 ng/ml for FL, and 10 ng/ml each for SCF, IL-3, IL-6, and IL-11. All cultures were performed at 37°C in a 5% CO_2 in air-humidified atmosphere, and samples were removed for analysis at the indicated time points. Viable cells were evaluated by trypan blue exclusion and cell counts were determined using a haemocytometer.

Megakaryocyte progenitor cell assay

A commercially available kit for evaluation of colony-forming units-megakaryocyte (CFU-Mk) was used (MegaCult-C, Stem Cell Technologies, Vancouver, Canada) following manufacturer's instructions. In short, 5×10^3 purified CD34$^+$ cells or suspension-culture expanded cells were seeded per double-chamber culture slide in serum-free medium containing thrombopoietin, IL-3, IL-6 and collagen. Cultures were incubated for 10 to 12 days followed by dehydration, fixation and immunocytochemical staining of the slides. Cultures were scored for the presence of pure megakaryocytic colonies consisting of at least 5 nucleated cells expressing Gp IIb/IIIa, without negatively stained cells (mixed and non-Mk colonies).

Flow cytometry

Regular two-colour flow cytometry analysis of suspension cultures was performed on a FACS Calibur flow cytometer and analysed using CellQuest software (Becton Dickinson, Mountain View CA). The percentage of antibody-specific stained cells was determined as compared with PE- or FITC-conjugated isotype controls. Thus, the number of stained cells in the culture was calculated using this percentage and the total cell counts determined by haemocytometer. Fold-expansion of stained cells after culture was expressed relative to the $CD34^+$ input level.

For sorting of cultured cells, 1×10^6 cells were labelled with CD34-PE and CD61–FITC antibodies, washed twice in PBS and filtered through a nylon sieve. Cells were separated by sorting on a FACStar flow cytometer (Becton Dickinson). The purity of the sorted cell fraction was greater than 950/0 as tested by re-analysis.

Migration assay

Expanded cells were washed and resuspended in serum-free medium without cytokines at a density of 10^6 cells/ml. Samples of 100 μl were placed on 8 μm pore Transwell plates (Costar, Cambridge, MA) and 600 μl of medium containing 100 ng/ml SDF-l (R&D Systems, Minneapolis, MN) was placed in the lower chamber. Controls for chemotaxis and chemokinesis were made by adding SDF-l to the upper chamber with cells and medium to the lower chamber (negative gradient), or by adding SDF-l to both chambers (zero gradient), respectively. After 3 h of incubation at 37°C in a $C0_2$ incubator migrated cells were collected and analysed.

Statistical analysis

Statistical evaluations were performed using a paired t-test. Values of P less than 0.05 were considered to be statistically significant.

Results

Expansion of $CD34^+$ and $CD61^+$ cells

The effect of different cytokine combinations on proliferation and differentiation of purified $CD34^+$ cells from mobilized peripheral blood to megakaryocytic ($CD61^+$) cells was determined. After 7 days incubation in serum-free suspension medium in the presence of MGDF plus SCF, the $CD61^+$ fraction was expanded 1.3 ± 0.5-fold over $CD34^+$ input number while the $CD34^+/CD61^-$ cell fraction was not expanded under these conditions (0.7 ± 0.l-fold expansion, mean ± SD, n=4). In order to obtain higher proliferation, additional cytokines known to stimulate proliferation of precursor cells and megakaryocytes were included in the cultures. In four paired studies we determined expansion after 7 days in the presence of MGDF plus SCF, compared to culture medium containing MGDF+ SCF+IL-3 and MGDF+SCF+IL-3+IL-6+IL-11+FL (six cytokines). Addition of IL-3 to serum-free cultures of MGDF+SCF resulted in a significant increase in proliferation of $CD61^+$ cells generating on average a three-fold expansion ($P<0.05$), while the number of $CD61^+$ cells was not further increased when the combination of six cytokines was provided. The $CD34^+/CD61^-$ cell fraction also showed a significantly higher proliferation in the presence of MGDF+SCF+IL-3 (2.7 ± 1(1-fold) compared to MGDF+SCF alone ($P<0.05$). The combination of six cytokines gave a further expansion of the $CD34^+/CD61^-$ cell fraction to 5.2 ± 1.9-fold over $CD34^+$ input values ($P<0.05$ compared to MGDF+SCF+IL-3).

In addition to megakaryocyte and progenitor cell proliferation we observed a 2.4 ± 1.4-fold expansion of $CD15^+$ cells in serum-free medium with MGDF+SCF +IL-3, and a 7.7 ± 3.3-fold increase in $CD15^+$ cells with the combination of six cytokines.

Expansion of clonogenic megakaryocyte progenitor cells

The starting population of purified $CD34^+$ cells contained on average a comparable number of megakaryocyte progenitors (CFU-Mk) with high and low proliferative capacity, as reflected in the number of large colonies (mean 48, with a range of 20-88 per 5000 $CD34^+$ cells plated, n=3) and small colonies (mean 45, range 32-52 per 5000 $CD34^+$ cells plated) formed in progenitor colony assay. We observed many loosely lying megakaryocytes besides colony formation in samples from cells that had been expanded in suspension culture. Therefore in this study, colonies were counted as CFU-Mk if they consisted of a minimum of five closely associated $CD41^+$ cells, and divided into megakaryocyte progenitors with low proliferative capacity when colonies contained 5-20 megakaryocytes, and progenitors with high proliferative capacity when CFU-Mk contained more than 2O megakaryocytes per colony.

The megakaryocyte progenitors that expanded in suspension culture consisted mainly of colonies with low proliferative capacity. The presence of MGDF plus SCF during culture resulted in 4.1 ± 1.6-fold expansion of CFU-Mk with 5-20 cells and maintained the high proliferative megakaryocyte progenitor population (1.3 ± 0.6-fold expansion). Addition of IL-3 besides MGDF+SCF and the combination of six cytokines increased the number of megakaryocyte progenitors with low proliferative capacity more than ten-fold over input number (12.1 ± 3.5-fold and l0.8 ± 3.2-fold, respectively). Culture in the combination of six cytokines lcd to a 4.3 ± 1.2-fold increase in megakaryocyte progenitors with high proliferative capacity, which was significantly more than in the presence of MGDF+SCF alone ($P<0.05$).

To evaluate the presence of CFU-Mk generating small and large colonies in subpopulations of suspension cultures depending on their expression of CD34 or CD61, day 7 cultured cells were divided into a $CD34^+/CD61^-$ fraction and a $CD61^+$ fraction by FACS sorting. The number and size of CFU-Mk in the subfractions were then determined, as shown in Table 1. The $CD61^+$ population contained progenitors with low proliferative capacity as this fraction gave rise mainly to colonies with <20 megakaryocytes. The $CD34^+/CD61^-$ population

Table 1. Distribution of megakaryocyte progenitors with high (>20 cells/colony) and low (5-20 cells/colony) proliferative capacity in day 7 suspension cultures generated in MGDF+SCF+IL-3. Values are given as the mean (range) per 5000 cells plated in colony assay for 3 independent experiments.

Size CFU-Mk	**$CD34^+/CD61^-$ fraction**	**$CD61^+$ fraction**
5-20 cells	158 (75-280)	41 (12-60)
>20 cells	23 (8-40)	1 (0-3)

contained megakaryocyte progenitors with differing proliferative capacity as demonstrated by the presence of both large and small colonies. In addition, the $CD34^+/CD61^-$ fraction of cultured cells showed a greater clonogenic potential than the $CD61^+$ fraction, as reflected in the number of colonies formed per cell plated (Table 1).

Kinetics of proliferation

In view of these results we were interested in the kinetics of $CD34^+$ and $CD61^+$ cell proliferation and megakaryocyte progenitor expansion in time. The $CD34^+/CD61^-$ cell population was two-fold increased by day 4, peaked at day 7, and decreased in cell number during prolonged culture. The $CD34^+/CD61^+$ cell population expanded 2.7-fold by 7 days of culture: this represented a 337-fold increase in double positive cells compared to the $CD34^+/CD61^+$ input values at the start of culture. By day 7, megakaryocytes with a more mature phenotype were detected as reflected by loss of the CD34 phenotype from $CD61^+$ cells. The proportion of CD61+ cells increased from approximately 1% $CD34^+/CD61^+$ cells in the starting population of purified cells to 26% double positive and 17% $CD34^-/CD61^+$ cells on day 7, and more than 60% $CD34^-/CD61^+$ cells on day 10 of culture. The number of CFU-Mk with low proliferative capacity expanded 13-fold and reached a maximum at day 7 of culture, while progenitors with a higher proliferative capacity showed a 4-fold expansion on day 4 and were not further increased during prolonged culture.

Maturation of megakaryocytes produced in suspension culture

To see if cells expanded in these suspension cultures were able to undergo differentiation into mature megakaryocytes, day 7 suspension cultures were enriched for megakaryocytic cells by FACS sorting of the $CD61^+$ cell population. Subsequently, cultured megakaryocytes were incubated in medium with 10% plasma and MGDF to induce proplatelet formation. Figure 1 shows cells which were initially cultured in MGDF+SCF+IL-3 at different stages of megakaryocyte maturation. Within 3 days after replating, megakaryocytes with a well developed demarcation membrane system (Fig lA), proplatelet formation and platelet-sized fragments (Fig lB) were observed . Similar results were found for $CD61^+$ cells generated in MGDF+SCF or in the presence of six cytokines, indicating that megakaryocytes expanded in the different cytokine combinations could develop into mature proplatelet-forming cells in vitro.

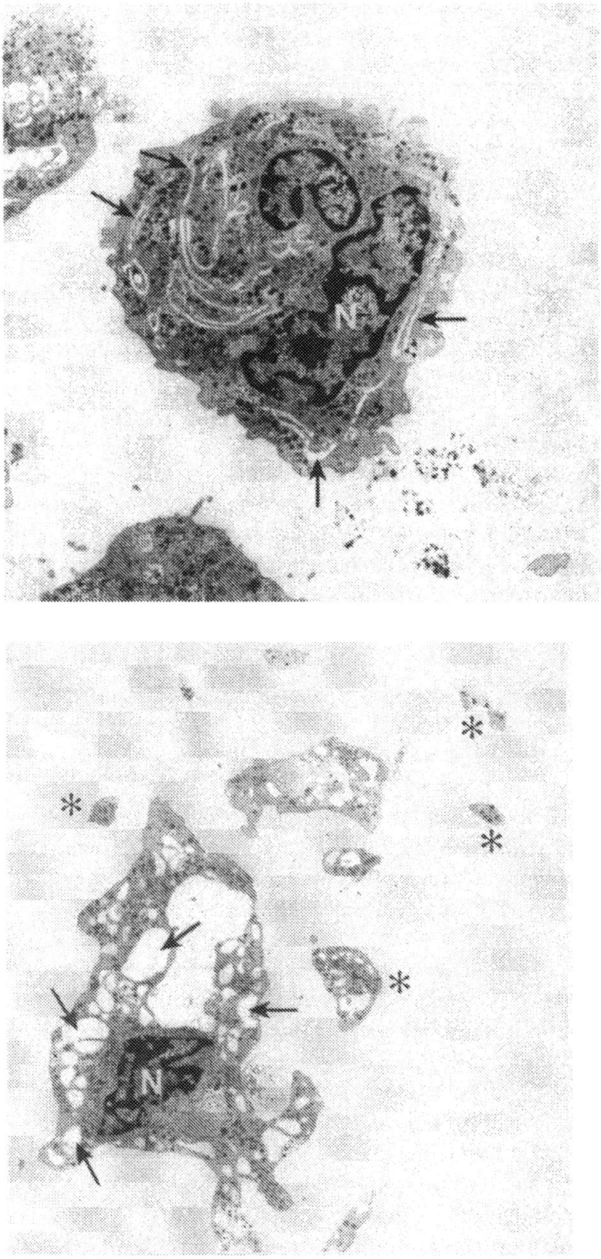

Figure 1. Electron microscopy of ex vivo expanded megakaryocytes. Top: Megakaryocyte demarcation membranes (arrows) and excentric-lying nucleus (N). Original magnification 5000×. Bottom: Megakaryocyte with well developed demarcation membrane system (arrows) and platelet-sized fragments (asterisks). Original magnification 3000×.

Expression of CXCR4 on expanded megakaryocytes

Previous experiments by others [26-29] demonstrated that the CXCR4 receptor is expressed on megakaryocytes derived from bone marrow platelets and on megakaryocytes generated *in vitro* from CD34$^+$ cells during of additional suspension culture with MGDF+SCF. We examined the cell-surface expression of the CXCR4 receptor by flow cytometry on megakaryocytes expanded in the presence growth factors. CXCR4 was expressed by the majority of CD61$^+$ cells generated after 7 days in suspension culture with MGDF+SCF (mean 82.3% ± 9.1%, n=8). Megakaryocytes cultured in the presence of additional cytokines showed a significantly reduced expression of CXCR4. A mean of 51.2 ± 14.3% and 49.0 ± 15.7% of CD61$^+$ cells were CXCR4+ when grown in medium containing MGDF+SCF+IL-3 or the combination of six cytokines, respectively ($P<0.01$ compared to MGDF+SCF).

Migration to SDF-1

We subsequently investigated the ability of megakaryocytes expanded *in vitro* to migrate to SDF-1 using Tran swell plates as described by Kim and Broxmayer [23]. Day 7 cultures were washed and resuspended in serum-free medium without cytokines prior to chemotaxis assays to exclude any effect of cytokines in the culture medium. In the absence of SDF-1, or when SDF-1 was added to both compartments of the Transwell plates, less than 2% of input cells migrated. The total number of cells that migrated during the 3 hour assay period was comparable for cells cultured in MGDF+SCF, MGDF+SCF+IL-3 or the combination of six cytokines. However, the ability of CD61+ cells to migrate to SDF-1 was consistently lower for cells grown in the presence of IL-3 or six cytokines. The migration potential of megakaryocytes cultured in MGDF+SCF+IL-3 was 4.2 ± 2.3-fold reduced compared to cells cultured in MGDF+SCF alone (mean ± SD, n=4, P=0.09) and showed a 4.3 ± 1.3-fold reduction for megakaryocytes cultured in MGDF+SCF+IL-3+IL-6+IL-11+FL (n=4; $P<0.05$ compared to MGDF+SCF). As we had found that megakaryocytes cultured in MGDF+SCF +IL-3 and the six cytokine combination expressed lower levels of CXCR4 than cells cultured in MGDF+SCF alone, we examined the migrated megakaryocytes for the presence of CXCR4. Double staining with anti-CD61 and CXCR4 antibodies demonstrated that approximately 90% of migrated megakaryocytes were CXCR4$^+$ for cells from all culture conditions. These results suggest that the lower expression of CXCR4 on megakaryocytes expanded in the presence of additional cytokines besides MGDF+SCF could, in part, account for the reduced migration towards SDF-1. No significant difference in the chemotactic potential of CD34+/CD61$^-$ cells generated in MGDF+SCF and in the additional presence of IL-3 or six cytokines was observed.

The effect of additional cytokines during growth on the ability of megakaryocyte progenitors to migrate to SDF-1 was evaluated by determining the number of CFU-Mk before migration and in the migrated fractions. 12.8 ± 1.7% of megakaryocyte progenitors with a low proliferative capacity and 26.2 ± 0.6% of progenitors generating larger CFU-Mk were able to migrate from cells cultured in MGDF+SCF. Migration from cultures expanded in MGDF+SCF+IL-3 and the six cytokine combinations was reduced for megakaryocyte progenitors with a low proliferative capacity (7.2 ± 1.7% and 8.7 ± 01% of CFU-Mk migrating, respectively) and was also reduced for progenitors with a higher proliferative capacity (11.3 ± 0.8% and 18.1 ± 2.6%, respectively).

Discussion

The number of CFU-Mk and $CD34^{+/}CD61^{+}$ cells in a graft are the most important factors predicting time to platelet engraftment after autologous stem cell transplantation [30-33]. These results suggest that by increasing the number of megakaryocyte progenitors in the stem cell transplant, a more rapid platelet recovery might be achieved. In the present study the effects of different growth factor combinations were tested for the *in vitro* generation of megakaryocyte progenitors from $CD34^{+}$ cells with special emphasis on the type of colony formation and the capacity of the generated megakaryocyte progenitors to migrate. The results demonstrate that addition of cytokines such as IL-3 or IL-6, IL-11 and FL to the combination of MGDF plus SCF significantly enhanced the degree of expansion of megakaryocytic progenitors. These results are comparable to the studies of Bertolini et al. [4], Dolzhansky et al. [18], Ratajczak et al. [34] and Van den Oudenrijn et al. [35] whereby high expansion of megakaryocyte progenitors was achieved after inclusion of additional cytokines to suspension cultures besides MGDF. However, in the present study we made a distinction between two types of megakaryocytic colonies: those, which consisted of 5-20 cells per colony and colonies of more than 20 megakaryocytic cells. The latter are the high proliferative colonies that are generated exclusively from the $CD34^{+}/CD61^{-}$ cell fraction, while the small colonies with a low proliferative potential originate from the $CD61^{+}$ and $CD34^{+}/CD61^{-}$ cell fractions. It appeared that the growth factor combination has a significant effect on the type of progenitor that is expanded. The addition of IL-6, IL-11, FL and/or IL-3 to the combination of MGDF plus SCF resulted especially in an expansion of progenitors with small colony formation as a consequence of the ongoing proliferation and differentiation of $CD34^{+}$ cells to $CD61^{+}$ cells. These data indicate that the degree of expansion has to be linked to the type of megakaryocyte progenitor expanded and cannot be the sum of the total number of colonies. These findings might indicate that inclusion of a combination of cytokines stimulating proliferation might propagate a cell population that especially affects the short repopulating potential post-transplantation. Despite the differences on the degree of expansion with the different growth factor combinations, no significant effect was observed on the maturation of the megakaryocytes from $CD61^{+}$ cells to platelets. Proplatelet formation was comparable for the different growth factor combinations used as demonstrated with electron microscopy studies.

Recent studies have indicated that the expression of CXCR4 on immature haematopoietic cells has important consequences for the degree of trans-endothelial chemotaxis and marrow homing [22,24,36]. In addition, Peled et al. [24] demonstrated that incubation with SCF and IL-6 induced surface expression of CXCR4 on $CD34^{+}$ cells and increased migration and bone marrow engraftment in NOD/SCID mice. In view of these results we questioned whether the different growth factors might affect the expression of CXCR4 and subsequently influence the degree of megakaryocyte migration. The present study shows that IL-3 or a combination of I L-3, IL-6, IL-II and FL significantly reduces the expression of CXCR4 on megakaryocytes cultured in the presence of MGDF plus SCF. Moreover, the reduced expression of CXCR4 did have functional consequences in view of the results of the cell migration assay. In relation to the reduced receptor expression, the chemotactic potential of CFU-Mk and CD61^{+} cells towards SDF-1 was found to be diminished, suggesting that IL-3 and additional growth factors besides MGDF+SCF significantly impair the capacity of megakaryocytic cells to migrate. However, it is unclear at the moment whether this finding has clinical consequences for engraftment of *ex-vivo* expanded stem cells. In conclusion, these findings indicate that the expansion of cell type and function is significantly affected by the growth factor combination, whereby expansion to high numbers of cells does not imply that the potential of the expanded cells to migrate is enhanced.

References

1. Beyer J, Schwella N, Zingsem J, et al. Hematopoietic rescue after high-dose chemotherapy using autologous peripheral-blood progenitor cells or bone marrow: A randomized comparison. J Clin Oncol 1995;13:1328-35.
2. Hartmann O, Le Corroller AG, Blaise D, et al. Peripheral blood stem cell and bone marrow transplantation for solid tumors and lymphomas: hematologic recovery and costs. A randomized, controlled trial. Ann Intern Med 1997;126:600-07.
3. 3. Kuter DJ, Cebon J, Harker LA, Petz LD, McCullough J. Platelet growth factors: potential impact on transfusion medicine (review). Transfusion 1999;39:321-132.
4. Bertolini F, Battaglia M, Pedrazzoli P, et al. Megakaryocytic progenitors can be generated *ex vivo* and safely administered to autologous peripheral blood progenitor cell transplant recipients. Blood 1997;89:2679-88.
5. Reiffers J, Cailliot C, Dazey B, Attal M, Caraux J , Boiron JM. Abrogation of post-myeloablative chemotherapy neutropenia by *ex-vivo* expanded autologous CD34-positive cells. Lancet 1999;354:1092-93.
6. De Sauvage FJ, Hass PE, Spencer SD, et al. Stimulation of megakaryocytopoiesis and thrombopoiesis by the c-mpl ligand. Nature 1994;369:533-38.
7. Lok S, Kaushansky K, Holly RD, et al. Cloning and expression of murine thrombopoietin cDNA and stimulation of platelet production *in vivo*. Nature 1994;369:565-68.
8. Wendling F, Marashovsky E, Debili N, et al. c-Mpl ligand is a humoral regulator of megakaryocytopoiesis. Nature 1994;369:571-74.
9. Kaushansky K, Lok S, Holly RD, et al. Promotion of megakaryocyte progenitor expansion and differentiation by the c-mpl ligand thrombopoietin. Nature 1994;369: 568-71.
10. Sitnicka E, Lin N, Priestley GV, et al. The effect of thrombopoietin on the proliferation and differentiation of murine hematopoietic stem cells. Blood 1996;87:4998-5005.
11. Ramsfjell V, Borge 0J, Veiby 0P, et al. Thrombopoietin, but not erythropoietin, directly stimulates multilineage growth of primitive murine bone marrow progenitor cells in xynergy with early cytokines: Distinct interactions with the ligands for c-kit and flt3. Blood 1996;88:4481-92.
12. Solar GP, Kerr WG, Zeigler FC, et al. Role of c-mpl in early hematopoiesis. Blood 1998;92:4-10.
13. Gurney AL, Carver-Moore K, de Sauvage FJ, Moore MW. Thrombocytopenia in c-mpl-deficient mice. Science 1994;265:1445-47.

14. Alexander WS, Roberts AW Nicola NA, Li R, Metcalf D. Deficiencies in progenitor cells of multiple hematopoietic lineages and defective megakaryocytopoiesis in mice lacking the thrombopoietin receptor c-mpl. Blood 1996;87:2162-70.
15. Carver-Moore K, Broxmeyer HE, Luoh SM, et al. Low levels of erythroid and myeloid progenitors in thrombopoietin- and c-mpl-deficient mice. Blood 1996;88: 803-08.
16. De Sauvage FJ, Carver-Moore K, Luoh SM, et al. Physiological regulation of early and late stages of megakaryocytopoiesis by thrombopoietin. J Exp Med 1996;183: 651-456.
17. Broudy VC, Lin NL, Kaushansky K. Thrombopoietin (c-mpl ligand) acts synergistically with erythropoietin, stem cell factor, and interleukin-11 to enhance murine megakaryocyte colony growth and increases megakaryocyte ploidy *in vitro*. Blood 1995;85:1719-26.
18. Dolzhansky A, Hirst,J, Basch RR, Karpatkin S. Complementary and antagonistic effects of IL-3 in the early development of human megakaryocytes in culture. Br J Haemat 1998;100:415-26.
19. Williams JL, Pipia, GG, Datta NS, Long MW. Thrombopoietin requires additional megakaryocyte active cytokines for optimal *ex vivo* expansion of megakaryocyte precursor cells. Blood 1998;91:4118-426.
20. Brugger W Heimfeld S, Berenson RJ, Mertelsmann R, Kanz L. Reconsitution of hematopoiesis after high-dose chemotherapy by autologous progenitor cells generated *ex vivo*. New Engl J Med 1995;333:283-87.
21. Williams SF, Lee WJ, Bender JG, et al. Selection and expansion of peripheral blood $CD34^+$ cells in autologous stem cell transplantation for breast cancer. Blood 1996; 87:1687-92.
22. Aiuti A, Webb IJ, Bleul C, Springer T Gutierrez-Ramos JC. The chemokine SDF-1 is a chemoattractant for human $CD34^+$ hematopoietic progenitor cells and provides a new mechanism to explain the mobilization of $CD34^+$ progenitors to peripheral blood. J Exp Med 1997;185:111-20.
23. Kim CH, Broxmeyer HE. *In vitro* behaviour of hematopoietic progenitor cells under the influence of chemoattractants: Stromal cell-derived factor-l, steel factor and the bone marrow environment. Blood 1998;91:100-10.
24. Peled A, Petit I, Kollet O, et al. Dependence of human stem cell engraftment and repopulation of NOD/SCID mice on CXCR4. Science 1999;283:845-48.
25. Nagasawa T, Hirota S, Takakura N, et al. Defects of B-cell lymphopoiesis and bone marrow myelopoiesis in mice lacking the CXC chemokine PBSF/SDF-l. Nature 1996;382:635-38.
26. Hamada T, Mohle R, Hesselgesser J, et al. Transendothelial migration of megakaryocytes in response to stromal cell-derived factor 1 (SDF-l) enhances platelet formation 1998;188:539-48.
27. Wang JF, Liu ZY, Groopman JE. The α-chemokine receptor CXCR4 is expressed on the magakaryocytic lineage from progenitor to platelets and modulates migration and adhesion. Blood 1998;92:756-64.
28. Kowalska MA, Ratajczak J, Hoxie J, et al. precursors, megakaryocytes and platelets express the HIV co-receptor CXCR4 on their surface: determination of respone to stromal-derived factor-1 by megakaryocytes and platelets. Br J Haemat 1999;104: 220-29.
29. Rivière C, Subra F, Cohen-Solal K, et al. Phenotypic and functional evidence for the expression of CXCR4 receptor during megakaryocytopoiesis. Blood 1999;93: 1511-23.
30. Dercksen MW, Weimar IS, Richel DJ, et al. The value of flow cytometric analysis of platelet glycoprotein expression on $CD34^+$ cells measured under conditions that prevent P-Selectin-mediated binding of platelets. Blood 1995;86:3771-82.

31. Takamatusu Y, Harada M,Teshima T, et al. Relationship of infused CFU-GM and CFU-Mk mobilized by chemotherapy with or without G-CSF to platelet recovery after autologous blood stem cell transplantation. Exp Hematol 1995;23:8-13.
32. Feng R, Shimazaki C, Inaba T, et al. $CD34^+/CD41a^+$ cells best predict platelet recovery after autologous peripheral blood stem cell transplantation. Bone Marrow Transplantation 1998;21:1217-22.
33. Gonzalez-Requejo A, Madero L, Diaz MA, et al. Progenitor cell subsets and engraftment kinetics in children undergoing autologous peripheral blood stem cell transplantation. Br J Haemat 1998;101:104-10.
34. Ratajczak MZ, Ratajczak J, Machalinski B, Mick R, Gewirtz AM. *In vitro* and *in vivo* evidence that *ex vivo* cytokine priming of donor marrow cells may ameliorate posttransplant thrombocytopenia. Blood 1998;1:353-59.
35. Van den Oudenrijn S, De Haas M, Calafat J, Van der Schoot CE. Von dem Borne AEGK. A combination of megakaryocyte growth and development factor and interleukin-1 is sufficient to culture large numbers of megakaryocytic progenitors and megakaryocytes for transfusion purposes. Br J Haemat 1999;106:553-63.
36. Zou YR, Kottmann AH, Kuroda M, Taniuchi I, Littman DR. Function of the chemokine receptor CXCR4 in haematopoiesis and in cerebellar development. Nature 1998;393:595-99.

MANIPULATION AND CONDITIONING OF T LYMPHOCYTES: DEVELOPMENT OF A CLOSED CULTURE SYSTEM FOR THE *EX VIVO* GENE TRANSFER INTO HUMAN T LYMPHOCYTES[1]

J.M. Robinet, Certoux, E., Rousseau, B. Petracca, I. Newton, P. Hervé, P. Tiberghien[2]

Introduction

Graft-versus-host disease (GvHD) is a severe complication following allogeneic haematopoietic stem cell (HSC) transplantation and is associated with significant morbidity and mortality [1]. GvHD results from the inappropriate activation toward the recipient's minor and/or major histocompatibility antigens of donor T cells contaminating the marrow graft and leading to host tissue injury [1]. T cell depletion of the HSC graft reduces the incidence and severity of GvHD, but results in higher rates of relapse and graft rejection [2]. An alternative to T cell depletion is the use of post-transplantation immunosuppression: this does not interfere with marrow engraftment and allows for a significant graft-versus-leukaemia (GVL) effect [3]. However, immunosuppression only partly controls GvHD occurrence and is associated with increased risks of secondary EBV-positive B lymphoma and late infections [4].

Thus, a major goal in HSC transplantation is to preserve the beneficial effects of donor T cells (engraftment-promoting properties and anti-leukaemic effects) and avoid their deleterious effects (GvHD-inducing activity). New approaches need to be developed in order to control more specifically GvHD-inducing donor T cells, while preserving the functions of other T cells and non-T cells (NK, B cells, monocytes, etc.). We have therefore recently initiated a clinical trial aimed at controlling GvHD occurrence using a gene therapy approach [5] which involves the introduction into donor T cells of a gene coding for a susceptibility factor ('suicide' gene) which renders them sensitive to a cytotoxic 'pro-drug' [6]. When administrated to the patient, this prodrug is transformed by gene-modified cells into a toxic compound, leading to the specific elimination of gene-modified, alloreactive donor T cells.

A suicide gene encoding the Herpes Simplex-thymidine kinase (HS-tk) enzyme is introduced *ex vivo* via a retroviral vector into activated donor T cells, together with a positive selection marker, the bacterial neomycin resistance gene (NeoR gene) encoding the neomycin phosphotransferase II enzyme. After a

---

1. This study was financed by the Association pour la Recherche sur le Cancer (ARC), the Ligue Nationale de Lutte contre le Cancer, Inserm/Agence Française du Sang, funds from the Programme Hospitalier de Recherche Clinique (PHRC) and the European Community (Biomed contracts).

2. ETS de Franche-Comté, Besançon, France.

selection process, based on the NeoR-mediated resistance of transduced cells to an analogue of neomycin toxic to eucaryotic cells (G418), gene-modified cells are reinfused into the patient, together with a T-cell depleted HSC graft. The aim is to obtain T-cell mediated beneficial effects. However, if GvHD occurs, ganciclovir (GCV), a thymidine analogue, is administered in order to induce the specific death of alloreactive, proliferating transduced T cells. GCV is specifically phosphorylated by HS-tk, after which GCV-monophosphate is further phosphorylated by cellular thymidine kinases. The DNA integration of the GCV-triphosphate during replication prevents DNA elongation by DNA polymerase, leading to growth arrest. The final result is the death of proliferating HS-tk-expressing T cells.

The conditions required for a stable transgene integration in target cells have been previously described [7]. Retroviral-mediated gene transfer requires preliminary T cell activation with OKT3 and IL-2 for 3 days, after which T cells are incubated for 24 hours with the retroviral-containing medium, then further rested for 24 hours. Transduced cells are then selected for 7 additional days in the presence of G418. Only transduced cells, expressing the NeoR gene, survive in the selection medium.

Cell therapy and *ex vivo* gene therapy represent growing fields of application for the treatment of various acquired, autoimmune or inherited diseases. Since gene therapy products are considered as drugs, gene-modified cells must be processed and handled in accordance with Good Manufacturing Practices (GMPs) standards. The *ex vivo* manipulation of blood products therefore requires a high level of technology in order to guarantee absolute safety of all products, both for patients and the environment. In the present study, we describe a closed culture system developed to generate and expand the gene modified T cells used in our clinical trial. We demonstrate that such a culture system makes it possible to reproducibly prepare gene-modified T cells for clinical use.

Materials and Methods

Selection of donors and recipients

Patients undergoing allogeneic HSC transplantation and presenting a high risk of GvHD were selected for the administration of gene-modified donor T cells. The rationale and description of this clinical trial have been previously published [5]. All donors were negative for HIV-1, HIV-2, HTLV, HBV, HCV. Informed consent was obtained from both donors and recipients.

Diposable sterile material

The following cell handling bags were provided by, or purchased from, Baxter (Deerfield, IL): Blood Collection Pack® bags (CP bags, ref. R7311), 400 ml Transfer Pack bags (400 ml TP bags, ref. R2074), Cell Wash Sets® (CWS, ref. ITX1015), LifeCell® bags (ref. R2111) and Cryocyte® bags (ref. R9951). The sampling bags (ref. VSE0000A) and the 150 ml Transfer Pack bags (150 ml TP bags, ref. VSE1001Q) were purchased from Maco-Pharma (Tourcoing, France). The interlink connection system and injection set were purchased from Baxter and Becton-Dickinson (Le Pont de Claix, France).

Serum source

Autologous serum. A 300-ml sample of peripheral blood was obtained from the HSC donor in a CP bag from which coagulant had been removed and left overnight at room temperature to allow blood clotting. The CP bag was then centrifuged (10min, 1850g, 20°C) using a semi-adapter (Baxter) in a KR4.22 centrifuge (Jouan, Saint-Herblain, France). The CP bag was then connected to a 150ml TP bag using a sterile connection device (Terumo, Elkton, MD) and the serum transferred into the TP bag using a plasma extractor. The TP bag was centrifuged to remove residual contaminating red blood cells. Serum was then collected in a CWS, decomplemented (30min, 56°C) and stored at 4°C until use.

Allogeneic secured AB serum. In some cases, AB serum (HIV-1, HIV-2, HTLV-1, HCV, HBV negative) from allogeneic donors was used instead of autologous serum. The serum was placed 4 months in quarantine and checked for the absence of anti-HLA antibodies, after which the serum donor was summoned again in order to redetermine her/his serological status before releasing the AB serum bags for clinical use.

Cell culture media and solutions

The following products were used for cell isolation: endotoxin-free Ficoll in glass bottles (Pharmacia Biotech, Uppsala, Sweden) and 0.9% NaCl solution bag (Macoflex: Maco-Pharma, Tourcoing, France). Cells were cultured in complete medium (CM) consisting of RPMI 1640 (BioWhittaker, Walkersville, MD) with L-glutamin, penicillin, streptomycin and 5 to 10% autologous or allogeneic AB serum. CM was supplemented, as indicated, with 500 or 1000U/ml Interleukin-2 (IL-2, Proleukin: Chiron, Amsterdam, Netherlands) and, as indicated, with 10ng/ml OKT3 (Cilag, Levallois-Perret, France) or 5µg/ml protamine sulfate (Choay, Gentilly, France) or 800µg/ml G418 (geneticin: Gibco BRL, Paisley, Scotland). Human serum was taken from its bag through an interlink injection site and added to the RPMI 1640 bottles using a syringe. Others supplements to the CM were introduced directly into the bags with a syringe connected to an interlink injection site. The retroviral vector G1TK1SVNa (Genetic Therapy, Inc., Gaithersburg, MD) derived from the Moloney murine leukaemia virus has been previously described [5].

Procedure for the generation of gene-modified cells

The generation of gene-modified T cells from the HSC donor required a 12-day culture period and involve 5 major steps:

Step 1: PBMC isolation and activation. 150ml of peripheral blood from the HSC donor were collected in an anticoagulated CP bag and diluted to 300ml final volume by adding 0.9% NaCl via a direct sterile connection to a NaCl solution bag, using a Terumo connection device. A volume of 100ml endotoxin-free Ficoll was transferred into a 400ml TP bag using a bag-to-bag transfer tube (Braun Medical, Boulogne, France). The CP bag was connected to this TP bag using a sterile connection device and the blood was gently layered over the Ficoll (Figure 1). PBMCs were then isolated by centrifugation (30min, 450g, 20°C) of the TP bag in a KR4.22 centrifuge using one or two semi-adapters

Figure 1. Blood deposit over Ficoll in a TP bag before centrifucation.

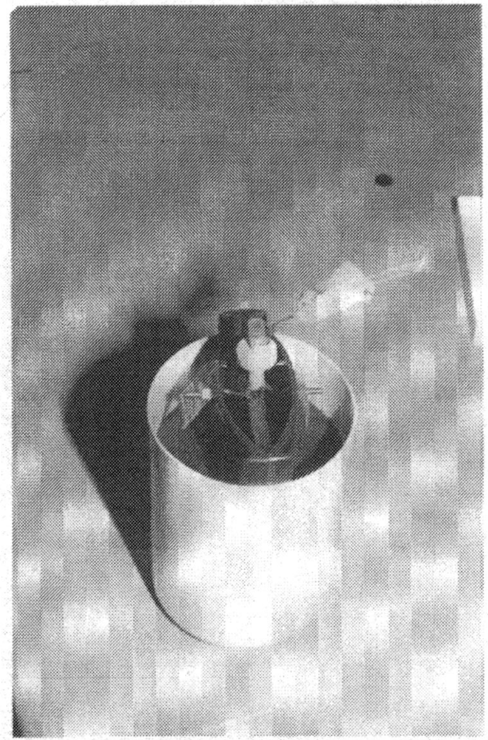

Figure 2. Centrifugation of the Ficull bag. The blue-coloured semi-adapters allow the Ficoll bag to be hung and to stand into the centrifuge burket.

(Baxter). These semi-adapters allow the bags to be hung and to stand into the centrifuge burket (Figure 2). The centrifuged TP bag (Figure 3) was placed on a plasma extractor, then connected to another TP bag into which the supernatant (plasma) was discarded. The PBMC layer was recovered in a third 400ml TP bag and resuspended in 200 to 300ml NaCl 0.9%. PBMCs were washed three times in this 400ml TP bag by successive 450g centrifugation processes (first wash: 20min, 4°C; second and third washes: 10min, 20°C). After the last wash, PBMCs were resuspended in CM, supplemented with 500U/ml IL-2 and 10ng/ml OKT3, then transferred into a LifeCell bag and cultured at 37°C and 5% CO_2 for 3 days. When numeration was required during the process, cell sampling was performed either by sterile connection of the LifeCell bag with a sampling bag or with a syringe connected to the interlink injection site.

Step 2: Retroviral-mediated PBMC transduction. After a 3-day culture, the activated cells were transferred in the retroviral vector-containing medium and supplemented with IL-2 (1000 U/ml) and 5µg/ml protamine sulfate. Cells were cultured for 24h with a multiplicity of infection of 3 in a LifeCell bag. A cell aliquot was cultured in parallel in CM with IL-2 without the retroviral vector (nontransduced control cells).

Step 3: Cell washing and resting period. On day 4, cells were pelleted in a CWS, then resuspended in CM with IL-2 (500U/ml) and further cultured in a LifeCell bag for an additional 24hrs.

Step 4: Selection of transduced cells. Transduced cell positive selection was performed by culturing them in CM containing 500U/ml IL-2 and 800µg/ml G418 from day 5 to day 12. The transduced and selected cells were termed «TK800» cells. An aliquot of transduced cells was cultured in parallel without G418 (transduced, unselected cells: TK0). Other controls included nontransduced, selected cells (C800: positive control for G418 toxicity) and nontransduced, unselected cells (C0) grown in parallel in CM+IL-2 with or without G418, respectively.

Step 5: Dead cell removal and cryopreservation of gene-modified cells until clinical use. At the end of the selection period (day 12), dead cells were removed by centrifugation (15min, 450g, 20°C) over Ficoll in a 400ml TP bag. Viable cells were transferred into a CWS, washed twice with 0.9% NaCl and cryopreserved in a suspension containing 4% human albumin (LFB, Les Ulis, France), 8 to 10% DMSO (Braun), then stored in liquid nitrogen in a Cryocyte bag until clinical use.

Quality controls

Transduction efficiency. Transduction efficiency was calculated on day 12 by comparing the proliferation rate of TK800 cells versus that of C0 control cells, or TK0 controls cells, according to the following formula:

$$\frac{\text{Fold increase in TK800 cell numbers from day 5 to day 12}}{\text{Fold increase in TK0 or C0 cell numbers from day 5 to day 12}} \times 100$$

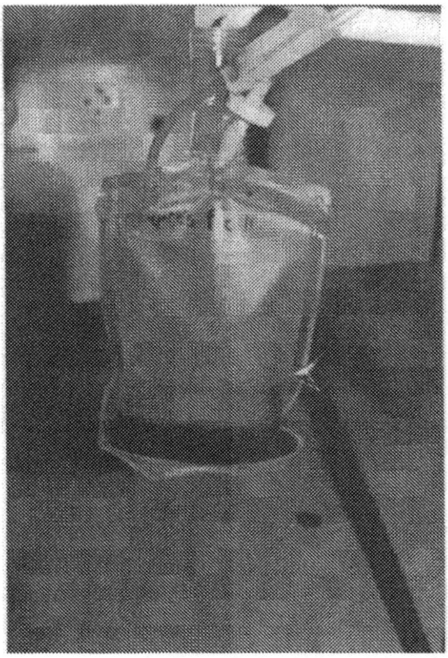

Figrue 3. TP bag after Ficoll centrifugation. Note the PBMC layer (arrow).

IL-2 dependence. After dead cell removal on day 12, TK800 cells were further cultured in CM in the presence or absence of IL-2 (500U/ml). Cell growth was assessed until complete death of TK800 cells cultured without IL-2.

Evaluation of ganciclovir sensitivity. Ganciclovir sensitivity of TK800 cells was evaluated as previously described [7] by ^{3}H-thymidine incorporation (^{3}HdT, 1μCi/well) (Amersham, Les Ulis, France) after 5 or 10 days culture in the presence or absence of ganciclovir (GCV: cymevan; Syntex, Puteaux, France). The assay was considered positive if inhibition of TK800 cell proliferation by 1μg/ml GCV was >80%.

Determination of CD3$^+$ cell rates. Cells were stained with FITC-labelled CD3 mAb (Becton-Dickinson) and analysed by flow cytometry using a FACScan (Becton-Dickinson) and CellQuest program.

Cell viability. Cell viability was assessed on cultured cells, or on frozen cells after thawing, by the trypan blue dye exclusion assay.

Safety controls

Sterility controls and endotoxin levels. Standard sterility controls were performed on the culture supernatant by mean of 14-day cultures in thiglycolate medium USP, Sabouraud liquid medium and tryptone soya broth, to search for the presence of aerobic and anaerobic bacteria, as well as fungi (Bacteriology Laboratory, CHU Minjoz, Besançon, France). Endotoxin levels were determined using a Limulus Amoebocyte Lysate (LAL) assay (sensitivity : 1EU/ml) by Euromedex (Souffelweyersheim, France).

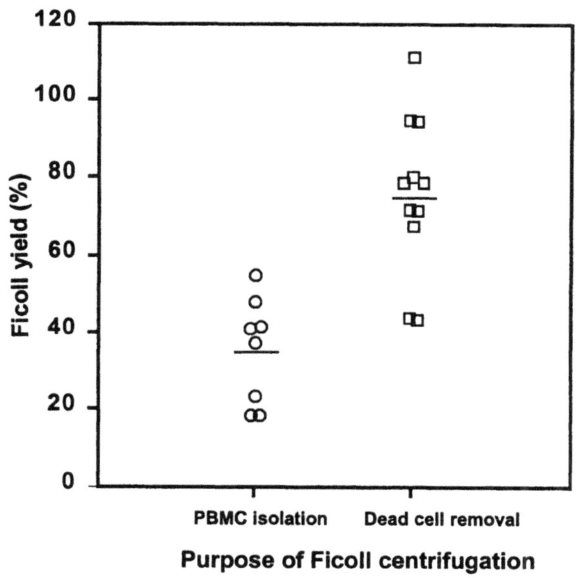

Figrue 4. Efficiency of Ficoll centrifugation performed in TP bags. Ficoll centrifugation was performed for the isolation of PBMCs from blood on day 0 (left) and for dead cell removal on day 12 (right).

Detection of mycoplasma contamination

Detection of mycoplasma contamination by PCR. Detection of mycoplasma contamination was performed on DNA from gene-modified cells. The nucleotide sequences of primers used in the mycoplasma group-specific PCR assay were as followed:

5' - GGCGAATGGGTGAGTAACACG - 3' and
5' - CGGATAACGCTTGCGACCTATG - 3'.

These primers allow detection of the most common species of mycoplasma: M. hyorinis, M. arginini, M. pneumoniae, M. fermentans, M. orale, M. pirum, Acholeplasma laidlawii, and Spiroplasma mirum.

Detection of mycoplasma by ELISA. Cell supernatants were screened for the most common mycoplasma species using a Mycoplasma detection ELISA kit (Boehringer Mannheim, Meylan, France).

Evaluation of replication competent recombinants (RCR)

Mus dunni assay. The *Mus dunni* assay [8, 9] was performed by Q-One Biotech LTD (Glasgow, Scotland) or by Genetic Therapy, Inc in order to amplify RCR particles which may have been present at low level in the test article (GMC or culture supernatant taken on day 12).

PCR Detection of recombinant virus. DNA from transduced T lymphocytes is screened by a nested PCR for recombinant virus genome detection. The prim-

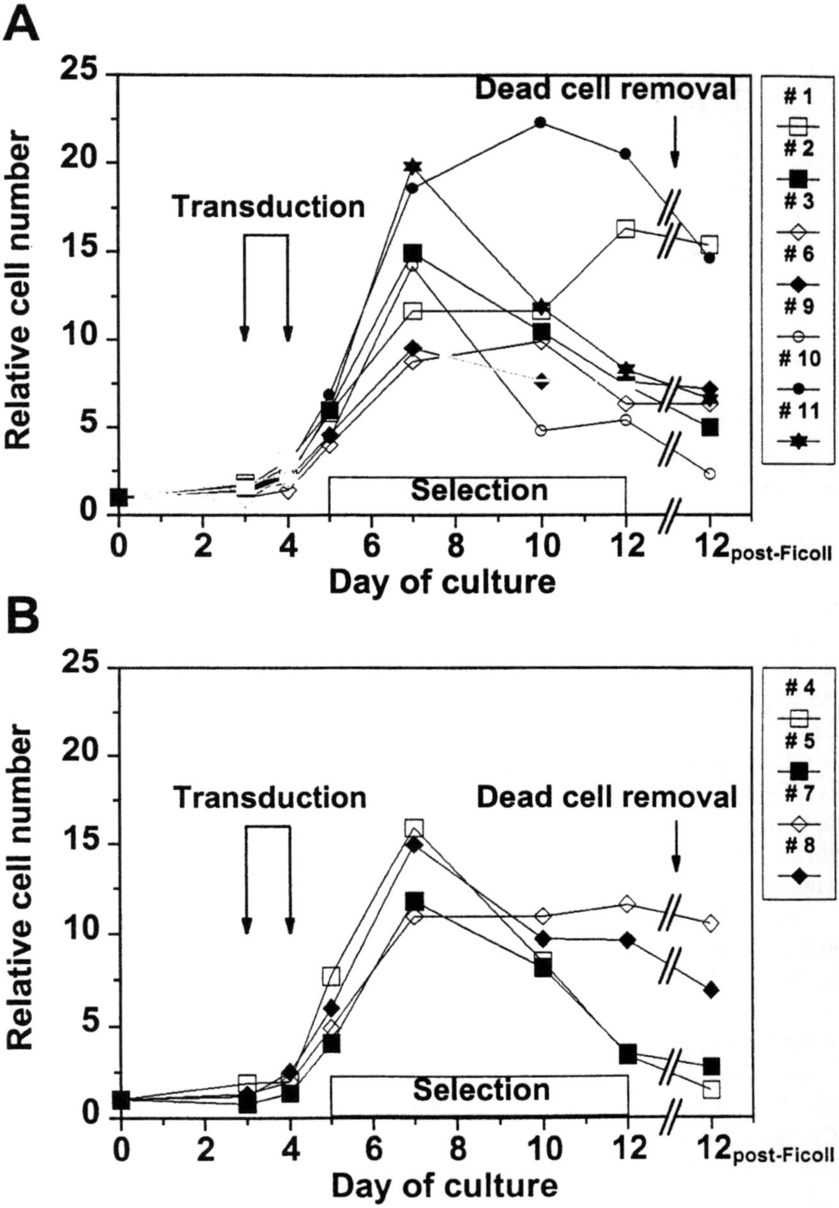

Figure 5. Proliferation of gene-modified cells cultured in autologous (A) or allogeneic AB serum (b). Cell growth is expressed as relative cell numbers.

ers used to amplify a segment of the amphotropic envelope gene are for the first PCR run:
5' - AAGTCAAAAGTCCGGGTCCGC - 3' and
5' - CCCGTGAAGCCCTTAATAGTC - 3';
for the second run, 2μl of PCR products from the first run are amplified with two internal primers:
5' - CTGTCTCCCTGCGGGTACTT-3' and
5' - GGAGTAGGGATGGCAGAGAGC-3'.
The sensitivity of this nested PCR, determined by amplification of diluted positive DNA is $2.5x10^{-5}$.

Results

PBMC5 recovery after blood centrifugation over a Ficoll layer

Twenty-three gene-modified cell suspensions have been produced for clinical purposes using our closed system, but analysis of cell growth are described only for the 11 first cell products, as previously reported in detail [10]. Similar results were obtained for the 12 remaining cell process (data not shown). For these 11 cell products, PBMCs were isolated from peripheral blood in 10 cases, and from a buffy coat in one case (donor # 7). Blood volume was 140 ± 8 ml (mean ± SE ; range: 100-180 ml). The Ficoll yield was 35.1 ± 4.9% (Figure 4); cell viability of the recovered PBMCs was always >95%.

Evaluation of gene-modified cell growth

Of the first 11 cell preparations performed for clinical use, 7 were performed in autologous serum (cell products # 1, 2, 3, 6, 9, 10 and 11) and 3 in allogeneic secured AB serum (cell products # 4, 5, and 7). Cell product # 8 was cultured in autologous serum from day 0 to day 3 and allogeneic AB serum subsequently. $92.7 \pm 7.2x10^6$ cells were activated on day 0 (range : 50 to $240x10^6$). Cell growth curves, expressed as relative cell numbers from day 0 to day 12 of culture, are depicted for cultures in autologous serum (Figure 5A) and in allogeneic AB serum (Figure 5B). Similar cell growth was observed for either autologous or AB serum. Cell numbers peaked either on day 7 (i.e. during the selection period) in 7 cultures or on day 10 in 2 cultures (# 3 and 10) and declined until day 12 (Figures 5A and 5B). No cell number decline was observed over the 12-day procedure for cultures # 1 and # 7. Maximum fold increase for all products was 14.7 ± 1.2 times (range: 9.6-22.2) and was similar in cultures using autologous serum (maximum fold increase: 15.3 ± 1.8) or allogeneic AB serum (maximum fold increase: 13.6 ± 1.1). Despite the loss of nontransduced cells during the selection period, the final overall cell number at the end of selection was 9.1 ± 1.6 fold the initial cell number (range: 3.4 to 20.4-fold increase from day 1 to day 12).

Cell growth for C0, TK0 and C800 control cells is shown for one representative donor (Figure 6). Growth of TK0 and C0 control cells were similar, demonstrating that the transduction medium does not affect the proliferative potential of the cells. All C800 cells died during the selection procedure. The difference

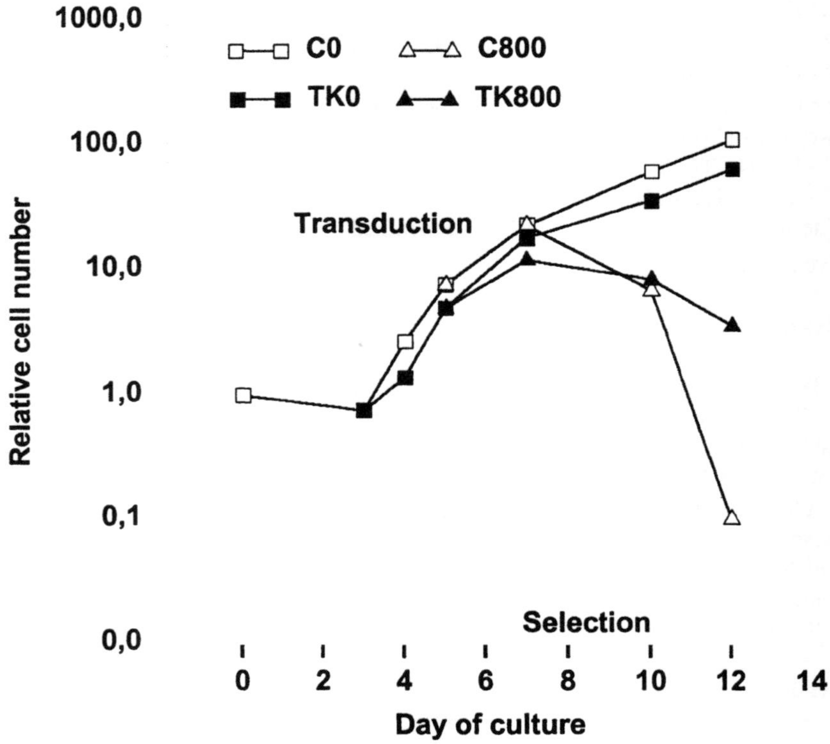

Figrue 6. Growth of transduced and selected cells (TK800) and control cells: nontransduced and unselected (C0), nontransduced and selected (C800), transduced and unselected (TK0) cells. All C800 cells died during the selection process, whereas only part of the cells were killed within the TK800 cell population. Data show one cell product (# 5) representative of 11.

between TK800 and C0 or TK0 cell numbers results from the elimination, during selection, of the nontransduced cell fraction in the TK800 population.

Efficiency of dead cell removal and viable cell recovery

Following dead cell removal at the end of selection, cell viability was 38.6 ± 3.8% (range : 16-52%) and 85.2 ± 2.5% (range : 72-93%) before and after Ficoll centrifugation, respectively. The Ficoll yield was 76.0 ± 5.8% (range 43.3-100% ; Figure 4) and was similar for cultures using autologous and AB serum (data not shown). On day 12, the total number of viable transduced cells after dead cell elimination was 7.2 ± 1.4 times higher than the initial PBMC number (range: 1.5-15.4-fold increase).

Quality controls

Twenty-three GMC products were produced, of which 14were reinfused to patients with a T-cell-depleted bone marrow from HLA-identical donors (n=12)

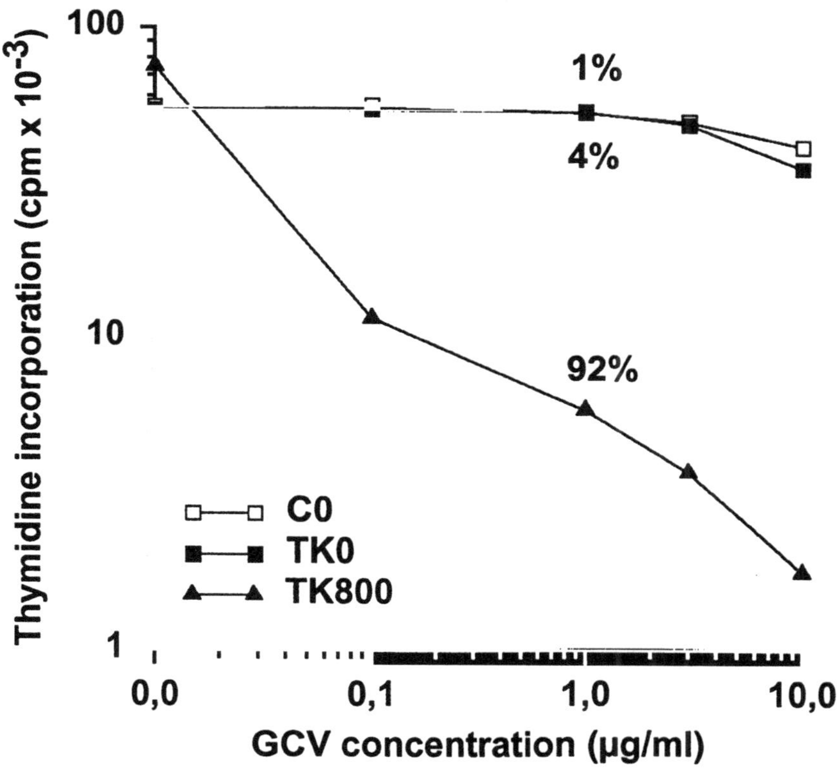

Figrue 7. Dose-dependent inhibition by GCV of Il-2-mediated TK800 cell proliferation. TK800 cells or C0 and TK0 control cells were cultured in flat-bottom 96-well microtiter plates (0.2×10^6 cells/well in triplicates) in the presence of IL-2 (500μ/ml) and the indicated GCV concentrations. Cell proliferation was assessed by ^{3}HdT incorporation after a 5-day culture. Percent inhibition of cell proliferation is indicated at 1μg/ml GCV for each cell type. Data show one cell product (# 5) representative of 11.

or from unrelated volunteer donors (n=2). The results of safety and quality controls are reported only for the GMC products that were reinfused to the 12 HLA-identical recipients (Table 1). Sterility controls (bacterial, endotoxin and mycoplasma contamination) were negative in all cell preparations for clinical use and no RCRs were detected by either of the assays (detection of Env gene by PCR and *Mus dunni* assay). The TK800 cells were IL-2-dependent as demonstrated by the the death of all cells in the absence of IL-2. The percentage of $CD3^+$ cells in the TK800 cell population was 85.9 ± 2.5% (n=12).

Efficient HS-tk gene expression was achieved in all cell preparations, as demonstrated by the GCV-induced inhibition of transduced, selected cells of 87 ± 1% (n=12) at 1μg/ml GCV (Table 1). Data on Figure 7 show one representative dose-dependent inhibition by GCV of IL-2-mediated TK800 cell proliferation. NeoR gene transfer was demonstrated by PCR in all TK800 cell popula-

Table 1. Results of safety and quality controls performed on cell preparations infused to HLA-identical recipients.

Control[1]	**Expected value**	**Observed values (Mean ±SE, n=12)**	**Observed values (Range)**
Sterility	Sterile	Sterile: 12/12	
Endotoxin level	<1 EU/ml	< EU/ml: 10/10	
Mycoplasma detection	Negative	Negative: 12/12	
PCR (*mus dunni* assay)	Negative	Negative: 12/12	
RCR (Env gene PCR)	Negative	Negative: 12/12	
Neo-R gene detection (PCR)	Positive	Positive: 12/12	
IL-2 dependence	Positive	Positive: 12/12	
% GCV inhibition	>80%	87 ± 1%	80 – 93%
% $CD3^+$ cells	>50%	85.9 ± 2.5	71 – 97%
Transduction efficiency		8.0 ± 1.6	1.5 – 19.7%
% viability after thawing	>60%	77.4 ± 1.2	60 – 92%

1. Not tested in two cell products.

tions tested. Overall TK800 transduction efficiency in infused cells was 8.0 ± 1.6% (Table 1). Direct assessment of transduction efficiency using competitive semi-quantitative PCR showed similar results (data not shown). Cell viability at the time of injection was 77.4 ± 1.2% (range 60-92%).

The 9 other cell products were not reinfused because of exclusion of the recipients for medical reasons (n=7) or, in two additional donors, for GMC preparation failure. Indeed, inadequate G418-induced selection of GMC was observed in one case, leading to a GCV sensitivity below 80% inhibition. A low transduction efficiency was observed in the last case, leading to insufficient GMC quantities.

Discussion

Cell therapy and *ex vivo* gene therapy are growing fields of application in the treatment of various acquired, autoimmune or inherited diseases. However, gene therapy products must be considered as drugs and must therefore be processed and handled in accordance with GMP standards. Thus, the *ex vivo* manipulation of blood products requires a high level of technology to ensure absolute safety of products for both the patients and the environment.

In the present study, we describe a method allowing the production of gene-modified cells in a standardised closed system. This includes the separation and culture of PBMCs, retroviral vector-mediated transduction, and the selection of gene-modified human T cells. Cell handling was always performed in bags and cell suspensions were transferred from one bag to another by sterile connections, thus avoiding opening the culture system. Because the cell suspensions were never directly in contact with the surrounding atmosphere, our method allowed for a drastic reduction in the risk of exogenous contamination (bacteria, mycoplasmas, viruses, endotoxins, etc.). Importantly, endotoxin levels were particularly low in all cell preparations tested (<1EU/ml). Among 23 «clinical grade» cell preparations tested for standard sterility, no bacterial, fungal or mycoplasma contamination was observed. Limiting the contamination risk is of major importance in cell therapy. In a gene therapy trial involving retroviral-mediated gene transfer using tumour infiltrating lymphocytes (TILs), one cell preparation out of 5 intended for clinical use was reported to display fungal contamination [11]. This contamination probably occurred during *ex vivo* culture, since a later culture of 'rescue' frozen cells was not contaminated. However, the use of this 'rescue' cell batch led to a lower proliferative potential and reduced transduction efficiency [11].

The closed culture system offers a safe procedure, not only for cell products, but also for the environment and the operator. It can be applied not only to the generation of gene-modified cells, but more broadly to any production and expansion of human T cells in immunotherapy. However, in the present study, some steps in the procedure were performed under a sterile flow hood (preparation and injection of OKT3, IL-2, protamin sulfate and G418 in culture bags). The use of a sterile flow hood could not be bypassed because of the conditioning of these agents. The connection of culture bags to an interlink injection site allowed their administration with a syringe without the need of a steel needle. This improves the operator's safety and reduces the risk of bag perforation.

One of the major improvements of our system is the development of the Ficoll centrifugation method in TP bags. The Ficoll solution is transferred from its bottle into a TP bag by sterile connection. However, it remains possible to introduce the Ficoll into the TP bag by directly connecting Ficoll bags to the 400ml TP bag, thus avoiding the use of bag-to-bag transfer tubes and Ficoll glass bottles. Similarly, the availability of RPMI 1640 medium in bags will soon allow us to avoid the opening of RPMI 1640 bottles under a flow hood, thus further reducing contamination risks.

Situations other than PBMC isolation or dead cell removal during *ex vivo* expansion in cell therapy protocols may benefit from Ficoll centrifugation in bags. Indeed, this procedure may also be applied to red blood cell removal after cytapheresis, when the hematocrit is too low to be removed by the machine, or similarly for the removal of red blood cells contaminating HSC samples in HSC transplantation in settings of donor/recipient ABO incompatibility. Ficoll centrifugation in TP bags was performed regardless of the hematocrit, with blood volumes as low as 30ml or with buffy coats. This method has however yet to be validated with HSC samples.

Ficoll efficiency, cell proliferation rate and transduction efficiency were reproducible and similar to those observed in conventional flask cell culture systems. It should be emphasised that the number of gene-modified cells recovered at the end of the process was higher than the initial cell number and was in large excess over the quantity required for infusion into patients. The number of cells obtained was sufficient to allow additional infusions in add-back settings performed in some of our patients (data not shown), as described by Bonini et al. [12].

Quality controls showed that all the results were within expected range values, both in terms of safety (sterility, absence of RCR particles) and phenotype ($CD3^+$ cell rate, IL-2-dependence, viability). Gene transfer-specific parameters (transduction efficiency, presence of NeoR transgene, G418 resistance and GCV sensitivity) were also in accordance with the data obtained in conventional culture systems and were within expected range values in all but two GMC products.

We have thus developed a closed culture system which makes it possible to perform routine large-scale preparation, expansion, transduction and selection of gene-modified PBMCs for clinical use. In particular, we have perfected a method for the isolation of PBMCs and removal of dead cells by Ficoll centrifugation in TP bags. This closed culture system is convenient, the method is reproducible and improves safety for both the operator and the cell product.

Acknowledgements

This study was financed in part by the Association pour la Recherche sur le Cancer (ARC), the Ligue Nationale de Lutte contre le Cancer, INSERM/Agence Française du Sang, the Programme Hospitalier de Recherche Clinique (PHRC), and the European Community (Biomed contracts).

References

1. Ferrara JLM, Deeg HJ, Burakoff SJ, eds. Graft-versus-host disease. 2nd ed. New York, NY : Marcel Dekker, Inc., 1997.
2. Maraninchi D, Gluckman E, Blaise D, et al. Impact of T-cell depletion on outcome of allogeneic bone marrow transplantation for standard risk leukemias. Lancet 1987;2:175-78.
3. Horowitz MM, Gale RP, Sondel PM, et al. Graft-versus-leukemia reactions after bone marrow transplantation. Blood 1990;75:555-62.
4. Crum Vander Woude A, Bierer BE. Immunosuppression and Immunophilin ligands : cyclosporin A, FK506, and Rapamycin In : Ferrara JML, Deeg HJ, Burakoff SJ eds. Graft-versus-host disease. 2nd ed. New York, NY : Marcel Dekker, Inc., 1997:111-49.
5. Tiberghien P, Brion A, Deconinck E, et al. Use of donor T-lymphocytes expressing herpes-simplex thymidine kinase in allogeneic bone marrow transplantation: a phase I-II study. Hum Gene Ther 1997;8:615-24.
6. Tiberghien P. ''Suicide'' gene for the control of graft-versus-host disease. Curr Opin Hematol 1998;5:478-82.
7. Tiberghien P, Reynolds CW, Keller J, et al. Ganciclovir treatment of herpes simplex thymidine kinase-transduced primary T-lymphocytes: an approach for specific in vivo donor T-cell depletion after bone marrow transplantation? Blood 1994;84: 1333-41.
8. Lander MA., Chattopadhyay SK. A Mus dunni cell line that lacks sequences closely related to endogenous murine leukemia virus and can be infected by ecotropic, amphotropic, xenotropic and mink cell focus forming viruses. J Virol 1984;52:695-98.
9. Haapala DK, Robey WG, Oroszalan SD, Tsai WP. Isolation from cats of an endogenous type C virus with a novel envelope glycoprotein. J Virol 1985;53: 827-33.
10. Robinet E, Certoux JM, Ferrand C, et al. A closed culture system for the ex vivo transduction and expansion of human T lymphocytes. J Hematother 1998;7:205-15.

11. Merrouche Y, Negrier S, Bain C, et al. Clinical application of retroviral gene transfer in oncology - Results of a French study with tumor-infiltrating lymphocytes transduced with the gene of resistance to neomycin. J Clin Oncol 1995;13:410-18.
12. Bonini C, Ferrari G, Verzeletti S, et al. HSV-tk gene transfer into donor lymphocytes for control of allogeneic graft-versus-leukemia. Science 1997;276: 1719-24.

RECOMBINANT PRODUCTS FOR HAEMOSTASIS

S.V. Seremetis[1]

Introduction

Plasma-derived factor VIII and factor IX concentrates were introduced in the mid-1960s, resulting in a dramatic increase in both the quality of life and life expectancy of patients with haemophilia A [1,2]. However, reliance on human plasma as the source of factor VIII resulted in the exposure of patients with haemophilia to foreign proteins as well as blood-borne viruses including hepatitis B, delta hepatitis, hepatitis C, hepatitis A, human parvovirus B19 and human immunodeficiency virus (HIV) [3-5]. Donor screening and viral inactivation processes dramatically reduced the risk of viral transmission; however, in the past few years certain plasma-derived products have been implicated in the transmission of hepatitis A [6-10], and concerns have been voiced regarding the potential for transmission of Creutzfeld-Jakob Disease (CJD). Therefore, the prospect of increased safety with clotting factor concentrates produced from recombinant DNA technology provides an attractive option for treatment of the haemophilias. Recombinant factor VIII (rFVIII), recombinant factor IX (rFIX), and recombinant factor VIIa (rFVIIa) are free from many of the risks associated with plasma-derived therapy and, as a product of a scalable manufacturing process, could theoretically be available in unlimited supply. Dramatic progress has been made in producing pure, safe, and effective FVIII and FIX and FVIIa from mammalian cell cultures transduced with the relevant vectors.

Recombinant Factor VIII

Production of rFVIII was made possible by the successful cloning and expression of the factor VIII gene, announced by two groups simultaneously in 1984 [11, 12]. The factor VIII gene is one of the largest genes ever cloned, and factor VIII is the largest recombinant protein licensed for use to date.

Two full length recombinant FVIII products are available – Kogenate and Recombinate. Kogenate was initially cloned and developed by Genentec, Inc. and is currently sold by Bayer. The product is prepared in a continuous cell culture process using baby hamster kidney (BHK) cells transfected with the

---

1. Dept Medicine, Haemophilia Treatment Center, Mount Sinai School of Medicine, New York, NY, USA.

Table 1. Recombinant clotting factor concentrates.

Product (manufacturer)	Specific procedure for viral inactivation/removal	Formulated with human albumin	Cells exposed to bovine protein
Recombinante (Baxter)	yes	yes	yes
Kogenate (Bayer)	no	yes	yes
Refacto (Pharmacia)	yes	no[1]	no
BeneFIX (Genetics Institute)	yes	no	no
Novo Seven® (Novo Nordisk)	no	no	yes

1. Cells grown in human albumin.

FVIII gene. Co-expression of recombinant FVIII with von Willebrand factor (vWf) in the culture media is not required for stabilization of the FVIII molecule, but a special production culture medium containing, for example, insulin, transferrin and albumin is used [13]. Recombinate was cloned and developed at Genetics Institute Inc. and is marketed by Baxter. It is produced in a cell culture process using Chinese hamster ovary (CHO) cells transfected with the FVIII gene. Co-expression of FVIII with vWf stabilizes FVIII. In the absence of vWf, the heavy chain of FVIII expressed in CHO cells is not associated with the light chain and both are degraded [14]. Both recombinant products are further purified using monoclonal antibody immunoaffinity and additional chromatographic steps. Following the development of large-scale techniques for fermentation and purification, pre-licensing trials began in 1987 with the rFVIII produced by Baxter (Recombinate) and in 1988 with the rFVIII produced by Bayer (Kogenate).

The results of the clinical trials of these two recombinant factor VIII concentrates were remarkably congruent. Factor VIII post-infusion recoveries were, on average, 2.5%/U/kg for both Kogenate and Recombinate, a result which is interpreted as analogous with the recovery of plasma-derived factor VIII. The circulating half-life of both of the recombinant factor VIII materials was, again, analogous to that of plasma-derived factor VIII and in the range of 12-14 hours. Efficacy of the transfused materials was also analogous to plasma-derived factor VIII in the context of both home treatment for mild as well as moderate bleeding episodes, including joint bleeding, muscle bleeding, urinary tract bleeding, as well as more severe bleeding episodes [15,16]. The total documented treatment episodes in the literature from these recombinant factor VIIIs is in excess of 20,000 treatment episodes at this point with over 100 million units having been used in the context of clinical trials since 1987. The remarkable data accumulated in the conduct of these trials represents the largest body of prospective data with regard to the treatment of haemophilia A, and, again, are data within the range of previous smaller studies using plasma-derived factor VIII. Adverse events were minor, and comparable to those observed in the use of plasma-derived factor VIII. Importantly, few allergic reactions were noted and no significant seroconversions were reported to any animal proteins which might have

Table 2. Preparation of recombinant clotting factor concentrates.

Products	Cells used for transfection	Formulation
Factor VIII		
Recombinate (Baxter)	CHO	human albumin
Kogenate (Bayer)	BHK	human albumin
Refacto (Pharmacia)	CHO	no albumin
Factor IX		
BeneFIX (Genetics Institute)	CHO	no albumin
Factor VIIa		
Novo Seven (Novo Nordisk)	BHK	no albumin

been introduced in the manufacturing process, including murine, bovine, and hamster proteins [15,16].

It was concluded from these trials that recombinant factor VIII concentrates were safe and clinically effective for the prevention and treatment of haemorrhage in haemophilia A, and that its behavior in vivo was similar to that of plasma-derived factor VIII concentrate. Both of these materials were licensed for general use in the early 1990s.

B-Domain Deleted Recombinant Factor VIII

The design of second-generation recombinant factor VIII is based on the knowledge of factor VIII structure and function. In plasma, factor VIII exists as a heterodimer consisting of a light chain (domains A3, C1 and C2) and the heavy chain derivatives (domains A1, A2 and B). The B domain is dispensable for the haemostatic activity of factor VIII [17]. The B-domain deleted molecule is much more stable than the intact molecule, and since it is less prone to proteolytic degradation, no human albumin is needed for stabilization.

As of January 1998, 113 patients (median age 26 years, range 8-73) had been followed in a worldwide study involving previously treated patients [18]. Overall, 30,045 infusions of 62,318,262 IU were administered over 28,056 exposure days. This was a median dose of 29.5 IU/kg given for 6,754 bleeding episodes. In this study, 71% of bleeding episodes resolved with one treatment, and 93% were resolved with three or less injections. The recovery was 2.7% per IU per kg infused and did not change over the duration of treatment [18]. Thus, it was concluded that r-factor VIII SQ was safe and efficacious in the treatment of previously treated patients with haemophilia A for on-demand and/or prophylactic treatment.

One previously treated patient developed an inhibitor after 93 exposure days and 3 years into the study with a peak level of 12.6 BU/ml. Although several other patients received r-factor VIII SQ from the same batch, they did not develop an inhibitor. The patient is still using r-factor VIII SQ successfully with increased doses.

Advantages of Recombinant Factor VIII SQ (rFVIII SQ)

A perceived advantage of r-factor VIII SQ is the fact that no human albumin is needed as a stabilizer and, therefore, it should be completely safe with regard to transmission of human blood borne viruses. However, it should be noted that albumin is used in the manufacturing process even though none is needed for a stabilizer [19]. It should be noted that albumin does have an excellent record of safety due to pasteurization.

Inhibitor Development with Recombinant Factor VIII

Perhaps the most significant contribution of the carefully conducted Kogenate PUP study was the careful, prospective monitoring of factor VIII inhibitors. In the initial report published in 1993, Lusher et al. reported the cumulative probability of inhibitor development as 24.8% by day 45.1 (6.5 weeks) after first treatment with r-factor VIII [20]. The Recombinate PUP study began in July 1990, approximately 18 months after the commencement of the Kogenate study. The occurrence of inhibitors in the Recombinate study mirrored that of the Kogenate study. Seventeen (23.9%) individuals developed inhibitors, five with peak titers > 10 BU and 12 with titers < 10 BU. The survival analysis showed that the probability of remaining inhibitor free in this group of patients was 88.4% after 8 days of exposure, 73.6% after 10 days of exposure, and 61.6% after 25 days of exposure to r-factor VIII. Inhibitors disappeared in five (29.4% inhibitor patients) on re-testing 2 to 16 months after the last positive inhibitor assay [15].

In addition, from October 1994 until January 1998, 101 severe, previously untreated haemophilia A patients were enrolled in a multinational safety and efficacy study of rFVIII SQ. In this study, 91% of all bleeding episodes were treated successfully with less than three treatments. Overall, 29% have developed an inhibitor with one-third having a titer of more than 10 BU and two-thirds with a titer less than 10 BU [16]. As of December 1998, the inhibitor rate was 16%, which is comparable to the rate of the other two full-length r-factor VIII concentrates.

The experience of the three r-factor VIII studies has shown that, in prospective trials that are carefully monitored with laboratory inhibitor assays at frequent intervals, the incidence of inhibitor development in infants and children with severe haemophilia A is much higher than previously thought. The high incidence of inhibitor occurrence that began to emerge in these studies prompted several retrospective reviews of the previous experience with plasma-derived concentrate. Reports appeared with an incidence of 20% to 30% following the administration of plasma-derived concentrates [21,22]. The highest reported incidence being an alarming 52% in a variety of plasma-derived products [23]. The cumulative incidence in patients receiving recombinant products is of the order of 25%, as discussed above.

Vermylen has reviewed how some haemophiliacs develop inhibitors [24]; clearly, it is a multifactorial process. However, there is probably enough evidence in the literature to suggest that repeated switching from one factor VIII product to another facilitates an immune response.

Thus, in promoting the use of r-factor VIII for the issues of viral safety, it is important to caution patients regarding the possible increased inhibitor risk. Furthermore, because of the high purity of recombinant products, they may increasingly be used for continuous infusion, and it is possible that the route of exposure, particularly for the patient with mild disease, may potentiate the risk of developing inhibitors [25].

Recombinant Factor IX

Factor IX presents unique challenges for recombinant technology. The factor IX gene and cDNA were cloned in 1982, [27,28] and the recombinant protein was expressed from cDNA in CHO cells [29]. The challenge in the development of r-factor IX for clinical use was the need for post-translational modifications, such as gamma carboxylation and propeptide cleavage [30]. For the manufacture of recombinant human (rh)-factor IX, the Genetics Institute uses a CHO cell line that has been co-transfected with the rh-factor IX cDNA expression plasmid and a cDNA expression plasmid that encodes an engineered form of the paired basic amino acid cleaving enzyme or PACE protease. PACE improves the processing efficiency of profactor IX expressed in CHO cells.

Another challenge has been to develop methods to eliminate foreign proteins from the production process. All cells used in the synthesis of rh-factor IX are grown in serum-free medium containing essential amino acids, salts, recombinant human insulin, and vitamin K, with no bovine serum or animal proteins. The master cell bank, working cell bank, and cells used to synthesise rh-factor IX undergo extensive testing, including analysis for adventitious and other rodent and human viruses [30]. This virtually eliminates the risk of transmission of blood borne viruses as well as spongiform agents.

Pharmacokinetic Studies of Recombinant Human Factor IX

A double-blind, randomized, crossover study was performed in 11 males with severe or moderately severe haemophilia B. Patients were randomized to receive 50 IU/kg of either rh-factor IX or monoclonal plasma-derived factor IX with a washout between infusions. The mean recovery was 37.81% or 0.84 IU/dl per IU/kg. This was only 72% of that observed with the monoclonal plasma-derived factor IX (52.61% and 1.71 IU/dl per IU/kg. This difference was statistically significant ($p<0.05$) (Figure 6). The elimination half-life of the two factor IX preparations was similar, 17.66 hours and 18.10 hours, respectively.

The 28% difference in recoveries is due to the difference in sulfation of Tyrl55 and phosphorylation of Ser158, suggesting that these residues are important in the clearance of factor IX. This may account for the difference in the volume of distribution, which may be the result of different intercompartmental clearance or binding or both.

Two multi-center studies evaluated the efficacy in treating bleeding episodes and surgical prophylaxis of recombinant human factor IX. More than 2,000 bleeding episodes have been reported, 80% of which required only one infusion for resolution. The average dose of these infusions has been in the range of 50 IU/kg for joint haemorrhage and 70 IU/kg for surgery, although the range of dosing was quite large. In general, investigators were aware of the decreased recovery noted in the pharmacokinetic studies and adjusted upwards in terms of routine dosing for joint haemorrhages. Typically, dosing for efficacy has been in the range of 1.3-1.5 times that which would normally be prescribed for plasma-derived factor IX. It is not certain whether lower doses would produce similar efficacies as this has not been carefully addressed. In fact, prescribing information currently available advocates for an upward adjustment of dosing based on historical data with plasma-derived factor IX.

In the trial of previously treated patients with recombinant human factor IX, a single patient developed a low titer inhibitor to factor IX after 39 exposure days to the material. The patient had had 500 previous exposure days to plasma-derived factor IX, but, interestingly, had received a single subcutaneous dose of monoclonal factor IX at 10 sites approximately two months prior to study entry. The peak inhibitor titer was 1 Bethesda unit with decay of the inhibitor after approximately one year. Although the inhibitor affected recovery when it was detected, it did not affect elimination half-life.

In the study of previously untreated patients or PUPs, only 1 of 40 in all PUPs who have had follow up inhibitor testing has developed inhibitors to factor IX with a peak of 30 Bethesda units noted.

To summarize the experience of clinical trials with recombinant factor IX, over 2,000 bleeding episodes were observed and analysed. The product appeared to be safe and effective in these episodes. The side effects of administration were minor and comparable to those reported with infusions of plasma-derived factor IX. In no case did any of these side effects result in cessation of the use of the recombinant factor IX. Very importantly, recovery of the recombinant factor IX was compromised with respect to plasma-derived factor IX with approximately a 30% lower recovery. This is consistent through all trials reported; and is thought to be as a consequence of differences in phosphorylation and sulfation of the molecule.

In contrast, the half-life of recombinant factor IX once in circulation is entirely analogous to that of previously reported half-lives of plasma-derived factor IX. It is thought to be important to conduct prospective controlled comparison trials of plasma-derived factor IX and recombinant factor IX in terms of pharmacokinetics. These trials are ongoing and data should be reported in the near future.

Recombinant Factor VIIa

Recombinant factor VIIa was developed by scientists at Novo Nordisk as a treatment for bleeding episodes in individuals with haemophilia complicated by inhibitor antibodies. Activated factor VII is not proteolytically active itself and does not, therefore, induce systemic activation of coagulation when infused. The factor VIIa complexes with tissue factor at the site of injury, and since it is not neutralized by circulating antithrombin, infused factor VIIa can reach the site of injury, where it complexes with tissue factor exposed at the site, inducing local haemostasis [34]. The process is independent of the presence of factors VIII or IX and is not affected by inhibitors to factors VIII or IX.

Human factor VIIa was cloned in the mid-1980s and was expressed in baby hamster kidney (BHK) cells [35]. The BHK cells are capable of gamma carboxylation, and the factor VII is secreted in a single chain form. The amino acid sequence has been shown to be identical to the human form.

The advantages of r-factor VIIa include viral safety, no systemic activation of coagulation, effectiveness independent of inhibitor titer, and an excellent overall safety profile allowing for home use and ease of administration. The main disadvantage is the short half-life and, therefore, the need for frequent administration. The high cost of this product may also be considered an issue, pending cost-impact analyses which take into account changes in duration of hospitalization and other sequelae.

References

1. Levine PH. Efficacy of self-therapy in hemophilia: a study of 72 patients with hemophilia A and B. N Engl J Med 1974;291:1381-84.
2. Aledort LM. Lessons from hemophilia. N Engl J Med 1982;306: 607-08.
3. Goldsmith MF. Hemophilia, beaten on one front, is beset on others. JAMA 1986;256:3200.
4. Pierce GF, Lusher JM, Brownstein SP, Goldsmith JC, Kessler CM. The use of purified clotting factor concentrates in hemophilia: influence of viral safety, cost and supply on therapy. JAMA 1989;261:3434-38.
5. Brettler DB, Levine PH. Factor concentrates for treatment of hemophilia: which one to choose? Blood 1989;73:2067-73.
6. Kasper CK, Lusher JM. Recent evolution of clotting factor concentrates for hemophilia A and B. Transfusion 1993;33:422-34.
7. Lusher JM, Kessler C, Laurian Y, Pierce G. Viral contamination of blood products. Lancet 1994;344:405-06.
8. Mannucci PM, Santagostino E, DiBona E, et al. The outbreak of hepatitis A in Italian patients with hemophilia: facts and fancies. Vox Sang 1994;67(Suppl 1):31-35.
9. Mannucci PM, Gdovin S, Gringeri A, et al. Transmission of hepatitis A to patients with hemophilia by factor VIII concentrates treated with organic solvent and detergent to inactivate viruses. Ann Intern Med 1994;120:1-7.
10. Centers for Disease Control. Hepatitis A among persons with hemophilia who received clotting factor concentrate – United States, September-December 1995. MMWR 1996;45: 29-32.
11. Wood WI, Capon DJ, Simonsen CC, et al. Expression of active human factor VIII from recombinant DNA clones. Nature 1984;312:330-37.
12. Toole JJ, Knopf JL, Wozney JM, et al. Molecular cloning of a cDNA encoding human antihaemophilic factor. Nature 1984;312:337-42.
13. Boedeker BGD. The manufacturing of the recombinant factor VIII, Kogenate. Transfus Med Rev 1992;6:256-60.
14. Gomperts E, Lundblad R, Adamson R, et al. The manufacturing process of recombinant factor VIII, Recombinate. Transfus Med Rev 1992;6:247-51.
15. Bray GL, Gomperts ED, Courter S, et al. A multicenter study of recombinant factor VIII (Recombinate): safety, efficacy and inhibitor risk in previously untreated patients with hemophilia A. Blood 1994;83:2428-35.
16. Seremetis S, Lusher JM, Abildgaard C, et al. Human recombinant DNA-derived antihemophilic factor (factor VIII) in the treatment of hemophilia A: conclusions of a 5-year study of home therapy. Hemophilia 1999;5:9-16.

17. Lusher JM, Spira J, Magill M. A four-year update of safety and efficacy of a second generation B-domain deleted factor VIII (R-VIII SQ) in previously untreated hemophilia A patients. Blood 1998;92(10 Suppl 1): 555a.
18. Toole JT, Pittman DD, Orr EC, Murtha P, Wasley LC, Kaufman RJ. A large region (=95kDa) of human factor VIII is dispensable for in vitro procoagulant activity. Proc Natl Acad Sci USA. 1986;83:5939-42.
19. Mikaelsson M, Eriksson B, Lind P, et al. Manufacturing and characterization of a new B-domain deleted recombinant factor VIII, r-VIII SQ. Thromb Haemost 1995;69:1205 (abstract 2350).
20. Lusher JM, Arkin S, Abildgaard CF, Schwartz RS. Recombinant factor VIII for the treatment of previously untreated patients with hemophilia A. N Engl J Med. 1993; 328:453-59.
21. Addiego J, Kasper C, Abildgaard C, et al. Frequency of inhibitor development in haemophiliacs treated with low-purity factor VIII. Lancet 1993;342:462-64.
22. Schwaab R, Brackmann HH, Meyer C, et al. Haemophilia A: mutation type determines risk of inhibitor formation. Thromb Haemost 1995;74:1402-06.
23. Ehrenforth S, Kreutz W, Scharrer I, et al. Incidence of development of factor VIII and factor IX inhibitors in haemophiliacs. Lancet 1992;339:594-98.
24. Vermylen J. How do some haemophiliacs develop inhibitors? Haemophilia 1998; 4:538-42.
25. Baglin T, Beacham E. Is a change of factor VIII product a risk factor for the development of a factor VIII inhibitor? Thromb Haemost 1998;80:1036-37.
26. Williams IJ, Peake IR, Goodeve AV. Recombinate PUP mutation study: relationship between factor VIII mutation and inhibitor development. Haemophilia 1998;4:228 (abstract 290).
27. Kurachi K, Davie EW. Isolation and characterization of a cDNA coding for human factor IX. Proc Natl Acad Sci USA 1982;79:6461-64.
28. Choo GH, Gould KG, Rees DJ, Brownlee GG. Molecular cloning of the gene for human anti-haemophilic factor IX. Nature 1982;299:178-80.
29. Kaufman RJ, Wasley IC, Furie BC, Furie B, Shoemaker CB. Expression, purification, and characterization of recombinant gamma-carboxylated factor IX synthesized in Chinese hamster ovary cells. J Biol Chem 1986;261:9622-28.
30. 30.White GC, Beebe A, Nielsen B. Recombinant factor IX. Thromb Haemost 1997;78:261-65.
31. Kim HC, McMillan CW, White GC, Bergman GE, Horton MW, Saidi P. Purified factor IX using monoclonal immunoaffinity technique: clinical trials in hemophilia B and comparison to prothrombin complex concentrates. Blood 1992;79:568-75.
32. Goldsmith JC, Kasper CK, Blatt PM, et al. Coagulation factor IX: successful surgical experience with a purified factor IX concentrate. Am J Hematol 1992;40:210-15.
33. Berntorp E, Bjorkman S, Carlsson M, Lethagen S, Nilsson IM. Biochemical and in vivo properties of high purity factor IX concentrates. Thromb Haemost 1993;70:768-73.
34. Hedner U, Glazer S. Management of hemophilia patients with inhibitors. Hematol Oncol Clin North Am 1992;6:1035-36.
35. Hagen FS, Gray CL, O'Hara P, et al. Characterization of a rDNA coding for human factor VII. Proc Natl Acad Sci USA 1986;83:2412.

APPLICATION OF MOLECULAR BIOLOGY TO THE ENGINEERING OF IV IMMUNOGLOBULINS[1]

D.L. Siegel[2]

Introduction

The use of passively-administered, plasma-derived, human immunoglobulin (Ig) preparations has many clinical indications particularly for the treatment of infectious, oncologic, immunodeficiency, and autoimmune diseases and for the prevention of unwanted immunity such as for Rh immune prophylaxis. Evolving concerns regarding the safety, supply, cost, and efficacy of such therapeutic reagents have created a need for developing methods for the production of immunoglobulins *in vitro*. Through the use of novel molecular technologies collectively referred to as antibody engineering, it is now possible to generate fully human antibodies in the laboratory which express precisely-controlled specificities and effector functions. Many of these approaches do so without the need for B-lymphocyte transformation, tissue culture, or even immunized individuals as a source of starting material. The following chapter will review these technologies with examples from the author's laboratory on their use in producing human antibodies directed against membrane-bound antigens expressed on human red blood cells (RBCs).

An Overview of Antibody Engineering Techniques

A main driving force for the development of *in vitro* methods for therapeutic antibody production is the need for an alternative to the use of rodent monoclonal antibodies in humans. As recognized early on, the therapeutic efficacy of murine-derived Ig preparations becomes severely limited by their induction of human anti-mouse antibody (HAMA) immune responses in the recipient. The most straight-forward way to avoid this phenomenon would be to produce human monoclonal antibodies using hybridoma technology analogous to that used for the mouse. Unfortunately, the absence of a suitable fusion partner for human B-cells and other technical difficulties introduced by EBV-transformation, made methods that rely on human B-cell immortalisation

1. Support: NIH grants R01-HL61844 and P50-HL54516 (SCOR in Transfusion Medicine and Biology Award).
2. Dept Pathol & Lab Med, Blood Bank/Transf Med Section, Univ Pennsylvania Med Center, Philadelphia, PA, USA.

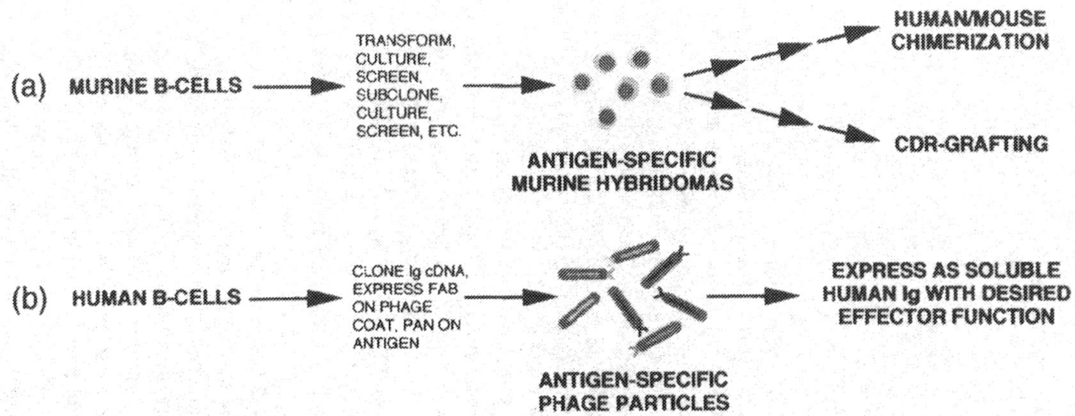

Figure 1. Cellular vs. molecular technologies for producing therapeutic immunoglobulins.

problematic [1]. Consequently, the field of antibody engineering can be roughly divided into those methods which are designed to humanize existing mouse monoclonal antibodies or those methods that produce human antibodies by cloning rearranged human Ig genes in bacteria.

To humanize murine antibodies (Figure 1a), chimeric antibodies can be created in which the mouse heavy and light chain constant region DNA are exchanged for their human equivalents [2]. After transformation into a suitable eukaryotic host cell, the expressed antibodies retain their murine variable regions and native antigen specificity. Murine-based therapeutic antibodies can be engineered to be even less immunogenic by grafting their heavy and light chain CDRs (complementary determining regions) onto a human antibody scaffold [3,4]. Despite the fact that many of these molecules can still lead to anti-idiotypic host responses, a number of humanized murine antibodies have made it to the clinical setting including the antibody c7E3 (abciximab, ReoPro®), an anti-thrombotic agent [5], 2B8 (rituximab, Rituxan®), an anti-CD20 antibody used to treat lymphoma [6], and daclizumab (Xenapax®), an anti-CD25 (IL2-receptor) antibody used to prevent renal allograft rejection [7].

An alternative to the use of murine antibodies as templates for the design of human-like antibodies, is the production of fully human antibodies from the outset by using molecular approaches which obviate the need for human B-cell transformation (Figure 1b). Collectively, these techniques, referred to as repertoire cloning or phage display, seek to immortalize Ig genes rather than the cells from which they were derived [8,9]. cDNA is prepared from the mRNA of the rearranged Ig genes from donor B cells, and heavy and light chain gene segments are amplified by the polymerase chain reaction (PCR) using sets of universal oligonucleotide primers. Heavy and light chain DNA are cloned into one of a number of available plasmid expression vectors [10], which, after transformation of *E. coli*, results in the accumulation of Ig Fab fragments in the bacteria's periplasmic space. After co-infection with M13 filamentous helper phage, phagemid particles are generated which contain the plasmid DNA and which incorporate ~1-3 copies of Ig Fab fragment as part of their viral coat protein. Libraries of such phage particles can then be panned against an antigen of interest to adsorb phage expressing the desired specificity. Fresh cultures of *E. coli* are infected with the affinity-purified phage for further propagation and enrichment utilizing several additional rounds of panning. Once monospecific phage are isolated, individual clones can be selected and expressed in bacteria as soluble antibody fragments unlinked to phage. These soluble molecules can be produced in a number of different forms [Fab, $F(ab')_2$, single-chain Fv fragments (scFv), etc.] and empowered with various effector functions as described later in this review. To date, a large number of human antibodies have been produced in this manner with specificities for autoantigens [11-20] and alloantigens [21-26] with potential therapeutic and/or diagnostic utility [9,27-33]. Compared to methods which rely on the isolation of murine hybridomas and subsequent humanization protocols, these molecular approaches offer a number of advantages as illustrated in the following examples from the author's laboratory.

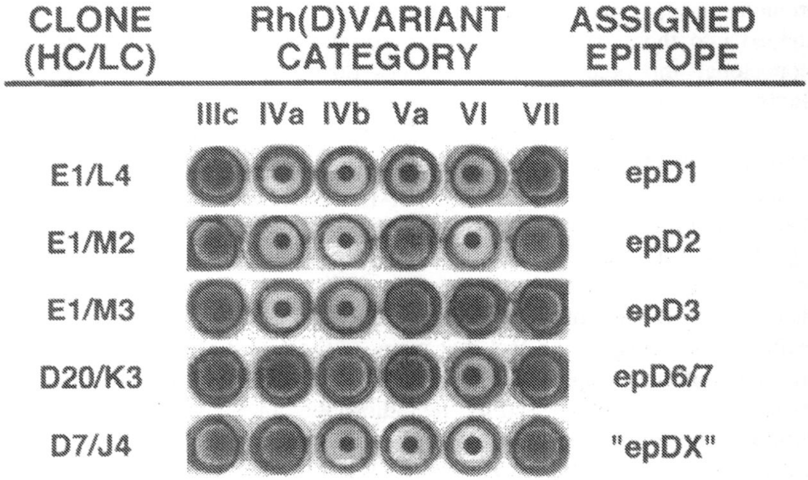

Figure 2. Determination of the Rh(D) binding epitope of anti-Rh(D) Fab/phage clones. The fine specificities of monoclonal Fab/phage clones were determined by their ability to agglutinate members of a panel of six Rh(D) variant RBCs. The figure illustrates five different agglutination patterns obtained from screening a cohort of 53 Fab/phage clones. The particular clones shown are identified by their unique heavy chain/light chain pairings as shown in Figure 4. For E1/M3, reactivity with additional Rh(D) variant cells would be required to distinguish its specificity for epD3 vs. epD9 [37]. Rh(D) epitope assignments are as per Lomas et al. [38]. An example of a novel Rh(D) epitope revealed by repertoire cloning is designated epDX. Reprinted with permission from Chang and Siegel [32].

Cloning Human Anti-RBC Ig Using Phage Display Technology

The administration of Rh(D)-immune globulin, a preparation of polyclonal anti-Rh(D) IgG derived from the sera of Rh(D)-sensitised individuals, is the standard of care for the prevention of Rh(D) all immunization from pregnancy. For over 30 years, the use of this drug has nearly eliminated haemolytic disease of the newborn (HDN), yet the mechanism by which it exerts its immunomodulatory effect is still not understood [34]. We believe that such an understanding could have wide application in the field of immunotherapy as one tries to develop specific therapeutic strategies to prevent or attenuate the production of pathogenic auto- and alloantibodies. To approach this problem, we wanted to develop an efficient method for the production of large repertoires of anti-Rh(D) antibodies from the B-cells of a given alloimmunized individual so that the human immune response to Rh(D) could be studied on a molecular level [35,36]. Such a method would also address a very practical issue, i.e. the fact that the supply of Rh(D)-immune globulin is dwindling ironically due to the drug's effectiveness in preventing alloimmunization to Rh(D) and the eventual unavailability of naturally-sensitised donors. Since the Rh(D) antigen is only immunogenic in humans, murine anti-Rh(D) monoclonals cannot be produced, humanized, and offered as a potential substitute. Therefore, having an endless

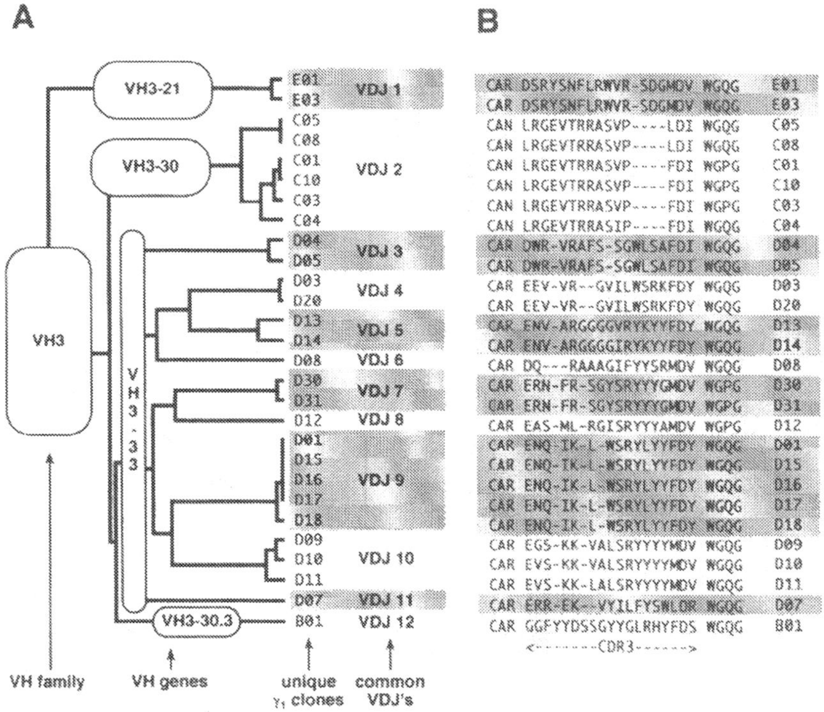

Figure 3. (A) Dendrogram and (B) CDR3 alignment of anti-Rh(D) heavy chains. The 28 unique heavy chain clones are organized by V_H family, V_H germline gene, and VDJ rearrangement. Each heavy chain clone is identified by a numeral preceded by a letter (B through E) which denotes its germline gene. The 28 heavy chains comprised 12 distinct VDJ regions, designated VDJ1-VDJ12. Clones with identical VDJ joins putatively result from intraclonal diversity of 12 original B lymphocytes from the donor. Reprinted with permission from Chang and Siegel [32].

source of Rh(D)-immune globulin relies on our ability to produce effective formulations of human anti-Rh(D) *in vitro*.

Because published reports describing the use of conventional B-cell transformation/culture techniques had succeeded in cloning at most ~8 anti-Rh(D) monoclonals from a given individual, we opted to utilize phage display technology. We constructed a pooled $\gamma_1\kappa/\gamma_1\lambda$ phage display library from the B-cells contained within ~30 ml of peripheral blood collected from an Rh(D)-negative individual previously sensitized to Rh(D). The libraries were panned for anti-Rh(D)-reactive Fab/phage using a novel cell-surface panning approach in which negative and positive selections were performed simultaneously [26]. Literally *millions* of anti-Rh(D) monoclonal antibodies were isolated from the library.

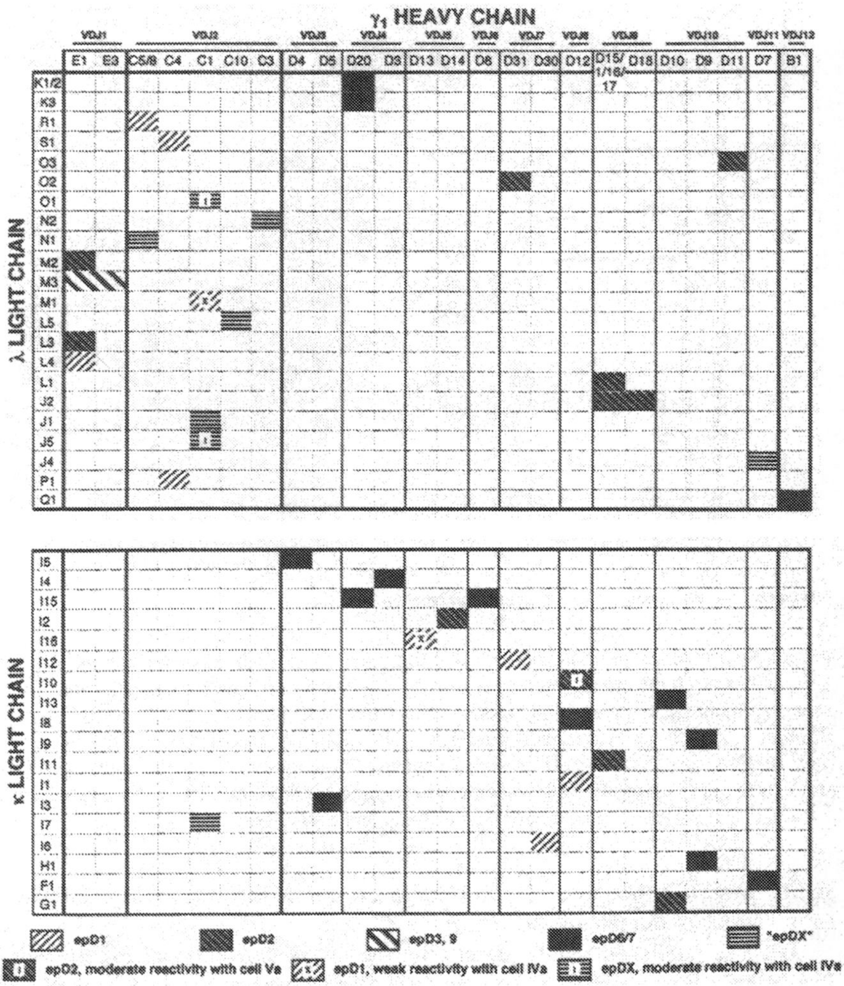

Figure 4. Matrix illustrating the genetic composition and epitope specificity of a cohort of anti-Rh(D) antibodies. The horizontal axis represents the unique γ_1 heavy chains and the vertical axis represents the unique λ and κ light chains (based on amino acid sequence). A shaded pattern at the intersection of a heavy chain/light chain pair indicates the Rh(D) epitope specificity observed for that Fab/phage antibody. A few clones gave mixed patterns of reactivity as illustrated. The order of heavy chains (left to right) and light chains (top to bottom) was determined by the multiple alignment of amino acid sequences. Reprinted with permission from Chang and Siegel [32].

Analysis of 83 randomly-selected positive clones revealed a total of 28 unique heavy and 41 unique light chains which had combined to produce 53 unique Fab antibodies that had specificity for at least half of the major Rh(D) epitopes (Figure 2). Surprisingly, despite this diversity, genetic analysis of this set of anti-Rh(D) antibodies revealed that they evolved from a remarkably small number of highly-related Ig germline heavy chain variable region genes (Figure 3) [32]. Several antibodies that differed in epitope specificity used identical heavy chains but different light chains. In particular, two such clones (E1/M2 and E1/M3) differed by only three amino acid residues in their light chains, which resulted in a change from the Rh(D) epitope D2 to epitope D3 specificity (Figure 4). Site-directed mutagenesis experiments revealed that only one of those amino acid differences was actually responsible for the epitope switch [39]. By mutating still other single amino acid residues in a given anti-Rh(D) antibody light chain, epitope specificities would migrate among epitopes D1, D2, and D3.

The data obtained from these studies, previously unobtainable without the ability to study dozens of antigen-specific antibodies cloned from one individual at one point in time, revealed new insights regarding the epitope topology of a clinically-significant human alloantigen as well as the human immune response raised against them. The observation that the epitope specificity of an anti-Rh(D) antibody can change during the course of somatic mutation within a sensitized individual led to a hypothesis that a process, which we termed epitope migration, may be a general feature of human immune responses to other closely-related blood group epitopes. From a practical standpoint, the generation of an extraordinarily large array of heavy and light chain variable region gene segments coding for human anti-Rh(D) antibodies provides a series of DNA cassettes that can now be subcloned into a eukaryotic expression vector that will convert the Fab constructs to full-length IgG suitable for clinical trials [40].

In a more recent series of experiments, we used phage display technology to clone a large repertoire of antibodies directed against human RBCs from peripheral blood mononuclear cells collected from a macaque previously hyperimmunized with human RBCs [41]. One of the goals of this study was to investigate whether the antibodies produced by non-human primates might serve as useful reagents for diagnostic and clinical use. We reasoned that development of such a system would obviate the need for alloimmunized human donors and might produce Ig that was relatively non-immunogenic when administered therapeutically (given that there is significant Ig sequence homology between monkey and man). In addition, one could theoretically obtain therapeutically-useful Ig specificities that would otherwise not be produced by healthy human subjects due to self-tolerance (e.g. antigens over-expressed on tumour cells). Through the use of our usual battery of *human* variable-region oligonucleotides [26], a Fab/phage display library was successfully constructed from the B-cells contained in ~50-ml of blood from a sensitized macaque. The library was panned on human RBCs and $>10^7$ anti-human RBC monoclonal antibody clones were obtained. Of 42 unique clones tested thus far, 34 display anti-Wr^b specificity, 3 display anti-Rh17 specificity, and the remaining 5 are undergoing immunoblotting to aid in their characterization. Remarkably, nucleotide analysis

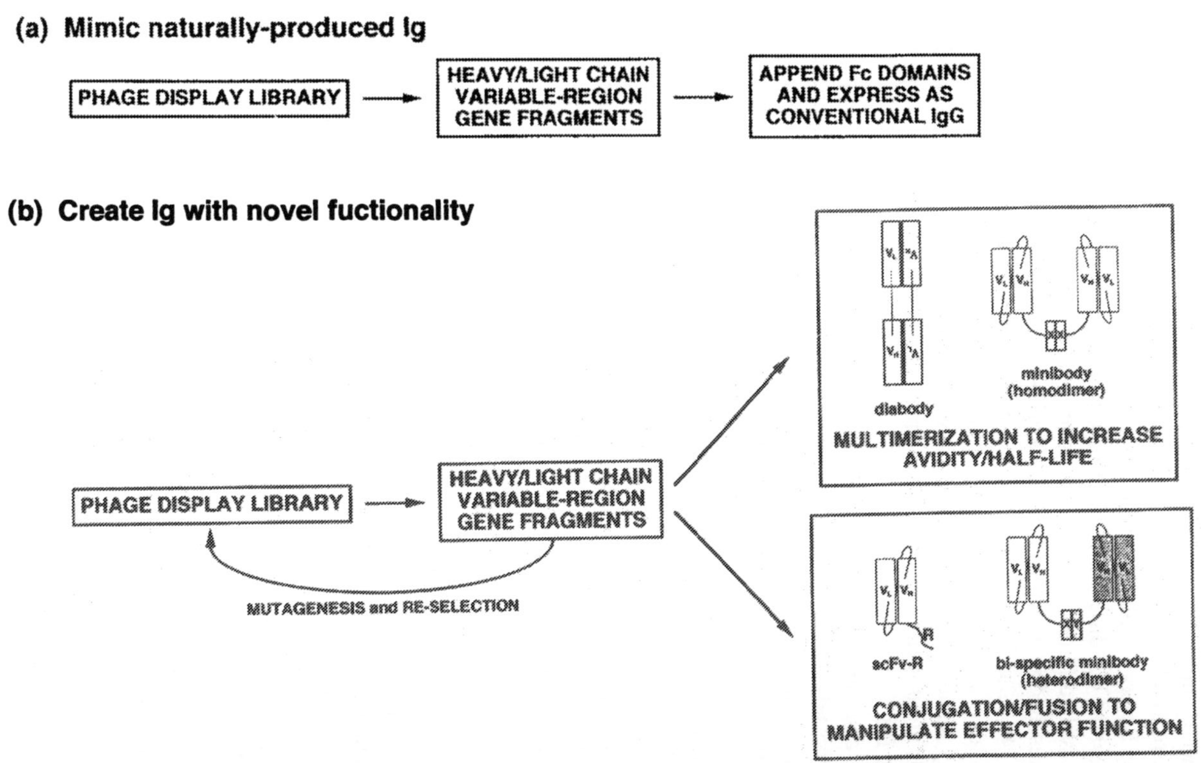

Figure 5. Strategies used to engineer recombinant therapeutic immunoglobulins. X and Y represent self-associating protein domains; "R" represents toxin, drug, enzyme, cytokine, etc. (see text for details).

revealed a restricted use of macaque Ig genes that were homologous to genes of the human VH4-family that humans use to encode allogeneic RBC Ig. Thus far, these results provide a proof of principle that phage display technology can be used to clone repertoires of non-human primate antibodies directed at high frequency human antigens that are difficult, if not impossible, to obtain from human sources.

Advantages of molecular methods for antibody production

As exemplified in the examples presented above, a clear advantage of phage display technology over conventional tissue culture methods is one of yield, especially when one is interested in the isolation of human monoclonal antibodies. Additional significant advantages are those of time and expense. The application of conventional tissue culture methods may require numerous cycles of cloning, screening, subcloning, re-screening, etc., over a period of many months. The construction of a phage display library can be accomplished in several weeks in experienced hands. Each round of selection can be accomplished in a day so that from start-to-finish, large arrays of antigen-specific Fab or scFv antibody fragments can be obtained in about a month. With the availability of pre-made generic phage display libraries constructed from naïve human repertoires [12,42,43] or synthetic human Ig gene segments [44-48], recombinant antibodies against a large variety of targets can be obtained by a single investigator in a few days [49].

The scheme outlined above for the production of human recombinant anti-Rh(D) – library construction, panning, selection of individual clones, conversion to full-length IgG – represents an attempt to recreate *in vitro* the types of molecules which are produced naturally (Figure 5a). However, phage display technology has the ability to create Ig with novel characteristics not found in nature (Figure 5b). For example, the affinity of a recombinant antibody can be improved *in vitro* through error-prone PCR [50], random mutagenesis of the antibody's CDRs [28], or by using mutator strains of *E. coli* [51]. After the introduction of mutations, antibodies are then re-expressed on phage and re-panned on antigen under conditions of increased stringency that select for higher affinity molecules. To increase the avidity and/or circulating half-life of recombinant molecules, scFv's can be made to dimerize into diabodies by shortening the V_H/V_L linker [52-54] or into minibodies [5] by expressing the antibody fragments as fusions to self-associating protein domains [56-59]. Designer effector functions can be engineered by appending cysteine residues to the carboxy termini of a recombinant Fabs or scFv's and conjugating them with drugs [60] or radionucleides such as ^{99m}Tc [61]. Alternatively, phage display-derived antibody fragments can be directly expressed as fusion proteins with bacterial toxins [62], cytotoxic enzymes [63], cytokines [64,65], or super-antigens [66]. A particularly interesting application of this technology is the design of bi-specific (or bi-functional) molecules in which heterodimers of recombinant antibody fragments are formed. As suggested by recent work [67], a bi-functional antibody comprising a tumour-specific antibody fragment and an antibody fragment that catalyzes the conversion of a chemotherapeutic pro-drug to its active form would serve as a magic bullet designed to release the toxic effect of a cytolytic agent only at its target site. Such a novel form of intravenous immune globulin, impossible to produce without the use of molecular biology, would prevent the non-specific systemic toxicity of chemotherapeutic agents.

Future Prospects

The field of antibody engineering has progressed rapidly over the past decade. With the development of phage display technology, large repertoires of human Ig molecules can be isolated without the use of time-consuming, expensive, low-yield cell culture methods and without the need for rodent antibody humanization. Improvements in the size and diversity of generic single-pot libraries created from non-immune sources [43] will not only eliminate the need to construct a new phage display library for each project, but will eventually obviate the need for immunized animals or human subjects. Further refinements in our ability to access such libraries by panning phage-displayed antibodies on cell-surfaces [26,68], tissue sections (69), or even by *in vivo* infusion [70], will provide an enormous number and variety of immunoglobulin reagents for therapeutic use in the future.

REFERENCES

1. Winter G, Milstein C. Man-made antibodies. Nature 1991;349:293-99.
2. Morrison SL. In vitro antibodies: strategies for production and application. Ann Rev Immunol 1992;10:239-65.
3. Jones PT, Dear PH, Foote J, Neiberger MS, Winter G. Replacing the complimentarity determining regions in a human antibody with those from a mouse. Nature 1986;321:522-25.
4. Riechmann L, Clark M, Waldmann H, Winter G. Reshaping human antibodies for therapy. Nature 1988;332:323-27.
5. Lincoff AM, Califf RM, Moliterno DJ, et al. Complementary clinical benefits of coronary-artery stenting and blockade of platelet glycoprotein IIb/IIIa receptors. Evaluation of Platelet IIb/IIIa Inhibition in Stenting Investigators. N Engl J Med 1999;341:319-27.
6. Reff ME, Carner K, Chambers KS, et al. Depletion of B cells in vivo by a chimeric mouse human monoclonal antibody to CD20. Blood 1994;83:435-45.
7. Vincenti F, Kirkman R, Light S, et al. Interleukin-2-receptor blockade with daclizumab to prevent acute rejection in renal transplantation. Daclizumab Triple Therapy Study Group. N Engl J Med 1998;338:161-65.
8. Burton DR, Barbas CF. Human antibodies from combinatorial libraries. Adv Immunol 1994;57:191-280.
9. Marks C, Marks JD. Phage libraries--a new route to clinically useful antibodies. N Engl J Med 1996;335:730-33.
10. Barbas CF, Kang AS, Lerner RA, Benkovic SJ. Assembly of combinatorial antibody libraries on phage surfaces: The gene III site. Proc Natl Acad Sci (USA) 1991;88: 7978-82.
11. Hexham JM, Furmaniak J, Pegg C, Burton DR, Smith BR. Cloning of a human autoimmune response: Preparation and sequencing of a human anti-thyroglobulin autoantibody using a combinatorial approach. Autoimmunity 1992;12:135-41.

12. Griffiths AD, Malmqvist M, Marks JD, et al. Human anti-self antibodies with high specificity from phage display libraries. EMBO J 1993;12:725-34.
13. Chazenbalk GD, Portolano S, Russo D, Hutchison JS, Rapoport B, McLachlan S. Human organ-specific autoimmune disease: Molecular cloning and expression of an autoantibody gene repertoire for a major autoantigen reveals an antigenic immunodominant region and restricted immunoglobulin gene usage in the target organ. J Clin Invest 1993;92:62-74.
14. Portolano S, McLachlan SM, Rapoport B. High affinity, thyroid-specific human autoantibodies displayed on the surface of filamentous phage use V genes similar to other autoantibodies. J Immunol 1993;151:2839-51.
15. Cha S, Leung PS, Gershwin ME, Fletcher MP, Ansari AA, Coppel RL. Combinatorial autoantibodies to dihydrolipoamide acetyltransferase, the major autoantigen of primary biliary cirrhosis. Proc Natl Acad Sci (USA) 1993;90:2527-31.
16. Barbas SM, Ditzel HJ, Salonen EM, Yang WP, Silverman GJ, Burton DR. Human autoantibody recognition of DNA. Proc Natl Acad Sci (USA) 1995;92:2529-33.
17. Eggena M, Targan SR, Iwanczyk L, Vidrich A, Gordon LK, Braun J. Phage display cloning and characterization of an immunogenetic marker (perinuclear anti-neutrophil cytoplasmic antibody) in ulcerative colitis. J Immunol 1996;156:4005-11.
18. Schier R, Bye J, Apell G, et al. Isolation of high-affinity monomeric human anti-c-erbB-2 single chain Fv using affinity-driven selection. J Mol Biol 1996;255:28-43.
19. Roben P, Barbas SM, Sandoval L, et al. Repertoire cloning of lupus anti-DNA autoantibodies. J Clin Invest 1996;98:2827-37.
20. Roark JH, Chang TY, Caton M, Siegel DL. Isolation of human IgG platelet autoantibodies using phage display [abstract]. Transfusion 1998;38:1S.
21. Burton DR, Barbas CF, Persson MMA, Koenig S, Chanock RM, Lerner RA. A large array of human monoclonal antibodies to type 1 human immunodeficiency virus from combinatorial libraries of asymptomatic seropositive individuals. Proc Natl Acad Sci (USA) 1991;88:10134-37.
22. Zebedee SL, Barbas CF, Hom Y-L, et al. Human combinatorial antibody libraries to hepatitis B surface antigen. Proc Natl Acad Sci (USA) 1992;89:3175-79.
23. Barbas CF, Crowe JEJ, Cababa D, et al. Human monoclonal Fab fragments derived from a combinatorial library bind to respiratory syncytial virus F glycoprotein and neutralize infectivity. Proc Natl Acad Sci (USA) 1992;89:10164-68.
24. Williamson RA, Burioni R, Sanna PP, Partridge LJ, Barbas CF, Burton D. Human monoclonal antibodies against a plethora of viral pathogens from single combinatorial libraries. Proc Natl Acad Sci (USA) 1993;90:4141-45.
25. Sasano M, Burton DR, Silverman GJ. Molecular selection of human antibodies with an unconventional bacterial B cell antigen. J Immunol 1993;151:5822-39.
26. Siegel DL, Chang TY, Russell SL, Bunya VY. Isolation of cell surface-specific human monoclonal antibodies using phage display and magnetically-activated cell sorting: Applications in immunohematology. J Immunol Meth 1997;206:73-85.
27. Crowe JE, Murphy BR, Chanock RM, Williamson RA, Barbas CF, Burton DR. Recombinant human respiratory syncytial virus (RSV) monoclonal antibody Fab is effective therapeutically when introduced directly into the lungs of RSV-infected mice. Proc Natl Acad Sci (USA) 1994;91:1386-90.
28. Barbas CF, Hu D, Dunlop N, et al. In vitro evolution of a neutralizing human antibody to human immunodeficiency virus type 1 to enhance affinity and broaden strain cross-reactivity. Proc Natl Acad Sci (USA) 1994;91:3809-13.
29. Cai X, Garen A. A melanoma-specific VH antibody cloned from a fusion phage library of a vaccinated melanoma patient. Proc Natl Acad Sci (USA) 1996;93:6280-85.
30. Barbas CF, Burton DR. Selection and evolution of high-affinity human anti-viral antibodies. Trends in Biotechnology 1996;14:230-34.

31. Pereira S, Maruyama H, Siegel DL, et al. A model system for detection and isolation of a tumor cell surface antigen using antibody phage display. J Immunol Meth 1997;203:11-24.
32. Chang TY, Siegel DL. Genetic and immunological properties of phage-displayed human anti-Rh(D) antibodies: Implications for Rh(D) epitope topology. Blood 1998;91:3066-78.
33. Huls GA, Heijnen IA, Cuomo ME, et al. A recombinant, fully human monoclonal antibody with antitumor activity constructed from phage-displayed antibody fragments. Nature Biotechnology 1999;17:276-81.
34. Mollison PL, Engelfriet CP, Contreras M. Blood Transfusion in Clinical Medicine. Oxford: Blackwell Scientific Publications, 1997.
35. Siegel DL. Isolation of human anti-red blood cell antibodies by repertoire cloning. Ann NY Acad Sci 1995;764:547-58.
36. Siegel DL, Silberstein LE. Expression and characterization of recombinant anti-Rh(D) antibodies on filamentous phage: A model system for isolating human red blood cell antibodies by repertoire cloning. Blood 1994;83:2334-44.
37. Scott M. Rh serology – Coordinator's report. Transfusion Clinique et Biologique 1996;3:333-37.
38. Lomas C, Tippett P, Thompson KM, Melamed MD, Hughes-Jones NC. Demonstration of seven epitopes on the Rh antigen D using human monoclonal anti-D antibodies and red cells from D categories. Vox Sang 1989;57:261-64.
39. Siegel DL, Chang TY. Epitope migration: Anti-Rh(D) antibodies as a model for human immunogenicity [abstract]. Blood 1998;92:671a.
40. Sanna PP, Samson ME, Moon JS, et al. pFab-CMV, a single vector system for the rapid conversion of recombinant Fabs into whole IgG1 antibodies. Immunotechnol 1999;4:185-88.
41. Siegel DL, Reid ME, Lee H, Blancher A. Production of large repertories of macaque mAbs to human RBCs using phage display [abstract]. Transfusion 1999;39:92S.
42. Marks JD, Hoogenboom HR, Bonnert TP, McCafferty J, Griffiths AD, Winter G. By-passing immunization. Human antibodies from V-gene libraries displayed on phage. J Mol Biol 1991;222:581-97.
43. Vaughan TJ, Williams AJ, Pritchard K, et al. Human antibodies with sub-nanomolar affinities isolated from a large non-immunized phage display library. Nature Biotechnology 1996;14:309-14.
44. Hoogenboom HR, Winter G. By-passing immunisation: Human antibodies from synthetic repertoires of germline V_H gene segments rearranged *in vitro*. J Mol Biol 1992;227:381-88.
45. Akamatsu Y, Cole MS, Tso JY, Tsurushita N. Construction of a human Ig combinatorial library from genomic V segments and synthetic CDR3 fragments. J Immunol 1993;151:4651-59.
46. Nissim A, Hoogenboom HR, Tomlinson IM, et al. Antibody fragments from a "single pot" phage display library as immunochemical reagents. EMBO J 1994;13 :692-98.
47. Griffiths AD, Williams SC, Hartley O, et al. Isolation of high affinity human antibodies directly from large synthetic repertoires. EMBO J 1994;13:3245-60.
48. de Kruif J, Boel E, Logtenberg T. Selection and application of human single chain Fv antibody fragments from a semi-synthetic phage antibody display library with designed CDR3 regions. J Mol Biol 1995;248:97-105.
49. Winter G, Griffiths AD, Hawkins RE, Hoogenboom HR. Making antibodies by phage display technology. Ann Rev Immunol 1994;12:433-55.
50. Hawkins RE, Russell SJ, Winter G. Selection of phage antibodies by binding affinity mimicking affinity maturation. J Mol Biol 1992;226:889-96.

51. Low NM, Holliger PH, Winter G. Mimicking somatic hypermutation: affinity maturation of antibodies displayed on bacteriophage using a bacterial mutator strain. J Mol Biol 1996;260:359-68.
52. Holliger P, Prospero T, Winter G. "Diabodies": small bivalent and bispecific antibody fragments. Proc Natl Acad Sci (USA) 1993;90:6444-48.
53. Whitlow M, Bell BA, Feng S-L, et al. An improved linker for single-chain Fv with reduced aggregation and enhanced proteolytic stability. Prot Eng 1993;6:989-95.
54. Wu AM, Chen W, Raubitschek AA, et al. Tumor localization of anti-CEA single chain Fvs: improved targeting by non-covalent dimers. Immunotechnol 1996;2:21-36.
55. Hu S, Shively L, Raubitschek A, et al. Minibody: A novel engineered anti-carcino-embryonic antigen antibody fragment (single-chain Fv-CH3) which exhibits rapid, high-level targeting of xenografts. Cancer Res 1996;56:3055-61.
56. Kostelny SA, Cole MS, Tso JY. Formation of a bispecific antibody by the use of leucine zippers. J Immunol 1992;148:1547-53.
57. Pack P, Pluckthun A. Miniantibodies: use of amphipathic helices to produce functional, flexibly linked dimeric Fv fragments with high avidity in Escherichia coli. Biochem 1992;31:1579-84.
58. Mallender WD, Voss EW. Construction, expression, and activity of a bivalent bi-specific single-chain antibody. J Biol Chem 1994;269:199-206.
59. Dubel S, Breitling F, Kontermann R, Schnidt T, Skerra A, Little M. Bifunctional and multimeric complexes of streptavidin fused to single chain antibodies (scFv). J Immunol Meth 1995;178:201-09.
60. Trail PA, Willner D, Lasch SJ, et al. Cure of xenografted human carcinomas by BR96-doxorubicin immunoconjugates. Science 1993;261:212-15.
61. Verhaar MJ, Keep PA, Hawkins RE, et al. Technetium-99m radiolabeling using a phage-derived single-chain Fv with a C-terminal cysteine. J Nucl Med 1996;37:868-72.
62. Chaudhary VK, Queen C, Junghans RP, Waldmann TA, FitzGerald DJ, Pastan I. A recombinant immunotoxin consisting of two antibody variable domains fused to Pseudomonas exotoxin. Nature 1989;339:394-97.
63. Youle RJ, Newton D, Wu YN, Gadina M, Rybak SM. Cytotoxic ribonucleases and chimeras in cancer therapy. Crit Rev Ther Drug Carrier Syst 1993;10:1-28.
64. Sabzevari H, Gillies SD, Mueller BM, Pancook JD, Reisfeld RA. A recombinant antibody-interleukin 2 fusion protein suppresses growth of hepatic human neuro-blastoma metastases in severe combined immunodeficiency mice. Proc Natl Acad Sci (USA) 1994;91:9626-30.
65. Boleti E, Deonarain MP, Spooner RA, Smith AJ, Epenetos AA, George AJ. Construction, expression and characterisation of a single chain anti-tumour antibody (scFv)-IL-2 fusion protein. Annals of Oncology 1995;6:945-47.
66. Dohlsten M, Abrahmsen L, Bjork P, et al. Monoclonal antibody-superantigen fusion proteins: tumor-specific agents for T-cell-based tumor therapy. Proc Natl Acad Sci (USA) 1994;91:8945-49.
67. Shabat D, Rader C, List B, Lerner RA, Barbas CF, 3rd. Multiple event activation of a generic prodrug trigger by antibody catalysis. Proc Natl Acad Sci (USA) 1999;96: 6925-30.
68. de Kruif J, Terstappen L, Boel E, Logtenberg T. Rapid selection of cell subpopu-lation-specific human monoclonal antibodies from a synthetic phage antibody library. Proc Natl Acad Sci (USA) 1995;92:3938-42.
69. Osbourn JK, Derbyshire EJ, Vaughan TJ, Field AW, Johnson KS. Pathfinder selection: in situ isolation of novel antibodies. Immunotechnol 1998;3:293-302.
70. Pasqualini R, Ruoslahti E. Organ targeting in vivo using phage display peptide libraries. Nature 1996;380:364-66.

DISCUSSION

Moderators: G.D.M. Beun and J.E. Wagner

J.E. Wagner (Minneapolis, MN, USA): GM-CSF plus SCF expanded the megakaryocyte progenitor population and increased CXCR4 expression 80%.

A.L. Drayer (Groningen, NL): Well, the megakaryocytes from these cultures showed CXCR4 expression, that is right.

J.E. Wagner: Yet they had reduced migration to SDF1.

A.L. Drayer: No, the cells which were grown in interleukin-3 showed the reduced expression.

J.E. Wagner: I missed that. With increased expression of CXCR4, I would have expected improved migration to SDF1.

A.L. Drayer: When cells are cultured in MDDF and stem cell factor we do see the CD61 positive cells migrating, but when cells are cultured in interleukin-3 or a combination with 6 growth factors they do not migrate. The CXCR4 expression is reduced under these conditions when we have reduced SDF1 migration. Yes, they track well.

J.E. Wagner: However, there are other things that account for this defect in homing which you have not addressed and migration is only one aspect. Adhesion factors and cell cycle status may also play a role.

G.D.M. Beun (Groningen, NL): Dr. Robinet, is your Ficoll indeed clinical grade?

E. Robinet (Besançon, F): Yes and it is possible now to have Ficoll in bags for bioexperiment.

J.E. Wagner: Dr. Robinet, you told us the percent that are PCR positive for the neoR gene after having been incubated in G418. What was the absolute number of cells recovered? I would not expect that a number of cells with the neoR gene may be lost because of lack of expression and were killed by the G418? What is the cell loss that occurs even though the gene was transferred.

E. Robinet: You mean what is the amount of cells, what was the percent that was lost in G418 even though the gene was transferred. But because of lack of expression there was no resistance. It is difficult to say as one part of the cells are growing that is the positive cells while part of the cells is eliminated.

J.E. Wagner: Can you estimate the number of cells that had the gene before G418 or is the frequency so low that you really cannot reliably estimate the number before and after G418.

E. Robinet: Yes, 10% of the cells are transfused and at the end of the process as far as can be evaluated by PCR all the cells are transfused and that is the most important part. I should say that even the non-transfused cells that have been incubated with G418 stem cells remain seems to be variable they are not blue in Trypan blue assay and when we try to culture them they do not appear to be killed.

M.K. Elias (Groningen, NL): Dr. Robinet, a very attractive approach that of using suicidal gene to selectively eliminate the undesirable effects of T cells while preserving their graft promoting properties and anti-leukaemic effects. We have adopted your protocol in an attempt to process the umbilical cord blood in closed system, but because as far as I know the Ficoll is not FDA approved we substituted it with Optiprep having with a different density. But I am afraid we could not reproduce your nice interface, probably also because we did not use the special insert for the centrifuge. You mentioned a 15% transduction efficiency and 85% positivity for for selected cells, which is comparable to gene transfer in the quiescent haematopoietic cells. Did you notice any difference between the AB serum and the autologous serum as far as transduction efficiency is concerned.

E. Robinet: Sofar we have not seen significant difference between autologous serum and AB serum in terms of transduction and efficiency or it seems not to be a major difference.

A.H. Lazarus (Toronto, C): Dr. Siegel, from what we know about monoclonal antibodies and the way they are monospecific I do not understand why you had some of the same for example heavy chains combining with a light chain recognising presumably a completely different amino space acid sequence. How do you know that these are in fact monoclonal and you do not have more than one phage heavy chain that may account now for two specificities?

D.L. Siegel (Philadelphia, PA, USA): The reason why we found the same anti-Rh(D) heavy chain with different light chains was because an original anti-Rh(D)-secreting B-cell clone in the donor preesumably underwent somatic mutaiton slightly altering the amino acid sequence of its light chain yet keeping teh heavy chain teh same. We succeeded in isolating such a series of clonally-related antibodies from this donor's immune repertoire using phage display technology. The reason why the group of antibodies – all with teh same heavy chain but slightly different light chains – recognise different Rh(D) epitopes has to do with the structural relatedness of at least some of the Rh(D) epitopes.

A.H. Lazarus: Do you have any idea what the affinity is of these antibodies?

D.L. Siegel: Yes, I made soluble Fabs (that is, Fabs unlinke_to phage particles) from some of the recombinant anti-Rh(D) antibodies, radio-iodinated them, and performed conventional binding assays with red blood cells. The affinities are in the order of 10^{-8} to 10^{-9} molar or greater so they are of very high affinity.

H.G. Klein (Bethesda, MD, USA): Dr. Robinet a few technical questions if I may. First I like to rephrase Dr. Wagner's question, just so I am sure I understand. You got 8-40% transduction efficiency of the genes of interest, but were you able to tell us what kind of expression you had from those transduced cells. Were they all expressing or only a small percentage or were you not able to measure that because of the growth of cells in your culture.

E. Robinet: Usually transduction efficiency is below 15% of transduction. The 40% was obtained only one time and we evaluated transduction efficiency by comparing the cell growth rate of not-transduced cells to the growth rate of transduced cells during the selection process and we have another approach by quantitative PCR of a neoR gene and we found the same kind of overreaction. Now, concerning the expression of the transgene; it is only the expression of the gene; the expression is assessed only by the functionality. So, we cannot exclude any infectious windows. There are some cells that have been transduced, that do express the neoR gene because they are going in resistance and do not express the gene. There is a run to run % of the cells that are resistant to the GCV. There are some retrovirus particles that have deleted the gene and there is no expression in these cells. There is a part of the cells that are resistant to GCV, but once again we never, there is never a independently of permanently. And it is true that we never reach 100% of inhibition. We are usually between 90-99.95 sometimes 99.95%, but we never reach 100%. Probably it is because some cells do not express the gene.

H.G. Klein: If I could ask you another small technical question; could you tell us the concentration of cells you had in the bags and I gather you did not have to change either the RPMI medium or the IL-2/CD3 during the course of the 12 day culture.

E. Robinet: Interleukin-3 is introduced only at the beginning and further the culture, IL-2 concentration is maintained at 500 units per ml and the cell concentration is maintained at 1 million per ml. That is once the concentration is over 1 million cells per ml, we just add RPMI medium to adjust the cell concentration to one million cells per ml; so this is until day 7. After the cell concentration begins to decline to become below one million per ml we do not concentrate the cells to maintain at 1 million per ml. We just add the IL-2 and the final concentration at the end of the culture is around 0.2 to 0.3 million cells per ml.

We concentrate the cells before the Ficoll centrifugation to remove the dead cells because we usually have on litre and a half of culture volume.

J.E. Wagner: The one reason that I asked that question was because in our experience using a retrovirus NGFR, neoR as well as the TK, we found that NGFR positive cells could be killed by G418 because it was not expressing neoR. Dr. Robinet, you may want to follow-up on what the clinical results were.

E. Robinet: Among the 12 patients we had 3 acute GvHDs of grade 2 and of grade 3. Two patients responded with a complete response to IVIg alone and one patient did respond partially and we tried in addition corticosteroids to have a complete response. There was a 4th patient who developed a chronic GvHD treated by administration of IVIg alone. So, it is possible to control this.

J.E. Wagner: So, the data suggest that at least in three patients the presence of neoR did not result in an immune reaction resulting in eradication of transduced cells infused. Also the data suggest that GCV is effective in stopping GvHD. Despite my concern cultured cells may not be functional, so this may be the first definite evidence that they are.

R. Atamian (Beirut, Lebanon): Dr. Seremetis, I noticed that the recombinant Factor VIIa has the same glycosylation as the natural Factor VIIa. If you are using single gene for recombinant Factor VIIa production from where is that glycosylation coming.

S.V. Seremetis (Besançon, F): The glycosylation is coming from the target baby hamster kidney cell, the usual mechanism of glycolysation.

R. Atamian: So, you are not only using single Factor VII gene for the collection, you are also using the genes that are responsible for the glycolysation.

S.V. Seremetis: No, the only transduced gene is the FVII gene everything else is the hamster kidney cell.

R. Atamian: The second question; if that porcine FVIII is associated with less immune antibody why do not we use porcine recombinant FVIII.

S.V. Seremetis: There is no experience with porcine recombinant FVIII and in fact I would actually contradict your statement about the immunogenicity of the porcine FVIII. In fact it is clinically useful in patients who have high titre, anti-human FVIII antibodies. It is generally transiently useful. The porcine factor VIII is functionally seen as FVIII in circulation. But typically very soon after infusion of the porcine factor VIII there is recognition of the exogenous porcine FVIII by whatever cause. Perhaps newly recruited clones of epitope specificity to porcine FVIII appear and in fact relatively high titre anti-porcine FVIII antibodies are seen very soon after use of the material. So, its clinical utility is typically only a week or so, I am not convinced that the use of porcine recombinant FVIII would be an important alternative strategy.

R. Atamian: Is there a change that we can get away from the immunogenicity to Factor VIII if we infuse the human being or the animal with the primary and/or secondary antibodies against Factor VIII before we transfuse the Factor VIII.

S.V. Seremetis: This is a very complicated area; the understanding of the genesis of the Factor VIII antibodies includes the observation that immune tolerance can be induced in some patients. By what mechanisms remains unclear and the timing of first introduction of Factor VIII may be important, perhaps as a role for continuous infusion very early on perhaps there is a role for pre-delivery infusion of Factor VIII. I mean there are many, many alternative strategies proposed but yours is certainly one that will be and should be considered.

D.J. Anstee (Bristol, UK): Dr. Seremetis, in the Factor VIIII where you engineered out the B domain, you used a peptide linker or you get a peptide linker in the product, if I understood you correctly. So, there is the potential that you might get antibodies to the linker.

S.V. Seremetis: No, it is not different from the usual linkage and plasma derived Factor VIII.

D.J. Anstee: So the sequence is based on the original.

S.V. Seremetis: Yes.

W.G. van Aken (Amsterdam, NL): Dr. Seremetis, would you like to comment on the virus inactivation which is currently used for the recombinant product, because I noticed that originally the products did not include the specific virus inactivation step and in the more recent ones I see the producers have changed the policies there.

S.V. Seremetis: Well, again we are well aware of the concerns with regards to potential for animal viruses and for other moieties in the manufacturing process. Certainly, it is that concern raised by the consuming public that has caused the manufacturers to introduce other steps in terms of viral inactivation. In fact, the original products are due to include viral inactivation steps. Both of the two original Factor VIIIs include a monoclonal antibody step which reduces in spiking assays about four logs of a variety of target viruses. So in fact there is some viral inactivation step early on. Whether this makes sense or not is a matter for an entire symposium; I do not think so and I also do not think that concerns should keep providers or patients from using the recombinant materials as they currently exist. However, currently people are choosing not to use these materials.

W.G. van Aken: So, would you agree with me that it is a bit of window dressing?

S.V. Seremetis: Absolutely and I also think that the new and ever more expensive products are for the window dressing, the new generations.

C.Th. Smit Sibinga (Groningen, NL): Dr. van der Schoot, I was very intrigued by the Taq/Man technology for the quantification of the PCR end products. What is the technical complexity of the technique and when do you foresee this technology more routinely applied in those centres which deal with the purging technology and minimal remaining tumour cells.

C.E. van der Schoot (Amsterdam, NL): I think we are already routinely applying Taq/Man technology for our routine diagnostics, so I think it is now for two years on the market and in the Netherlands now most university hospitals have a Taq/Man. The only disadvantage so far in our opinion are the costs; it is very expensive, but it safes workload. So, in the end I think it is superior already for routine not for research settings, but if you also can use it as an endpoint measurement, then already it is very simple. The software together with the apparatus is very well developed. So, formerly we were trying to develop a PCR with a lot of experiments and now a technician is working for one day at the computer and then there is a perfect PCR reaction. It directly works; you have to buy your probe and at the moment I think it will be 350 dollars for each probe. So, you have to use it in a routine setting, if you have a lot of samples then you can use that probe I think for 10,000 reactions.

C.Th. Smit Sibinga: Thank you very much. Dr. Siegel, what would you foresee in the near future for a more routine application of the panning and the phage display technology for the production of specific immunoglobulins.

D.L. Siegel: Well, I think certainly in the area of diagnostics it can be used to make blood group reagents particularly for those specificities that are in short supply. Once you isolate the antibodies, which really takes very little time with this technology, I can foresee the phage-displayed antibodies used as diagnostic reagents themselves using anti-M13 phage antibody as the "Coombs" reagent. Alternatively, they can be converted into immunoglobulins and expressed recombinantly. They can be converted into IgMs so as to create direct agglutinins or into IgGs for the production of, for example, synthetic Rh(D)-immune globulin. I did not mention it, but these phage-displayed antibodies can be used to type red blood cells using the gel card technique. I took the gel card and placed the Coombs reagent in the gel with the anti-M13 antibody. If you incubate cells with the phage and put them on top of a column and spin, you get the appropriate sedimentation patterns as with conventional typing sera. In terms of therapeutics, there are a number of groups who have already been using phage display to get antibodies to a number of infectious agents like RSV, for example, and in that case have incorporated antibodies made by the bacteria into aerosols that can be inhaled. There is a growing literature on using this technique to get tumour specific antibodies. There is a recent paper in Nature Biotechnology[1] on this. One can also express the recombinant anti-tumour antibodies as fusion proteins to certain toxins like Pseudomonas exotoxin which will facilitate target cell destruction. So, I think in the very near future, this technology will provide useful reagents for diagnostics and for therapeutics.

C.Th. Smit Sibinga: Great, a short comment to dr. Seremetis. You mentioned as one of the other indications for the use of recombinant VIIa liver disease. I am sure you are aware of the indications at this point of time under investigation clinically in liver transplants, where it has been shown already in the initial data to reduce tremendously the blood need and demand in liver transplant. I think there it is of real benefit.

D.J. Anstee: Dr. Drayer, do you think that the current viability of cytokines and combinations that you could make from those that are available, now will allow the scale up of megakaryocyte cultures to the level which will enable them to be used in a clinical context of transplantation you talked about. Do we know enough about it to get enough in culture to have an effect or is there another step at the research level.

A.L. Drayer: I think it will be possible seeing the high expansions we and others have seen. I think it would be at the progenitor level expending those early megakaryocyte progenitor cells and then seeing if these will engraft. But I think if you further develop the cells into more mature cells then you might have a difficulty in the engraftment. So it would be the expansion of the progenitor cells and then the assumption that having enough of these would speed the recovery.

C.E. van der Schoot: Maybe I can add, because in a recent issue of the Lancet[2] there was a paper about expanded progenitor cells not only directed to megakaryocytes. They were culturing the cells with stem cell factor, TPO and G-CSF and they had a neutropenic period of 0-2 days so it is incredible and also the low platelet count I think only for 5 days. So, it is already shown that it is working in vivo.

J.E. Wagner: That is an interesting combination, because that is the combination that the group in the Netherlands has previously reported to give marked decrease and homing. So, it actually goes against the idea that homing would markedly be impaired by that combination used.

C.E. van der Schoot: I think this is a combination driven by Amgen.

1. Huls GA, Heijnen LA, Cuomo ME et al. A recombinant, fully human monoclonal antibody with antitumour activity constructed from phage-displayed antibody fragments. Nature Biotechnol 1999; 17:276-81.
2. Reiffers J, Caillot C, Dazey B et al. Abrogation of post-myeloablative chemotherapy neutropenia by ex-vivo expanded autologous CD34-positive cells. Lancet 1999;345: 1092-93.

J.E. Wagner: One part of your question was whether or not from the practical point of view that we think that we would be able to do this in a clinical setting, because of the combinations of certain cytokines. I was wondering if that was part of your question, because the problem has always been, can you get the different companies to work with one another, should you cross companies. At least one thing though, this is not something perhaps for this group, because you are all very interested in GMP, but on the other hand from a laboratory point of view, which we plan in doing clinical testing with, is actually using cytokines from R&D systems. So, not for clinical use. On the other hand the FDA is allowing us to use them, so we get around this barrier of working with different companies.

IV. NEW HORIZONS IN CLINICAL MEDICINE

TARGETED IMMUNOTHERAPY- THE USE OF DENDRITIC CELLS IN ANTI-TUMOR VACCINES

G.A.P Hospers[1], J van Zanten[1], N.H. Mulder[1], L.F.M.H. de Leij[2]

Introduction

Vaccination with foreign antigens is one of the most efficacious medical interventions used to prevent infectious disease. The application of this concept to cancer treatment has been one of the first goals of the pioneers of vaccination. However to achieve efficacious anti-tumour vaccination in patients with tumours has been elusive until now. This might change with the recognition of the central role of dendritic cells (DC) in vaccination.

In 1973 Steinman and Cohen [1] have identified DC as a critical element of the immune system. DC have been characterized by their ability to process and present antigen, express co-stimulatory molecules and stimulate naive T cells. It appears that DC targeted peptide vaccines were effective in promoting the expansion of antigen-specific T cells in vitro and more importantly also in vivo. The resulting T cell responses are capable of either treating or preventing the outgrowth of a tumour challenge in experimental models.

It is now clear that tumours of various histologic origins and etiologies (viral, oncogene etc.) are susceptible to DC based therapies.

In phase I and II clinical trials more than 200 patients with cancer have been treated and anti-tumour effects in vivo have been described.

This review deals with the different phenotypes of DC, the effect of maturation of the DC on their ability to promote immune responses, the methods of loading antigens on DC, large scale production of DC for clinical use, the effect of the site of administration on immune responses and the first clinical results.

Classification of Dendritic Cells

Dendritic cells (DC) constitute a heterogenous population of leukocytes with specialized antigen-presenting cell (APC) function [2]. Typical features of DC are their dendritic morphology and the abundant expression of the major histocompatibility complex (MHC) antigens and various co-stimulatory molecules. DC are regarded as the professional antigen-presenting cell for the initia-

1. Dept Clinical Oncology Univ Hospital Groningen, NL.
2. Dept Immunology, Univ Hospital Groningen, NL.

tion of T cell responses. Three derived DC populations have been indentified in vivo; the epithelial DC (Langerhans cells), the non-epithelial DC (interstitial DC), and the activated mature DC. DC do not express classical T cell, B cell or NK cell markers. The expression of classical monocyte markers on precursor and mature DCs is controversial, possibly due to different culture conditions. The lack of a DC-specific marker has hampered the research in this field for years. However, to date the combination of the expression of the following surface molecules is thought to be characteristic for DC; CD1, high levels of MHC class I and II, the co-stimulatory molecules CD80, CD86, CD40, the adhesion molecules ICAM1, LFA3, ICAM3, and CD83 as a marker for mature DC. The expression of CD14 decreases during the differentiation from monocytes into DC.

The function of many of these surface markers is not known. For example the CMRF-35 antigen shows considerable similarity to the Fc receptor for polymeric IgA and IgM and may be a target for antigen loading for immunotherapy [3]. The understanding of the function of the DC specific antigens may contribute to our insight into the development of more effective therapeutic interventions strategies.

Maturation of DC

In 1980 Steinman established that DC may exist in different stages of differentiation which appear to relate to their function [4]. The immature DC are present in nonlymphoid tissues, where their function is to capture antigen. Subsequently these cells migrate to secondary lymphoid organs where they process these antigens and start to stimulate naive T cells present in these compartments. These lymph node DC (mature DC) lose antigen-capturing capacity while acquiring the ability to stimulate T cells.

Immature DC do have high levels of endocytic activity. They use different pathways for uptake; macropinocytose, receptor-mediated endocytose e.g via the mannose receptor, Fc receptor and of engulfment of apoptotic bodies. The high constitutive level of macropinocytosis allows DC to take up large volumes of fluid (as much as the cell's own volume in 2 hours) and concentrate the particles and soluble antigens in the endocytic compartment. Antigen capture is optimal for immature DC generated form peripheral blood monocytes after 5-7 days culturing with GM-CSF and IL4. DC are able to directly stimulate CD8 positive T cells because immature DC can process exogenous soluble proteins captured by macropinocytose for MHC class I presentation. Upon maturation in vitro as induced by e.g. LPS, TNF or CD40L, DC starts to express CD83, a marker which is restricted to mature DC. Schuler (1991) shows that mature DC, although highly stimulatory for allogeneic T cells, are unable to take up and present soluble antigens, a fact that is explained by the rapid downregulation of both endocytic activity and MHC class II biosynthesis upon maturation [5]. The knowledge that only immature DC can capture antigens may be important for the therapeutic application of DC based vaccines.

A dramatic effect of maturation is seen in the peptide MHC complexes. In immature DC, class II molecules have a relatively short half life of about 10 h, which shifts to more that 100 h in mature DC. As a consequence of these coordinate changes in class II biosynthesis and stability, DC soon after induction of maturation, assemble a larger number of peptide-MHC complexes that are retained in stable form for long periods of time in the absence of further class II synthesis. Such a stable form of peptide-MHC complexes allows the dendritic cells to home to secondary lymphoid organs to activate antigen specific T lymphocytes.

Large Scale Production of DC for DC Based Vaccins

DC acquired for vaccination studies can be used from three sources. Firstly, DC can be purified from peripheral blood. However the low frequency of DC in blood limits their use for vaccination. Secondly, CD34 progenitors can be isolated form bone marrow or peripheral blood. If cultured in the presence of GM-CSF and TNFα, these cells differentiate within 14 days into mature DC (CD1a, CD83, HLA-DR +DC). The addition of IL4 blocks the differentiation into macrophages and SCF enhances the yield of DC. This procedure results in a high rate of cell proliferation which might be very useful in the case of retroviral transfection of DC. A disadvantage is the use of TNF, which is added from the beginning since the phase of immature and mature DC is difficult to regulate. Thirdly, monocytes derived from peripheral blood can be differentiated in the presence of GM-CSF and IL4 in vitro, in immature DC. Only after the addition of TNF or CD40L or monocyte conditioned medium the DC matures. The distinct phases of antigen capture and antigen presentation can be regulated in this way. Using the latter method, up to 3×10^6 cells can be obtained from 40 ml blood. For large scale production, the use of apheresis material is preferred after in vivo treatment with G-CSF or GM-CSF because it enhances the DC yield greatly .

Production in concordance with GMP guidelines might require several adaptions in culture methods. For example, the use of serum free medium is recommended supplemented with 1-5% autologous serum instead of culture medium with FCS.

The methods in loading DC with peptides, can be divided in an ex-vivo and in-vivo approach. Both approaches will be discussed.

Methods of Loading Antigens on DC

Ex-vivo

The production and isolation of DC makes it possible to load DC ex-vivo

There are many ways to deliver antigens ex-vivo to DC. An easy and often used method is pulsing the DC with known peptides such as an Ig derived idiotype peptide in case of patients with Multiple Melanoma or MART 1/Melan A peptides in patients with malignant melanoma. The drawback of this method is the fact that synthetically made peptides which are recognized by certain T cells may not be representative for the tumour presented peptides and T cells prestimulated with such a peptide, do not recognize the tumour cells. Therefore, it might be better to use longer peptides or complete proteins which allows the

DC to process its own peptides and present them in context of MHC class I or II.

In case of unknown antigens, a variety of alternative approaches for loading DC have been used, such as tumour lysates, mRNA derived from the tumour, fusion with tumour cells or apoptotic bodies derived from the tumour.

Also DC can be genetically modified with genes coding for antigens.

Both non-viral and viral vectors have been used to deliver genes to DC. As expected the efficiency of a nonviral approach is low (1-5%). The viral vectors are more efficient in delivering genes to DC. It is quite likely that the viral vectors also coat viral proteins which can stimulate the immune response and can serve as an adjuvans in DC based vaccines.

In-vivo

Different methods can be used to load DC in-vivo

First DC can be delivered directly into the tumor. To counter act the immune privileged site of the tumor, the injected DC can be transfected with cytokine genes, as for example IL-12. Another possibility is genetic immunization (inoculation of plasmid DNA encoding a polypeptide antigen). DC are thought to play a central role in vivo in presenting the antigens delivered by plasmid DNA. The loading of these antigens can be through the MHC class II restricted presentation of antigen secreted by neighbouring transfected cells, MHC class I restricted "cross" presentation of antigens released by neighbouring transfected cells and direct presentation of antigens by transfected DC themselves.

Finally, one may choose to target the gene expression to DC in-vivo at the level of transcription [6]. Recently a DC specific transcriptional unit termed the dectin-2 promoter has been characterized and has been found to be as efficient as the conventional CMV promoter.

The Site of Administration of the DC Based Vaccin

The site of antigen delivery in "classical vaccination" is an important determinant of outcome of anti-tumor DC based vaccination. Appropriate targeting of an antigen to the immune system can affect the type of response (local/systemic, humoral/cellular, class I and class II restricted, Th1/Th2) and can also be effective with low antigen dosis. Eggert et al. [7] showed in a B16 mouse melanoma tumor model an increased anti-tumor and immune response after the DC have been administered subcutaneous compared to the intravenous administration. Delivery of the DC based vaccin directly into the tumor can counteract the activation of the immune response due to the effect of the tumor on the environment and CTL function.

The Ideal DC Based Anti-Tumor Vaccin

The ultimate goal for vaccination is an antigen specific stimulation of naive T cells against tumor cells to achieve a therapeutic tumor response. Although there is an extensive knowledge of the biology and types of DC, this did not yet result in a perfect DC based anti-tumor vaccin or criteria for such a vaccin. In 1995

Mayordomo demonstrated that GM-CSF plus IL-4 cultured DC were superior in anti-tumor effects to those derived from GM-CSF alone or GM-CSF plus TNF. In fact, anti-tumor responses in this mouse model were only observed in the cells that had been cultured with GM-CSF and IL-4, and exhibited an immature phenotype [8]. In addition, the importance of using terminally maturing DC in vivo as vaccin is still hypothetical, and has to be determined in either murine of human studies.

Clinical Result

Based on the different animal models showing the capacity of DC to elicit an antitumour immune response clinical trials were started.

The first reported clinical trial of DC-based cancer therapy was initiated in 1994 [9]. In this protocol, patients underwent a 48-hour DC purification procedure with precursors obtained from peripheral blood mononuclear cells after a series of gradient density centrifugations. Isolated DC were pulsed with idiotype protein and administered intravenously in six patients with B cell lymphoma. A total of four therapeutic vaccinations were applied. In all vaccinated patients evidence of immune responsiveness in the form of lymphocytic proliferation and/or cytolytic activity against the lymphocytic targets was seen. One patient had a complete tumor regression, one SD and two patients showed progression. The group of Murphy (1999) treated in phase I and II studies more than 100 patients with prostate cancer with DC based vaccins [10]. The DCs were pulsed with the HLA-A2 specific PSMA peptides (PSM-P1 and P2). These DC cells were monocyte-derived cells cultured in vitro with IL4 and GM-CSF and given intravenously. About 30 % of the patients responded (NPCP criteria) and in about 50% of the patients a reduction in the prostate-specific antigen was found. The observed responses observed in the phase I and II trials were clinical relevant for an average duration of over 370 days.

A third study concerns patients with melanoma who received 1-3 x 10^6 DC injected directly into the inguinal lymph nodes. In this trial, 16 patients were treated with DC pulsed with autologous tumor lysate or synthetic HLA-A1 or HLA-A2 presented melanoma associated peptides for tyrosinase [11]. A complete response was seen in two patients and partial response in three patients.

Holtl (1999) used mature DC (CD83+) loaded with tumor cell lysate in four patients. During the antigen pulse dendritic cells were activated with a combination of tumor necrosis factor α and prostaglandin E2 [12]. A substantial proliferative response of peripheral blood mononuclear cells against tumor and normal lysate was observed after 2 vaccinations. In one patient a 60% reduction in tumor volume was observed after vaccination.

Reichardt (1999) used idiotype determinant on the multiple myeloma protein that can be regarded as a tumor specific marker [13]. The precursor DC were collected from peripheral blood by leukapheresis. DC were identified by high MHC class II and the absence of B and T cell markers. Two of the 12 patients developed an Id specific cellular proliferative immune response and one of three patients studied, developed a transient but Id specific CTL response. The two patients who developed a cellular Id specific response remain in complete remission.

Morse (1999) used DC loaded with an HLA-A2 restricted fragment of the tumor antigen carcinoembryonic antigen (CAP-1 peptide), for the treatment of patients with advanced CEA expressing malignancies [14]. Up to 1 x 10^8 cells (66% non mature DC) could be produced under serum free conditions, loaded in-vitro and cryopreservation was possible without a significant loss of viability of the DC. The dose levels up 1×10^8 cells could be given i.v. weekly of biweekly. During treatment in three patients a grade 3/4 liver toxicity was detected in 21 of these treated patients and no responses were found in 19 evaluable patients.

In conclusion DC based anti-tumor vaccins can result in an immunological anti-tumor response and in clinical responses. The clinical results are difficult to compare due to the different origin of the DCs, different antigens used, different ways of loading and administration. Ten ongoing DC based clinical trials are listed on the PDQ [15]. As far as we know, no phase III study are initiated.

At this moment there is no registered DC based anti-tumor vaccin worldwide.

Conclusion

In the last years, the ontogeny and the phenotype of the DC as well as its capacity as an antigen presenting cell have been extensively explored in vitro. With these new insights of the DC this cell is exploited as a source for vaccination protocols. The clinical results of a relatively non-toxic DC based vaccin in patients with prostate cancer is promising.

The antitumour effect of the different types, maturation state, method of obtaining and loading of DCs have not yet been adequately investigated in animal and human trials. Also differences in the effect of these vaccins on various tumour types has to be studied. These studies will eventually teach us what the most effective DC based antitumour therapy is and where this approach fits in the anti-cancer strategy.

References

1. Steinman RM, Cohn ZA. Identification of a novel cell type in peripheral lympoid organs of mice. Morphology, quantitation, tissue distribution. J Exp Med 1973;137: 1142-45.
2. Steinman RM, Bancherau J. Introduction. In: Lotze MT and Thomson AW, eds. Dendritic cells, biology and clinical applications, Academic Press, San Diego, USA, 1999.
3. Jackson DG, Hart DNJ, Starling GC, Bell JI. Molecular cloning of a novel member of theimmunoglobulin gene superfamily homologous to the polymeric immunoglobulin receptor. Eur J Immunol 1992;22:1157-63.
4. Steinman RM, Nussenzweig MC. Dendritic cells: features and functions. Immunol Rev 1980;53:127-47.
5. Schuler G, Koch F. Enrichment of epidermal Langerhans cells. In: Schuler G (ed). Epidermal Langerhans Cells. CRC Press, Boca Raton FL, 1991:139-57.

6. Takashima A, Morita A. Dendritic cells in genetic immunization. J Leukoc Biol 1999;66:350-56.
7. Eggert AA, Schreurs MW, Boerman OC, et al. Biodistribution and vaccine efficiency of murine dendritic cells are dependent on the route of administration. Cancer Res 1999;59: 3340-45.
8. Mayordomo JI, Zorina T, Storkus WJ, et al. Bone marrow-derived dendritic cells pulsed with synthetic tumour peptides elicit protective and therapeutic antitumour immunity. Nat Med 1995;1:1297-302.
9. Hsu FJ, Benike C, Fagnoni F, Liles TM, et al. Vaccination of patients with B-cell lymphoma using autologous antigen-pulsed dendritic cells. Nat Med 1996;2:52-58.
10. Murphy GP, Tjoa BA, Simmons SJ, et al. Infusion of dendritic cells pulsed with HLA-A2-specific prostate specific membrane antigen peptides: a phase II prostate cancer vaccine trial involving patients with hormone-refractory disease. Prostate 1999;38:73-78.
11. Nestle FO, Alijagic S, Gilliet M, et al. Vaccination of melanoma patients with peptide- or tumor lysate-pulsed dendritic cells. Nat Med 1998;4328-32.
12. Holtl L, Rieser C, Papesh C, Ramoner R, et al. Cellular and humoral immune responses in patients with metastatic renal cell carcinoma after vaccination with antigen pulsed dendritic cells. J Urol 1999;161:777-82.
13. Reichardt VL, Okada CY, Liso A, et al. Idiotype vaccination using Dendritic cells after autologous peripheral blood stem cell transplantation for multiple myeloma- a feasibility study. Blood 1999;93:2411-19.
14. 14. Morse MA, Deng Y, Coleman D, et al. A phase I study of active immunotherapy with carcinoembryonic antigen peptide (CAP-1) pulsed, autologous human cultured dendritic cells in patients with metastatic malignancies expressing carcinoembryonic antigen. Clin Canc Res 1999;5:1331-38.
15. PDQ Clinical Trail Abstract; http://cancernet.nci.nih.gov.

USE OF *IN VITRO* AND *IN VIVO* GENETICALLY MANIPULATED CELLS FOR IMMUNOTHERAPY OF CANCER

M. Mincheff, S. Zoubak, H.T. Meryman[1]

Two recent discoveries in basic immunology have made the prospect of immunotherapy of cancer less distant. The first discovery relates to the nature of the antigens that are recognized by T cells, the likely effectors in cancer immunotherapy [1-5]. T cell receptors react with antigen-derived peptides bound to major histocompatibility (MHC) molecules. Specialized antigen presenting cells (APCs) express peptides from extracellular antigens in association with class II molecules, while parenchymal (and cancer) cells express MHC class I molecules displaying peptides from intracellularly synthesized proteins [6,7]. Different peptides may bind to one and the same MHC molecule thus creating different "bulging away" antigenic specificities [4,5].

Tumor-Associated Antigens

Immunological destruction of tumours is expected to be mediated primarily by lymphocytes [8]. CD8+ CTLs play a dominant role in most animal models [9,10], but CD4+ T helper cells [11], particularly of the Th1 type, are also important. For tumor destruction by CTLs, tumor-cell antigens must be processed by the tumor cells into peptides, which are then associated with MHC class I molecules and expressed on the cell membrane [11,12]. All parenchymal cells, as well as most tumor cells, express membrane-bound, peptide-occupied MHC class I molecules as a constant display of their most recent biochemical activity [13-16]. Based on their pattern of expression, these tumor-associated antigens can be classified into several major groups (Table 1).

Tumor transformation is associated with the activation of genes that are completely silent in most normal tissues. The products of such genes can be considered as tumor antigens and have been used as targets for immunotherapy of established tumours [17,18]. Prototype antigens of this group are encoded by the human Melanoma Antigen Encoding (*MAGE)* genes [17] and by the Renal Antigen Encoding (*RAGE*) genes [18].

Differentiation antigens such as the melanocyte tyrosinase [19] or the carcinoembryonic antigen (CEA) [20] constitute a second major group of tumor-associated antigens. Unlike MAGE and RAGE, or mutated antigens (see below), differentiation antigens are also expressed in a number of normal tissues and this

1. Biomedical Research Institute, Rockville, MD, USA.

Table 1. Tumor-associated antigens.

Type of Antigens	Products	Reference
Products of "silent" genes	MAGE; RAGE	17,18
Differentiation antigens	Melanocyte Tyrosinase, CEA	19,20
Mutated antigens, products of chromosomal translocation	CDK 4 , β-catenin, *CASP-8*, bcr-abl	21,22,23,24
Oncogene amplification	as HER-2/neu (C-erbB-2), p53	25,26,27
Viral antigens	E7	28
Tissue-specific antigens	PSA, PSMA	29

fact has raised concerns about possible autoimmune side effects if they are used as targets for immunotherapy.

Mutations that occur during malignant transformation also give rise to tumor-associated antigens. Among these are mutations affecting the human genes encoding for cyclin-dependent kinase (CDK) 4 [21], β-catenin [22] and *CASP-8* [23]. Although the specificity of these antigens is strictly anti-tumor, they can not be used as generic cancer vaccines because they are unique to each patient. A subgroup of mutated antigens includes chimeric proteins encoded at the site of chromosomal translocations. For example, myeloid leukemias often express a chimeric protein, bcr-abl, which results from a t(9;22) chromosomal translocation [24].

A fourth group of tumor-associated antigens consists of over-expressed ubiquitous products. Such products of oncogene amplification, as the HER-2/neu (C-erbB-2) antigen, are a manifestation of genetic instability that is often associated with the genesis and progression of cancer [25]. The HER-2/neu oncogene is the second member of the epidermal growth factor receptor family. It is overexpressed in many different types of human cancers, including breast, ovarian, lung, gastric, and oral cancers [25]. Overexpression of HER-2/neu in breast cancer has been associated with poor overall survival and has been shown pre-clinically to enhance malignancy and the metastatic phenotypes. HER-2/neu-encoded p185 protein is a receptor tyrosine kinase that can be associated with multiple signal transduction pathways. However, it is not yet clear how a specific signal pathway may correspond to a specific biological response. A peptide, derived from HER-2/neu was found to be the target of lymphocytes, infiltrating some HLA-A2 ovarian carcinomas [25,26]. Similarly, CTL have been successfully raised against a tumor cell overexpressing p53 [27].

An additional group of tumor-associated antigens comes from oncogenic viruses. Products encoded by viral genetic material such as the E7 oncoprotein from the human papilloma virus (HPV)16, are expressed in tumor cells and can successfully be used for stimulation of tumor-specific CTLs [28].

Finally, in certain cases, tissue specific antigens can be used as targets for cancer immunotherapy. Prostate cells provide a particularly useful model for the testing of this strategy since the application to prostate cancer does not mandate discrimination between normal and malignant cells [29].

T cells would not recognize a tumor that lacked MHC molecules, or lacked molecules involved in antigen processing and transport to MHC class I [30,31]. This implies that, in the clinical application of gene therapy for cancer, strategies not solely dependent on MHC class I-mediated tumor cell recognition by cytotoxic T lymphocytes (CTL) may need to be developed. For example, a complete loss of HLA class I has been reported to occur in 34% of primary prostate cancers and up to 80% of lymph node metastases [30]. When individual allelic expression was assessed, the minimum estimate of down regulation was up to 85% in the primary prostate cancers and almost 100% in the metastases [30]. A de-differentiation of tumor cells with loss of HLA molecules may be part of the natural course of the disease. Such cells, however, continue to express proteins on their membrane. For example, de-differentiated prostate tumor cells continue to express PSMA [31]. Other cancer cells may also express membrane bound tumor-associated antigens that can serve as potential targets. Use of the native molecule as target, rather than peptides derived from it, however, requires involvement of the humoral (antibody-mediated) arm of the immune system. In this respect, stimulating T cell immunity by naked DNA or viral immunization may be only the initial step. Experimental evidence from animals suggests that prior DNA immunization followed by boosts with soluble protein can result in the production of high titer, cytotoxic, antigen-specific antibodies [32] such as would be required for tumor lysis and destruction.

Lymphocyte Activation

Despite the fact that many peripheral solid tumours such as sarcomas and carcinomas express tumor-specific antigens that can serve as targets for immune effector T cells, overall immune surveillance against such tumours seems relatively inefficient. A generally accepted model in T cell immunology postulates that T cells require two distinct signals for activation. The first comes from the ligation of the T cell receptor with the MHC-peptide complex, while a second, often called co-stimulatory, originates from the binding of the T cell to accessory molecules on the antigen-presenting cell. [33,34]. Antigen recognition in the absence of signal two, as might occur in peripheral tissues and in most neoplastic tumours, renders T and B lymphocytes non-reactive (anergic) or deletes them [34]. Parenchymal cells and tumor cells are ineffective APCs because they do not normally express the second, co-stimulatory signal. Furthermore, to function properly, lymphocytes, macrophages and APCs need to collaborate within organized lymphoid tissues such as lymph nodes and spleen [35-40]. In fact, translocation of antigens from the periphery to secondary lymphoid organs appears to be a key step in the initiation of immunity. T cells do not react against antigens that are immunologically sequestered [34] or antigens that do not reach a certain concentration level in lymphoid organs [41]. Whether an immune response is mounted against a tumor-associated antigen will therefore depend on the presentation of this antigen by professional APCs in sufficient amounts and over an adequate time in secondary lymphoid tissues.

Antigen transport to secondary lymphoid tissues and primary sensitisation of T lymphocytes is mainly mediated by dendritic cells (DCs) which reside in peripheral non-lymphoid tissues. Appropriate and repeated activation of cytotoxic T lymphocytes by DC, in concert with local inflammatory processes that lead to the formation of organized lymphoid tissue in the target organ, is likely to be crucial in the development of destructive immunity. The success of a therapeutic intervention destined to elicit tumor specific immunity will therefore depend on 1) recruitment of immature DC to tumor(s), 2) optimisation of DC loading with tumor antigens and 3) enhanced DC maturation and migration to regional lymph nodes.

Factors Influencing Dendritic Cell Function

Immature dendritic cells residing in the periphery, e.g. Langerhans cells in the epidermis or mucous membranes, form a network of sentinel cells that, upon certain stimuli (Table 2), take up and process antigen, mature and migrate to the regional lymph node.

Table 2. Growth factors and membrane-bound receptors influencing DC function.

Growth Factors and Membrane-Bound Receptors	Effect on DC Function	References
Heat shock proteins (hsps)	"Danger signals"	42-51
Flt3-ligand	Increase production of myeloid and lymphoid DC	52,53
GM-CSF	Maturation of DC precursors	54,56
CD40	Production of IL-12; Th1 bias	57
TNF-alpha	DC maturation and migration	58
CD80; CD86	Ability of DC to bias T cell development towards Th1 versus Th2 phenotype	61

Immature dendritic cells are in constant contact with auto-antigens in the periphery. Whether an immune response is initiated against these antigens depends on their transfer and presentation in the regional lymphatic tissue. What triggers this transport and immunogenic presentation still remains an unresolved issue. One of the major line of thinking suggests that the immune system essentially makes a value judgement of what is dangerous, be it self or non-self, and it is the cells of the dendritic cell lineage that make this distinction [42-49]. The danger model proposes that the most likely signals of danger are the passive signals of cell distress, such as the functionally essential cytoplasmic proteins released only when cells undergo a violent death - necrosis. It has been shown, however, that DCs are capable of inducing CTLs by picking up antigens from virally infected, apoptotic cells [48]. Danger signals, therefore, might not be restricted to the passive signals released during necrosis, but could include active signals produced by the innate immunity's response to infection [49]. The most likely group of proteins that could serve such a function is the group encompassing the heat shock proteins (HSPs), which are induced by stress and by inflammation [50,51]. The HSPs are natural chaperones of peptides that reflect the immunologic composition of the cell. Immunization with tumor-derived HSP-peptide complexes (HSP-PC) induces a potent antigen-specific immune response against both the primary tumor and pre-established metastases and the response is restricted to the tumor from which the HSPs are purified [52,53].

Certain growth factors such as Flt3-Ligand (FL), granulocyte-macrophage colony-stimulating factor (GM-CSF) and tumor-necrosis factor alpha (TNF-alpha) also influence DC development and migration. Injection of mice with the cytokine Flt3-ligand (FL), for example, dramatically expands mature lymphoid and myeloid-related DC subsets [54,55]. In contrast, injection of a polyethylene glycol-modified form of granulocyte/macrophage colony-stimulating factor (GM-CSF) into mice only expands the myeloid-related DC subset [56]. These DC subsets differ in the cytokine profiles they induce in T cells *in vivo*. The lymphoid-related subset induces high levels of the Th1 cytokines interferon gamma and interleukin (IL)-2 but little or no Th2 cytokines. In contrast, the myeloid-related subset induces large amounts of the Th2 cytokines IL-4 and IL-10, in addition to interferon gamma and IL-2. FL- or GM-CSF-treated mice injected with soluble ovalbumin display dramatic increases in antigen-specific antibody titers, but the isotype profiles seem critically dependent on the cytokine used. Although FL treatment induces up to a 10, 000-fold increase in ovalbumin-specific IgG2a and a more modest increase in IgG1 titers, GM-CSF treatment favours a predominantly IgG1 response with little increase in IgG2a levels. These data suggest that distinct DC subsets have strikingly different influences on the type of immune response generated *in vivo* and may thus be targets for pharmacological intervention.

Recently, it has been shown that ligation of CD40 on dendritic cells triggers production of IL-12 [57]. TNF-alpha also provides an important signal for the migration of epidermal Langerhans cells (LC) from the skin [58]. Intradermal administration of homologous recombinant TNF-alpha to mice induces both a rapid reduction in the frequency of LC local to the site of exposure and, somewhat later, an accumulation of dendritic cells in draining lymph nodes. The influence of TNF-alpha on LC function, however, may be dose-dependent, with lower concentrations inducing migration, but higher concentrations immobilizing LC in the epidermis [58].

A number of recent studies have suggested an important role for the CD28 (B7) co-stimulatory pathway in influencing T cell differentiation. Following initial activation, CD4 + T cells can differentiate along either of two pathways, giving rise to distinct sub-populations as characterized by patterns of cytokine production and effector function. Th1 differentiation results in production of IL-2 and gamma interferon (gamma IFN), macrophage activation and induction of delayed-type hypersensitivity (DTH) responses. Th2 differentiation leads to secretion of IL-4, IL-5 and IL-10 (allergic type of response) and may be associated with the suppression of cell-mediated immunity [59,60]. Among the two known natural ligands for CD28, expressed by DCs, CD80 (B7-1) has been found to primarily bias the response towards Th1, while CD86 supports preferential Th2 differentiation [61]. This issue, however, is quite controversial since others [62] have found that co-administration of CD86 cDNA along with DNA encoding HIV-1 Ags intramuscularly dramatically increased Ag-specific CTL responses.

Strategies to Increase DC Loading with Tumor Antigens Based on Genetic Manipulation of Tumours or DC

Loading of DC with tumor antigens can be achieved by increasing the traffic of such cells through tumor tissue, combined with simultaneous release of "danger" signals. This has been accomplished by transfecting of tumor cell, either *in vivo* or *in vitro*, with those growth factors known to facilitate dendritic cell function (Table 2).

In vivo transfection of tumor cells has already undergone a phase I trial [63]. Evidence of anti-tumor activity was "suggested in patients with advanced malignancy that received multiple injections" with interferon gamma retroviral vector. This, and other studies have also suggested that transgene protein expression was sufficient to induce strong and prolonged immunity and that recombinant adenovirus injected directly into the tumor is a highly efficient vector for immunizing patients against a transgene protein [64]. Administration of adenovirus vectors (Ad) to immunologically naive experimental animals almost invariably has resulted in the induction of systemic anti-Ad neutralizing antibodies. Unlike the case for experimental animals, administration of such vectors to humans does not invariably evoke a systemic anti-Ad neutralizing antibody response [65]. Furthermore, the extent of the response, in cases where it occurred, is dictated by pre-existing antibody titers and modified by the route of administration but does not seem to be dose dependent.

Tumor cells can also be transfected *in vitro* and used as tumor-cell vaccines. For example, injection of tumor cells that have been transduced with the gene encoding the cytokine GM-CSF results in local accumulation of dendritic cells in and around the vaccination site [66]. Despite the therapeutic promise of this vaccine strategy, however, clinical development has been limited by the need to establish in culture the tumor of each patient and to perform individualized gene transfer. To circumvent these issues, we have selected to work at the dendritic cell level.

The myeloid DCs originate from a bone marrow CD34+ precursor common to granulocytes and macrophages [67]. A CD 14+ intermediate has been described in peripheral blood with the potential to differentiate along the DC or the macrophage pathway under distinct cytokine conditions [68,69]. Depending on a concentration gradient in the milieu, peptides derived from tumor-associated antigens can be loaded onto the MHC molecules of dendritic cells simply by incubation *in vitro* [70,71] and recently, trials using this approach have been initiated [72, 73].

Alternatively, DC can be genetically engineered to produce tumor-associated antigens and, respectively, peptides derived from them [74]. Such genetic engineering can be accomplished either *in vitro* (transfection with plasmids or viral vectors) [74] or *in vivo* ("naked" DNA immunization, immunization with viral vectors) [76-77]. While *in vitro* techniques have uniformly produced a rather low efficiency of DC transfection, DNA vaccination has proven to be an effective means of immunization against various noxi, including malaria [78], influenza [79], rabies [80], herpes simplex [81], leishmaniasis [82], tuberculosis [83] and borrelia infection [84], and may have the potential for prevention or treatment of autoimmune diseases [85] or allergies [86]. The immunogenicity of plasmid DNA is enhanced by short-term sequences that contain CpG dinucleotides in particular base contexts [87-89]. Several reports have shown that injection of naked DNA can induce both humoral and cellular immune responses directed against the encoded immunogenic proteins [90]. The development of cytotoxic (Tc1) T lymphocyte (CTL) responses is particularly important against virally infected cells or tumor cells. Experimental work in animals has shown that the intensity of the immune response can be increased by the introduction of sequences encoding for costimulatory molecules (CD86) [91] or of soluble lymphokines (GM-CSF and IL-12) [92,93].

Several groups have recently reported on the safety and biologic effectiveness of DNA vaccines for immunizations against HIV and malaria [78,94]. The use of PSA-recombinant vaccinia virus was also reported to be safe and to lead to an immunologic response [95]. Using PSMA as a target antigen, we have initiated a clinical trial for immunotherapy of prostate cancer. Prostate-specific membrane antigen is a hydrolase with the substrate and pharmacological characteristics of a neuropeptidase [96]. PSMA was originally considered to be restricted almost exclusively to prostate epithelial cells [97]. It is an integral, type II membrane protein and it is highly expressed in both normal and neoplastic prostate tissue and in prostate cancer metastases [98].

The extracellular portion of the human PSMA (XC-PSMA) was cloned into the pCR2.1 vector (Invitrogen, Carlsbad, CA) by RT-PCR of total mRNA, that was obtained from the human prostate cancer cell line LNCaP (CRL 1740, ATCC). The complete human CD86 cDNA was cloned into the pCR2.1 vector by RT-PCR of total mRNA, isolated from human monocytes of a healthy donor.

Both the cDNAs were sub-cloned into a modified mammalian expression vector pcDNA 3.1 (Invitrogen, Carlsbad, CA) after deletion of the neomycin resistance gene. Both the XC-PSMA and CD86 plasmids contain the corresponding cDNAs under the regulation of a CMV promoter and a Bovine Growth Hormone polyadenylation signal. The combined PSMA/CD86 plasmid contains both the XC-PSMA and CD86 cDNAs, each under a separate CMV promoter and a polyadenylation signal.

The entire expression cassette from the XC-PSMA plasmid described above, was inserted in a replication deficient (E1, E3 deletions) adenoviral (Ad5) vector (Quantum Biotechnologies Inc., Toronto, Canada). The resulting adenovirus Ad5-PSMA is a replication deficient recombinant adenovirus in which the replication-essential genes E1 and E3 are replaced with the expression cassette containing a coding sequence for the extracellular portion of the human PSMA.

All patients were immunized intradermally, around the naval area. Prior to injection, both of the plasmids and the Ad5-PSMA were diluted with sterile pyrogen-free saline. We also used soluble GM-CSF (sGM-CSF. Leukine. Sargramostim), which is a recombinant human granulocyte-macrophage colony-stimulating factor produced by recombinant DNA technology in a yeast expression system (Immunex, Seattle, WA). The 24-hour DTH reaction was recorded following each re-immunization, the aim being to maintain an intense response following each application. Depending on the intensity of the DTH response patients' doses for the following boost have varied between 100 to 800 μg of DNA. The viral dose, whenever virus was applied, has always been 5×10^8 PFUs. Standard laboratory tests included CBC, urinalysis, liver enzymes. antinuclear antibodies, erythrocyte sedimentation rate, PSA. Each patient had a pelvic CAT scan, chest radiograph and a cardiograph on entry and on week 20 (week 10 for the 10 patients immunized with virus only). Additionally. analysis of HLA DR^+; $CD4^+$; $CD8^+$; $CD3^-/CD16^+CD56^+$; $CD3^+$; $CD11b^+$; $CD25^-$: $CD19^-$ cells as well as CD4/CD8 ratio, prior to, and following immunotherapy. were performed by flow cytometry. Safety was defined as lack of untoward clinical or laboratory events, with particular attention to local and systemic reactions. as well as evidence of anti-nuclear or anti-double-stranded DNA antibody.

Ours is the third report on DNA vaccination in humans, but probably the first one that involves patients who were repeatedly immunized for more than one year. The primary objective of the study was to determine the safety of the PSMA vaccine after repeated intradermal injections. Recently, detectable PSMA levels have been found in tumor vascular endothelium, primary renal tumours. and, although at much lower density, in some normal tissues such as duodenal mucosa and a subset of proximal renal tubules [98]. Using PSMA as a target for immunotherapy could be seriously offset by the development of serious toxicity such as autoimmunity. However, our trial, as well as other trials that target PSMA, has shown that the evoking of a T-cell immune response against PSMA does not lead to autoimmune reactions in other organs.

Fifty five patients were included in the study. So far, almost one and a half years since the study began, no patient has experienced any short- or long-term side effects, including anti-DNA antibody. The only exception is the develop-

Table 3. Distribution of patients in the phase I clinical study according to presence of metastases and response to therapy

Patients with detectable LN or bone metastases				**Patients with no detectable metastases**			
Combined hormone and immune therapy		**Immune therapy only**		**Combined hormone and immune therapy**		**Immune therapy only**	
Res-ponder	Non-responder	Res-ponder	Non-responder	Res-ponder	Non-responder	Res-ponder	Non-responder
9	11	2	4	9	1	12	7

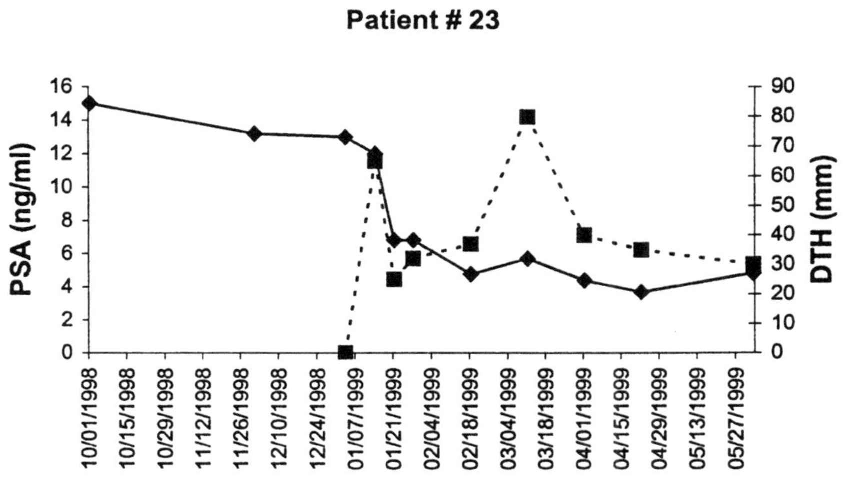

Figure 1. Patient # 23. Patient # 23 was diagnosed in June of 1998 (T2bNoMo, Gleason pattern 3-4). Immunotherapy initiated in January 1999. Each black square represents a DTH value (in mm) following immunization. DTH values are normalized to 100 ug DNA. Diamonds represent PSA values. No other therapy but immune therapy. Currently the prostate has decreased in volume by 50%, no tumor is palpable and he has no obstruc-obstructive voiding symptoms.

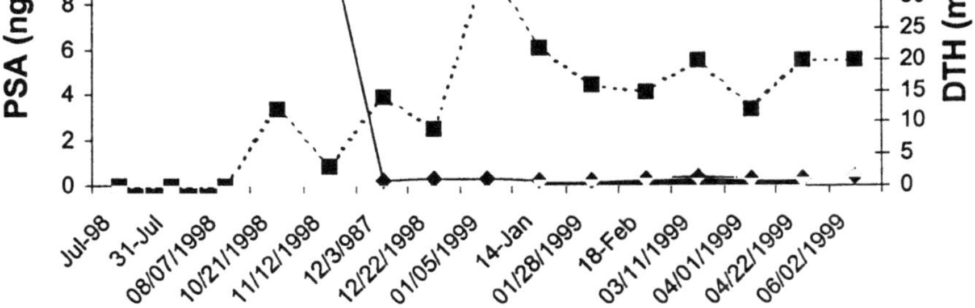

Figure 2. Patient # 10 was diagnosed with prostate cancer with bone metastases on May of 1998 (T3cNxM1b, Gleason pattern 2-3). Immunotherapy initiated in July 1998. Each black square represents a DTH value (in mm) following immunization. DTH values are normalized to 100 ug DNA. Diamonds represent PSA values. No other therapy but immune therapy. Currently the patient has no obstructive voiding symptoms and no bone pains.

ment of fragile vasculature in the immediate vicinity of the prostate in three of the four patients who had immunotherapy prior to surgery. Additional studies are needed before the significance of this observation can be determined. We found that repeated local injection of rHuGM-CSF (Sargramostim) is a safe procedure and that *in vivo* transfection with CD86 is also well tolerated. On the other hand soluble recombinant GM-CSF, as well as co-expression of CD86 with the target antigen increases the likelihood for successful immunization. Use of a cocktail of a complex PSMA/CD86 plasmid and sGM-CSF leads to uniform immunizations in all patients.

Uniform and safe anti-PSMA immunization was also achieved with the use of the recombinant adenoviral vector. However, an important consideration is whether a pre-existing adenovirus-specific immunity might compromise the ability of this vector to deliver antigen. This concern, in view of the effectiveness of the complex plasmid plus sGM-CSF in immunizing 100% of the patients, has led us to the conclusion that the best priming for induction of anti-PSMA T cell immunity requires only the use of the PSMA/CD86 plasmid-soluble GM-CSF cocktail.

All patients became immunized against PSMA as demonstrated by the development of positive DTH reaction. The patients were heterogeneous with regard to local advancement of disease, presence of distant metastases, of hormone treatment and refractoriness, which does not permit unequivocal interpretation of the results. Nevertheless, several responders to the immunotherapy could be identified.

The response rate was clearly dependent on the stage of the disease. (Table 3).

Twenty one of the twenty nine patients who had no evidence of distant metastases, including twelve of the nineteen patients who were on immunotherapy only, responded to the therapy with a drop in PSA levels and, whenever applicable, with improvement of local disease (Figure 1).

In comparison, eleven of the twenty six patients with detectable distant metastases showed improvement, including two of the 6 who were solely on immunotherapy (a representative case is shown in Figure 2).

Conclusion

Tumor immunotherapy can successfully be achieved by *in vivo* manipulation of patient's dendritic cells by naked DNA and adenoviral immunization. The procedure is safe and well tolerated. Autoimmunity against PSMA can be induced and maintained by repeated immunizations with DNA. No side effects have been registered one and a half-year following the onset of immunizations against a self-antigen.

References

1. Bjorkman PJ, Saper MA, Samraoui B, Bennett WS, Strominger JL, Wiley DC. Structure of the Human Class I Histocompatibility Antigen, HLA-2. Nature, 1987; 329: 506-12.

2. Bjorkman PJ, Saper MA, Samraoui B, Bennett WS, Strominger JL, Wiley DC. The Foreign Antigen Binding Site and the T Cell Recognition Regions of Class I Histocompatibility Antigens. Nature, 1987;329:512-18.
3. Brown J, Jardetzky T, Saper MA, Samraoui B, Bjorkman PJ, Wiley DC. A Hypothetical Model of the Foreign Antigen Binding Site of Class II Histocompatibility Molecules. Nature, 1987;332:845-50.
4. Guo HC, Jardetzky TS, Garrett TPJ, Lane SL, Strominger JL, Wiley DC. Different Length Peptides Bind to HLA-Aw68 Similarly at Their Ends But Bulge Out In The Middle. Nature, 1992;360:364-66.
5. Padlan EA, Margulies DH. Feeling out the complex. Curr Biol 1997;1:R17-20.
6. Allen PM, Beller DI, Braun J, Unanue ER. The handling of Listeria monocytogenes by macrophages: the search for an immunogenic molecule in antigen presentation. J Immunol 1984;132:323-31.
7. Monaco JJ. A molecular model of MHC class-I-restricted antigen processing. Immunol Today 1992;178:173-79.
8. Hellstrom KE, Hellstrom I. Cellular immunity against tumor specific antigens. Adv Canc Res 1969;12:167-223.
9. Boon T, Cerottini JC, Van den Eynde B, van der Bruggen P, Van Pel A. Tumor antigens recognized by T lymphocytes. Ann Rev Immunol 1994;12:337-65.
10. Melief CJ, Kast WM. T-cell immunotherapy of tumours by adoptive transfer of cytotoxic T lymphocytes and by vaccination with minimal essential epitopes. Immunol Rev 1995;145: 167-77.
11. Greenberg PD. Adoptive T cell therapy of tumours: mechanisms operative in the recognition and elimination of tumor cells. Adv Immunol 1991;49:281-355.
12. Chiodoni C, Paglia P, Stoppacciaro A, Rodolfo M, Parenza M, Colombo MP. Dendritic cells infiltrating tumours cotransduced with granulocyte/macrophage colony-stimulating factor (GM-CSF) and CD40 ligand genes take up and present endogenous tumor-associated antigens, and prime naive mice for a cytotoxic T lymphocyte response. J Exp Med 1999;190:125-133.
13. Boon T, Old LJ. Cancer tumor antigens. Curr Opinion Immunol 1997;9:681-83.
14. Armstrong TD, Pulaski BA, Ostrand-Rosenberg S. Tumor antigen presentation: changing the rules. Cancer Immunol Immunother 1998;46:70-74.
15. Imaeda S. The evolution of anti-tumor immunotherapy. J Invest dermatol 1997;108: 695-97.
16. Van den Eynde BJ, van der Bruggen P. T cell defined tumor antigens. Curr Opinion Immunol 1997;9:684-93.
17. De Plaen E, Arden K, Traversi C, Gafori JJ, Szikora J-P, De Smet C, Brasseur F, van der Bruggen P, Lethe B, Lurquin C et al. Structure, chromosomal localization and expression of twelve genes of the MAGE family. Immunogenetics 1994;40:360-69.
18. Gaugler B, Brouwenstijn N, Vantomme V, et al. A new gene coding for an antigen recognized by autologous cytolytic T lymphocytes on a human renal carcinoma. Immunogenetics 1996;44:323-30.
19. Topalian SL, Gonzales MI, Parkhurst M, et al. Melanoma-specific CD4+ T cells recognize non-mutated HLA DR-restricted tyrosinase4epitopes. J Exp Med 1996; 183:1965-71.
20. Hodge JW. Carcinoembryonic antigen as a target for cancer vaccines. Cancer Immunol Immunother 1996;43:1127-34.
21. Wolfel T, Hauer M, Schneider J, et al. A $p16^{INK4a}$-insensitive CDK4 mutant targeted by cytolytic T lymphocytes in a human melanoma. Science 1995;269:1281-84.
22. Robbins PF, El-Gamil M, Li YF, et al. A mutated β-catenin gene encodes a melanoma-specific antigen recognized by tumor-infiltrating lymphocytes. J Exp Med 1996;183:1185-92.

23. Mandruzzato S, Brasseur F, Andry G, Boon T, van der Bruggen P. A CASP-8 mutation recognized by cytolytic lymphocytes on a human head and neck carcinoma. J Exp Med 1997;186:785-93.
24. ten Bosch GJA, Joosten AM, Kessler JH, Melief CJM, Leeksma OC. Recognition of BCR-ABL positive leukemic blasts by human CD4+ T cells elicited by primary *in vitro* immunization with a BCR-ABL breakpoint peptide. Blood 1996;88:3522-27.
25. Fisk B, Blevins TL, Wharton JT, Ioannides CG. Identification of an immunodominant peptide of HER-2/neu protooncogene recognized by ovarian tumor-specific cytotoxic T lymphocyte lines. J Exp Med 1995;181:2109-17.
26. Lustgarten J, Theobald M, Labadie C, et al. Identification of HER-2/neu CTL epitopes using double transgenic mice expressing HLA-A2.1 and human CD8. Hum Immunol 1997;52:109-18.
27. Ropke M, Hald J, Guldberg P, et al. Spontaneous human squamous cell carcinomas are killed by a human cytotoxic T lymphocyte clone recognizing a wild-type p53-derived peptide. Proc Natl Acad Sci USA 1996;93:14704-07.
28. Ressing ME, Sette, A Brandt RMP, et al. Human CTL epitopes encoded by human papillomavirus type 16 E 6 and E7 identified through *in vivo* and *in vitro* immunogenicity studies of HLA A*0201-binding peptides. J Immunol 1995;154:5934-43.
29. Mincheff M, Tchakarov S, Zoubak S, et al. Naked DNA and adenoviral immunizations for immunotherapy of prostate cancer. A Phase I/II Clinical Trial. Eur Urol 2000;38:208-17.
30. Blades RA, Keating PJ, McWilliam LJ, George NJ, Stern PL. Loss of HLA class I expression in prostate cancer: implications for immunotherapy. Urology 1995; 46:681-86.
31. Sweat SD, Pacelli A, Murphy GP, Bostwick DG. Prostate-specific membrane antigen expression is greatest in prostate adenocarcinoma and lymph node metastases. Urology 1998;52:637-40.
32. Letvin NL, Montefiori DC, Yasutomi Y, et al. Potent, protective anti-HIV immune responses generated by bimodal HIV envelope DNA plus protein vaccination. Proc Natl Acad Sci USA 1997;94:9378-83.
33. Bretscher P, Cohn M. A theory of self-nonself discrimination. Science 169:1042-1049,1970
34. Schwartz RH. A cell culture model for T lymphocyte clonal anergy. Science 1990; 248:1349-56.
35. Falkenberg U, Leenen PJM, Falkenberg FW. Characterization of mouse macrophage differentiation antigens by monoclonal antibodies. Cell Immunol 1989;124:77-94.
36. Ford WL. Lymphocyte migration and immune responses. Progr Allergy 1975;19: 1-59.
37. Kraal G, Ter Hart H, Meelhuizen C, Venneker G, Claasen E. Marginal zone macrophages and their role in the immune response against T-independent type 2 antigens: modulation of the cells with specific antibody. Eur J Immunol 1989;19:675-80.
38. Mackay CR. T-cell memory: the connection between function, phenotype and migration pathways. Immunol Today 1991;12:189-92.
39. Tew JG, Kosco MH, Burton GF, Szakal AK. Follicular dendritic cells as accessory cells. Immunol Rev 1990;117:185-212.
40. Van Vliet E, Melis M, van Ewijk W. Marginal zone macrophages in the mouse spleen identified by a monoclonal antibody. Anatomical correlation with a B-cell subpopulation. J Histochem Cytochem 1985;33:40-44.
41. Kurts C, Miller JF, Subramaniam RM, Carbone FR, Heath WR. Major histocompatibility complex class I-restricted cross-presentation is biased towards high dose antigens and those released during cellular destruction. J Exp Med 1998; 188:409-14.
42. Janeway CA. The immune system has evolved to discriminate infectious non-self from non-infectious self. Immunol Today 1992;13:11-15.

43. Ibrahim MAA, Chain BM, Katz DR. The injured cell: the role of the dendritic cell system as a sentinel receptor pathway. Immunol Today 1995;16:181-86.
44. Matzinger P. Tolerance, danger and the extended family. Annu Rev Immunol 1994; 12;991-1045.
45. Medzhitov R, Janeway CA. Innate immunity: impact on the adaptive immune response. Curr Opinion Immunol 1997;9:4-6.
46. Fearon DT. Seeking wisdom in innate immunity. Nature 1997;388:323-24.
47. Reid CDL. The biology and clinical application of dendritic cells. Transf Med 1998;8:77-86.
48. Albert ML, Sauer B, Bhardwaj N. Dendritic cells acquire antigen from apoptotic cells and induce class-I-restricted CTLs. Nature 1998;392:86-89.
49. Colaco C. Towards a unified theory of immunity: antigen capture and the immune response. Cell Mol Biol 1998;6:883-90.
50. Polla BS, Kantengwa S. Heat shock proteins and inflammation. Curr Top Microbiol Immunol 1991;167:93-125.
51. Morimoto RI, Kline MP, Bimston DN, Cotto JJ. The heat-shock response: regulation and function of heat-shock proteins and molecular chaperones. Essays Biochem 1997;32:17-29.
52. Tamura Y, Peng P, Liu K, Daou M, Srivastava PK. Immunotherapy of tumours with autologous tumor-derived heat shock protein preparations. Science 1997;278:117-20.
53. Przepiorka D, Srivastava PK. Heat-shock proteins-peptide complexes: potentials cancer immunotherapeutic agents. Mol Med Today 1998;4:478-84.
54. Tarr PE. Granulocyte-macrophage colony-stimulating factor and the immune system. Med Oncol 1996;13:133-40.
55. Pulendran B, Smith JL, Caspary G, et al. Distinct dendritic cell subsets differentially regulate the class of immune response *in vivo*. Proc Natl Acad Sci USA 1999; 96:1036-41.
56. Pulendran B, Smith JL, Jenkins M, Schoenborn M, Maraskovsky E, Maliszewski CR. Prevention of peripheral tolerance by a dendritic cell growth factor: flt3 ligand as an adjuvant. J Exp Med 1998;188:2075-82.
57. Bianchi R, Grohmann U, Vacca C, Belladonna ML, Fioretti MC, Puccetti P. Autocrine IL-12 is involved in dendritic cell modulation via CD40 ligation. J Immunol 1999;63:2517-21.
58. Cumberbatch M, Dearman RJ, Kimber I. Induction by tumour necrosis factor alpha of dose-related changes in Langerhans cell frequency in mice. Arch Dermatol Res 1999;291:453-58.
59. Mosmann TR, Coffman RL. Heterogeneity of cytokine secretion patterns and functions of helper T cells. Adv Immunol 1989;46:111-47.
60. Mosmann TR, Li L, Hengartner H, Kagi D, Fu W, Sad S. Differentiation and functions of T cell subsets. Ciba Found Symp 1997;204:148-54.
61. Soos JM, Ashley TA, Morrow J, Patarroyo JC, Szente BE, Zamvil SS. Differential expression of B7 co-stimulatory molecules by astrocytes correlates with T cell activation and cytokine production. Int Immunol 1999;11:1169-79.
62. Agadjanyan MG, Kim JJ, Trivedi N, et al. CD86 (B7-2) can function to drive MHC-restricted antigen-specific CTL responses *in vivo*. J Immunol 1999;162:3417-27.
63. Nemunaitis J, Fong T, Burrows F, et al. Phase I trial of interferon gamma retroviral vector administered intratumorally with multiple courses in patients with metastatic melanoma. Hum Gene Ther 1999;10:1289-98.
64. Gahery-Segard H, Molinier-Frenkel V, Le Boulaire C, et al. Phase I trial of recombinant adenovirus gene transfer in lung cancer. Longitudinal study of the immune responses to transgene and viral products. J Clin Invest 1997;100:2218-26.
65. Harvey BG, Hackett NR, El-Sawy T, et al. Variability of human systemic humoral immune responses to adenovirus gene transfer vectors administered to different organs. J Virol 1999;73:6729-42.

66. van Elsas A, Hurwitz AA, Allison JP. Combination immunotherapy of B16 melanoma using anti-cytotoxic T lymphocyte-associated antigen 4 (CTLA-4) and granulocyte/macrophage colony-stimulating factor (GM-CSF)-producing vaccines induces rejection of subcutaneous and metastatic tumours accompanied by autoimmune depigmentation. J Exp Med 1999;190: 355-66.
67. Banchereau J, Steinman RM. Dendritic cells and the control of immunity. Nature 1998;392: 245-52.
68. Cella M, Sallusto F, Lanzavecchia A. Origin, Maturation and Antigen Presenting Function of Dendritic Cells. Curr Op Immunol 1997;9:10-16.
69. Thurner B, Roder C, Dieckmann D, et al. Generation of large numbers of fully mature and stable dendritic cells from leukapheresis products for clinical application. J Immunol Methods 1999;223:1-15.
70. Grey HM, Ruppert J, Vitiello A, et al. Class I MHC-peptide interactions: structural requirements and functional implications. Cancer Surv 1995;22:37-49.
71. Hermans IF, Daish A, Moroni-Rawson P, Ronchese F. Tumor-peptide-pulsed dendritic cells isolated from spleen or cultured *in vitro* from bone marrow precursors can provide protection against tumor challenge. Cancer Immunol Immunother 1997;44:341-47.
72. Murphy G, Tjoa B, Ragde H, Kenny G, Boynton A. Phase I clinical trial: T-cell therapy for prostate cancer using autologous dendritic cells pulsed with HLA-A0201-specific peptides from prostate-specific membrane antigen. Prostate 1996;29:371-80.
73. Tjoa BA, Simmons SJ, Elgamal A, et al. Follow-up evaluation of a phase II prostate cancer vaccine trial. Prostate 1999;40:125-29.
74. Alijagic S, Moller P, Artuc M, Jurgovsky K, Czarnetzki BM, Schadendorf D. Dendritic cells generated from peripheral blood transfected with human tyrosinase induce specific T cell activation. Eur J Immunol 1995;25:3100-07.
75. Robinson HL, Torres CA. DNA vaccines. Semin Immunol 1997;9:271-83.
76. Liu MA, Fu TM, Donnelly JJ, Caulfield MJ, Ulmer JB. DNA vaccines. Mechanisms for generation of immune responses. Adv Exp Med Biol 1998;452:187-91.
77. Lai WC, Bennett M. DNA vaccines. Crit Rev Immunol 1998;18:449-84.
78. Wang R, Doolan DL, Le TP, et al. Induction of antigen-specific cytotoxic T lymphocytes in humans by a malaria DNA vaccine. Science 1998;282:476-80.
79. Pertmer TM, Robinson HL. Studies on antibody responses following neonatal immunization with influenza hemagglutinin DNA or protein. Virology 1999;257: 406-14.
80. Osorio JE, Tomlinson CC, Frank RS, et al. Immunization of dogs and cats with a DNA vaccine against rabies virus. Vaccine 1999;17:1109-16.
81. Bernstein DI, Tepe ER, Mester JC, Arnold RL, Stanberry LR, Higgins T. Effects of DNA immunization formulated with bupivacaine in murine and guinea pig models of genital herpes simplex virus infection. Vaccine 1999;17:1964-69.
82. Paillard F. DNA vaccination for leishmaniasis. Hum Gene Ther 1998;9:1849-50.
83. Ulmer JB, Montgomery DL, Tang A, et al. DNA vaccines against tuberculosis. Novartis Found Symp 1998;217:239-53.
84. Luke CJ, Carner K, Liang X, Barbour AG. An OspA-based DNA vaccine protects mice against infection with Borrelia burgdorferi. J Infect Dis 1997;175:91-97.
85. Ruiz PJ, Garren H, Ruiz IU, et al. Suppressive immunization with DNA encoding a self-peptide prevents autoimmune disease: modulation of T cell costimulation. J Immunol 1999;162:3336-41.
86. Broide D, Raz E. DNA-Based immunization for asthma. Int Arch Allergy Immunol 1999;118: 453-56.
87. Bird AP. CpG-rich islands and the function of DNA methylation. Nature 1986;321: 209-13.
88. Klinman DM, Barnhart KM, Conover J. CpG motifs as immune adjuvants. Vaccine 1999;17: 19-25.
89. Roman M, Martin-Orozco E, Goodman JS, et al. Immunostimulatory DNA sequences function as T helper-1-promoting adjuvants. Nat Med 1997;3:849-54.

90. Oliveira SC, Rosinha GM, de-Brito CF, et al. Immunological properties of gene vaccines delivered by different routes. Braz J Med Biol Res 1999;32:207-14.
91. Moro M, Gasparri AM, Pagano S, et al. Induction of therapeutic T-cell immunity by tumor targeting with soluble recombinant B7-immunoglobulin costimulatory molecules. Cancer Res 1999;59:2650-56.
92. Armitage JO. Emerging applications of recombinant human granulocyte-macrophage colony-stimulating factor. Blood 1998;92:4491-508.
93. Trinchieri G, Scott P. Interleukin-12: basic principles and clinical applications. Curr Top Microbiol Immunol 1999;238:57-78.
94. MacGregor RR, Boyer JD, Ugen KE, et al. First human trial of a DNA-based vaccine for treatment of human immunodeficiency virus type 1 infection: safety and host response. J Infect Dis 1998;178:92-100.
95. Sanda MG, Smith DC, Charles LG, et al. Recombinant vaccinia-PSA (PROSTVAC) can induce a prostate-specific immune response in androgen-modulated human prostate cancer. Urology 1999;53:260-66.
96. Carter RE, Feldman AR, Coyle JT. Prostate-specific membrane antigen is a hydrolase with substrate and pharmacologic characteristics of a neuropeptidase. Proc Natl Acad Sci USA 1996;93:749-53.
97. Sweat SD, Pacelli A, Murphy GP, Bostwick DG. Prostate-specific membrane antigen expression is greatest in prostate adenocarcinoma and lymph node metastases. Urology 1998;52:637-40.
98. Troyer JK, Beckett ML, Wright GL Jr. Detection and characterization of the prostate-specific membrane antigen (PSMA) in tissue extracts and body fluids. Int J Cancer 1995;62:552-58.

TRANSPLANTATION OF UNRELATED DONOR UMBILICAL CORD BLOOD: CURRENT RESULTS AND FUTURE DIRECTIONS

J. E. Wagner[1], T. DeFor[1,2] and J. Kurtzberg[3]

Introduction

Transplantation of haematopoietic stem cells (HSC) derived from bone marrow and umbilical cord blood (UCB) of HLA-identical sibling donors has been successfully utilized in the treatment of patients with high-risk or recurrent haematological malignancies, bone marrow failure syndromes, selected hereditary immunodeficiency states and metabolic disorders [1-3]. However, limited availability of suitable HLA matched donors and the high risk of graft-versus-host disease (GvHD), has limited the usefulness of HSC transplantation. In an attempt to remedy the shortage of suitable donors and reduce the risk of GvHD, banked UCB from unrelated neonatal donors has been used as an alternative source of HSC for transplantation since 1994. In 1996, Kurtzberg et al. [4] and Wagner et al. [5] reported preliminary clinical results of unrelated donor UCB transplantation in 25 and 18 patients at Duke University and University of Minnesota, respectively. Early clinical data suggested that unrelated donor UCB might contained sufficient numbers of HSC to achieve engraftment in most pediatric recipients with lower than anticipated risk of severe acute GvHD. Since the publication of these two preliminary reports in 1996, more than 200 patients have been transplanted with unrelated donor UCB at Duke University and University of Minnesota. This report is an update of the combined experiences at these two institutions.

Patients and Methods

Patients with acute leukemia, bone marrow failure syndromes, immunodeficiency states or inborn errors of metabolism were eligible for unrelated donor UCB transplantation if: 1) an HLA-compatible related or unrelated bone marrow donor was not immediately available at the time needed, and 2) the subject/parent(s) consented to the transplant procedure. Protocols for myeloablative

1. Dept Pediatrics, Div Bone Marrow Transplantation.
2. Dept Pediatrics, Divs Bone Marrow Transplantation and Epidemiol and Clin Research, Univ of Minnesota School of Med, Minneapolis, MN, USA.
3. Dept Pediatrics, Div Hematol/Oncol/Bone Transplantation, Duke Univ Medical Center, Durham, NC, USA.

therapy and use of unrelated donor UCB for transplantation were reviewed and approved by the respective institutional review boards at Duke University Medical Center and the University of Minnesota. Median age of the patients was 7.2 years (range, 0.2-58) and median weight was 21.6 kg (range, 4.8-92).

Haematopoietic recovery and GvHD

Haematological recovery was defined as time to absolute neutrophil count (ANC) $\geq 5 \times 10^8$/L (first of three consecutive laboratory measurements on different days) and platelet count $\geq 5 \times 10^{10}$/L (first of seven consecutive laboratory measurements on different days without transfusion support. Patients were evaluated for acute GvHD daily during initial hospitalisation, at least once weekly after initial discharge during the first 100 days, and at routine follow-up evaluations at the transplant centers at 6 months, 1 year and at least yearly thereafter. Diagnosis of acute GvHD was based on clinical criteria with histopathologic confirmation when possible. Overall staging was based on published criteria (16) and assigned by the medical staff at each institution.

Statistical analysis

Data regarding transplant patient characteristics, post-transplant complications and outcomes were prospectively collected at Duke University and University of Minnesota. The major statistical end points of this study were the Kaplan-Meier estimates of engraftment, grade II-IV and grade III-IV acute GvHD, chronic GvHD, non-relapse mortality, relapse (for patients with malignant diseases), and survival. Event times for neutrophil engraftment were measured from the date of transplantation to date of neutrophil recovery with censoring for early death (i.e., before day 21, n=8) or evidence of persistent malignant disease (n=4). Patients who had very slow engraftment (i.e., achieved an ANC $\geq 5 \times 10^8$/L after day 42) or failed to have marrow reconstitution of donor origin were scored as graft failures. Event times for GvHD and relapse were measured from date of transplantation to date of event with censoring of patients at death. Univariate comparisons of the major endpoints were completed with 95% confidence intervals using the log-rank statistic. Multiple regression analysis was performed using a stratified Cox proportional hazards model.

Results

Haematopoietic recovery and engraftment

Overall probability of neutrophil recovery by day 42 and platelet recovery by 6 months were 0.87 (0.82-0.92) and 0.73 (0.64-0.82), respectively. The median time required to achieve an ANC $\geq 5 \times 10^8$/L and a platelet count $\geq 5 \times 10^{10}$/L are 26 days (10-59) and 3.4 months (0.9-8.2), respectively. While time to neutrophil recovery strongly correlated with the dose of cryopreserved nucleated cells and CFU-GM (correlation coefficient, -0.46 [$p<0.01$]), probabilities of neutrophil and platelet recovery were not affected by degree of HLA disparity. In multiple regression analysis, among continuous variables the nucleated cell dose and CFU-GM dose (determined before cryopreservation) and the CD34 cell dose (determined after thawing) were the only factors significant predicting neutrophil and platelet recovery.

Acute and Chronic Graft-versus-Host Disease

The overall probabilities of grade II-IV and grade III-IV acute GvHD for the entire group of patients was 0.37 (0.29-0.45) and 0.14 (0.08-0.20) by day 100 after unrelated donor UCB transplantation, respectively. In univariate analysis, no factor was associated with risk of acute GvHD, including CD3 cell dose or degree of HLA disparity. The overall probability of chronic GvHD was 0.11 (0.02-0.20).

Survival

With a median follow-up of 1.1 years, the probabilities of survival at 1 year and 2 years after unrelated donor UCB transplantation are 0.48 (0.40-0.56) and 0.44 (0.35-0.53), respectively. Survival was superior in patients with non-malignant disease (0.61 [0.46-0.76]) compared to that in patients with malignant disease (0.36 [0.25-47]). Increasing degree of HLA disparity did not significantly alter the probability of survival after unrelated donor UCB transplantation. In multiple regression analysis, only cell dose was found to impact upon survival.

Discussion

At the outset, it was clear that banked unrelated donor UCB had four positive attributes, namely: 1) rapid availability, 2) absence of risk to the HSC donor, 3) absence of donor attrition, and 4) a very low risk of transmissible infectious diseases, such as cytomegalovirus (CMV) and Epstein-Barr virus (EBV). Until recently, little data has been available on the use of unrelated donor UCB to know whether UCB offered additional risks and benefits in terms of engraftment, risk of GvHD, relapse and survival.

The principal objective of this analysis was to determine whether specific factors, such as degree of HLA disparity, impacted engraftment, risk of acute or chronic GvHD, non-relapse mortality, relapse and survival. It is notable that graft cell dose (nucleated cell, CD34 cell and/or CFU-GM dose) and not HLA disparity was consistently identified as the most important variable for predicting engraftment, non-relapse mortality and survival.

An important clinical question is whether unrelated donor UCB is less likely to generate severe acute GvHD and extensive chronic GvHD reaction as compared to marrow from adult unrelated donors. Kernan et al. [6] reported an incidence of grade III-IV GvHD of 0.47 ± 0.06 for a heterogeneous population transplanted with unrelated donor marrow, including both younger and older recipients and recipients of unmanipulated and T cell depleted marrow. Davies et al. [7] reported an incidence of grade III-IV GvHD of 32-49% depending upon the degree of HLA disparity (0 versus 1). In both instances, the probability of severe GvHD was influenced by recipient age and degree of HLA disparity with a lower risk observed in patients ≤ 18 years of age or in patients with HLA matched donors. In a study reported by Balduzzi et al. [8] in a pediatric population, 42-60% of children had grade III-IV acute GvHD depending upon degree of HLA disparity (0 versus 1) and 37% risk of extensive chronic GvHD. As the frequency of grades III-IV acute GvHD in the present study was 14%, these data suggest that unrelated UCB may be associated with a reduced risk of severe acute and extensive chronic GvHD despite greater HLA disparity even when taking into consideration the age of the recipients. However, formal comparisons between recipients of unrelated donor marrow and UCB using identical GvHD grading schemes are required before any conclusions can be made. The absence of extensive chronic GvHD after unrelated donor UCB transplantation is particularly striking. While decreased GvHD raises the concern that the graft-versus-leukemia (GVL) may also be decreased, relapse after unrelated donor UCB transplantation does not appear to be increased thus far.

In summary, we and others [3,9] have demonstrated that cryopreserved UCB from HLA 0-3 antigen mismatched unrelated donors contains sufficient numbers of transplantable haematopoietic stem and progenitor cells for most patients. In addition, the data presented indicate that the probabilities of grade III-IV acute GvHD and extensive chronic GvHD are low. Moreover, the results of this statistical analysis demonstrate the importance of graft cell dose in determining outcome after unrelated donor UCB transplantation. These data suggest that the choice of UCB graft should be based primarily on cell dose rather than degree HLA disparity for patients with more than one HLA 1-2 antigen disparate UCB units. The importance of cell dose on transplant outcomes provide the most compelling argument for focusing on the collection of larger UCB grafts and for investigating *ex vivo* HSC expansion for future clinical trials.

References

1. Bone Marrow Transplantation, edited by Forman SJ, Blume KG, Thomas ED, Blackwell Scientific Publications, Cambridge, MA, 1994.
2. Wagner JE, Kernan NA, Steinbuch M, et al. Allogeneic sibling donor umbilical-cord-blood transplantation in children with malignant and non-malignant disease. Lancet 1995;346:214-19.
3. Gluckman E, Rocha V, Boyer-Chammard A, et al. Outcome of cord-blood transplantation from related and unrelated donors. N Engl J Med 1997;337:373-81.
4. Kurtzberg J, Laughlin M, Graham ML, et al.. Placental blood as a source of hematopoietic stem cells for transplantation in unrelated recipients. N Engl J Med 1996;335: 157-66.
5. Wagner JE, Rosenthal J, Sweetman R, et al. Successful transplantation of HLA-matched and HLA-mismatched umbilical cord blood from unrelated donors: Analysis of engraftment and acute graft-versus-host disease. Blood 1996;88:795-802.
6. Kernan NA, Bartsch G, Ash RC, et al. Analysis of 462 transplantations from unrelated donors facilitated by the National Marrow Donor Program. N Engl J Med 1993;328: 593-601.
7. Davies SM, Shu XO, Blazar BR, et al. Unrelated donor bone marrow transplant: Influence of HLA-A and B incompatibility on outcome. Blood 1995;86:1636-42.
8. Balduzzi A, Gooley T, Anasetti C et al. Unrelated donor bone marrow transplantation in children. Blood 1995;86:3247-56.
9. Rubinstein P, Carrier C, Scaradavou A, et al. Outcomes among 562 recipients of placental-blood transplants from unrelated donors. N Engl J Med 1998;339:1565-77.

CLOSING REMARKS: STRATEGIES AND FUTURE DEVELOPMENTS IN APPLIED MOLECULAR BIOLOGY

H.G. Klein, Bethesda, MD, USA

Despite the theme it is particularly challenging to summarise a meeting on molecular biology – I hope that none felt they had to jump from a speeding car.

We learned from every speaker the power of the tools of molecular biology.

The discoveries of microbiologist and microbial geneticists were of questionable relevance to most of us in 1950's and 60's, but proved to revolutionise medicine and Transfusion Medicine is the 1980's and 90's –

a. structure of DNA
b. discovery of bacterial restriction enzymes like polymerases, ligases, episomes, spliceosomes, plasmids and bacteriophages
c. Watsen-Crick rules of replication – DNA → RNA → protein
d. exception that tests the rule – reverse transcriptase for diagnostic tests and cDNA libraries
e. amplification technologies like PCR, rtPCR, TMA (Transcription Mediated Amplification) and all their variations
f. applying these basic science discoveries to chip technology and coupling with computers will allow complex diagnostic procedures and turn from gene structure to functional genomics, immunity/autoimmunity in real time basic science → clinical application → sophisticated technology → answer basic questions. Knock in mice and knock out mice models → precursors for genetic therapies.

Application of RBC genotyping – 21 systems cloned and sequenced. Practical application for multitransfused patients and fetus at risk for HDN. But also genes that are present but not expressed – e.g. GATA .x mutations for Duffy. Early in the understanding of correlation between genotype and phenotype. Also continuing need for medical history – marrow transplant/artificial insemination. Future will bring us soluble RBC antigen – although we can remove antigens with galactosidases from coffee beans and chicken livers, hide them with PEG, but none of the chemical modification is yet practical – perhaps better to try to engineer genetically. Future will learn more about structure of antigen and more about epitopes and structure/function.

The HLA system is more complex yet. Learning more and more about inevitable heterogeneity at the molecular level with chip technology, we shall literally be able to sequence these areas in a matter of minutes using technology described as 'Walt Disney Molecular Biology'. We have almost foolproof individual identification methodology – one real challenge is how to guard and store this information for donors and patients and the paramount question how to relate this high resolution genotype information to the chemical problems of transplantation, immune responsiveness, vaccine production and tolerance. While RBC genotyping has not eliminated agglutination, HLA sequencing is still complemental by functional assays. But functional genomics and chip technology will undoubtedly replace radioactive thymidine uptake and MLC's.

The progressive/dramatic improvement in blood safety were reviewed and the power of amplification technologies in detecting a very few viral particles appreciated. Transcription mediated amplification (TMA) can yield a million fold amplifications in a matter of minutes and we can develop multiplex systems for simultaneous detection of several agents. The major problems with such testing today relates to logistic issues, not technical ones – yet 2 important principles are obvious. Reductions of the window period by 50% (HIV) or 85% (HCV) still gives a limited benefit because of the current low residual risk. Most importantly, kinetics of viral replication – time course and number of circulating virions guarantee that even the most sensitive amplification technology will not replace serology. If the virion is not in the aliquot to be tested, it cannot be amplified [also emerging variances] – you cannot fool Mother Nature. The real benefits of testing advances should apply to developing countries or Resource Limited Regions of the World. Prevalence and incidence are high, blood is scarce, and testing is too costly, so often not performed. Simple walk-away tests – like lateral flow tests are possible using recombinant antigens, we ought to be able to optimise sensitivity and specificity – part of the challenge is discovering the appropriate composite antigen – the other may be dealing with patent issues and the wealthy corporations that control them.

The elegant studies of platelet recovery and survival in baboons taught an important principle: one can use sophisticated diagnostic techniques to demonstrate platelet activation and disappearance of the activation marker *in vivo*. The conventional wisdom that this marker is related to platelet sequestration proves to be incorrect. Sophisticated technology cannot substitute for careful experimentation and correct interpretation of results. Activation may yet prove to be the storage lesion, but degranulation and p-selectin membrane expression do not appear to be causative. Similarly, conventional wisdom concerning donor mechanisms of antigen presentation may not explain entirely the allloimmune response to platelet transfusers. A recipient mechanism mediated at the molecular level by inducible nitric oxide may explain why 10-20% of recipients of leukocyte reduced platelets still develop alloimmunity.

Molecular techniques have revolutionised stem cell transplantation. We have developed and standardised methodology to identify progenitor cells that have been mobilised by recombinant cytokines. Now a new molecule may identify a mere primitive (stem-like) cell. Clearly the future will teach us more about mobilisation, identification, trafficking and engraftment of haematolymphoid progenitors. Molecular markers for neoplastic cells help identify and quantify their numbers? autografts and tumour recurrence in transplanted patients. Only 19% of AML patients do not have a chromosomal translocation that can be used as a diagnostic target for other tumours, such as lymphoid malignancies, antigen receptor gene rearrangements provide diagnostic target and in the future perhaps for idiotypic vaccines.

Recombinant cytokines give hope for expanding progenitor cells such as megakaryocytes for management of the aplasias encountered in cancer chemotherapy and transplantation. They also provide the milieu for retroviral insertion into lymphoid and stem cells. Perhaps expansion technology will help us apply small volume collections to need large volume recipients. Markedly less GvHD, a donor autoimmune disease. We need to understand molecular reasons for these differences. The prospects are exciting. Recombinant proteins for treatment of coagulation disorders are here – additional improvement to FVIII and FIX may not be cost effective. Recombinant VIIa on the other hand is just beginning to find its role(s). The elegant application of the phage display technology provides a novel note for engineering human immunoglobulins for standard diagnostic and therapeutic use, but also for identifying new epitopes on red cells and answering most basic questions about the human immune response such as the restriction of genes mediating Rh antibody response.

Finally molecular biology has helped define the professional antigen presenting cell, dendritic cell and show the way to produce candidate tumour vaccines. Count me among those who believe that the ability to insert and modify normal human genes and to use DNA and dendritic cells to produce idiotype vaccines, will revolutionise the treatment of cancer and viral infections to a degree comparable to the antibiotic revolution a generation ago. I fully expect to be present at this revolution. Perhaps we shall hear more about these advances at the 30th International on Blood Transfusion Symposium in Groningen.

DISCUSSION

Moderators: H.G. Klein and C.Th. Smit Sibinga

H.G. Klein (Bethesda, MD, USA): Dr. Hospers, have you any data on the use of, or comparison of peptide pulsing of dendritic cells with transduction with the same antigen; for example using retroviral vectors and measuring the tumour response in animal models?

G.A.P. Hospers (Groningen, NL): No comparative studies have been done, also not in the animal studies. The site of the administration has been documented, but there are a lot of possibilities you have to check in the animal model too to be sure which kind of vaccine you need. There are only theoretical thoughts about what is the best way, but it has not been proven until now.

M.K. Elias (Groningen, NL): Dr. Mincheff, it was a very nice presentation, thank you very much. You mentioned the types of tumour-associated antigens. Some tumour-associated antigens are expressed on a large array of tumour cells like MUC-1 which is expressed on breast, prostate, pancreatic and renal cell carcinomas. Is it, talking about future applications, theoretically possible to use such a universal tumour-associated antigen to pulse the dendritic cells in order to produce, let me call it a broad spectrum vaccine for this variety of tumour cells or should you load the dendritic cells with the tumour specific antigen or at least an antigen which is integral to the malignant process?

M.S. Mincheff (Bethesda, MD, USA): It is a very good question and I do not know whether anybody has the answer. It is a great idea, however, the more antigens you immunise against the greater chance of developing an untoward autoimmunity. The same is true when you use whole tumour cells for immunisation.

M.K. Elias: It may be less specific but it is definitely more immunogenic.

M.S. Mincheff: When you immunise with whole tumour cells you may develop response against unknown tumour antigens – you may end up even discovering new tumour- associated antigens. Furthermore, the response will be broader and more potent. However, you always risk the development of autoimmunity and we have selected to work with a signal antigen. However, in two of our non-responding patients we additionally injected tumour cells transfected with GM-CSF. We have not seen any signs of autoimmunity so far. Whether this has anything to do with the fact that those two patients had prior treatment with DNA vaccines – I do not know. But I agree with you that the response against a vaccination with tumour cell may be more potent and at the same time I would like to raise the possibility for autoimmunity.

W.G. Hughes (Sydney, AUS): Dr. Wagner, I have been to many talks, listening to these sort of presentations for ten years and I think this is one of the few areas of transplantation that was theoretically proposed and now actually been proven by your hard data. What I like to ask is how long in your estimation will it be before this technique would be securely available for adult transplantation.

J.E. Wagner (Minneapolis, MN, USA): Well, the results specifically in adults overall in our institution were approximately 35% survival between the two groups. It is feasible but clearly the limitation of cell dose is the prohibiting factor. I would say that this is a guess, but I would say probably we can find an acceptable corporal graft only for about 40% of the adults if you use the cut-off of 2 x 10^7. Clearly of our 12 patients 10 patients had less than 2 x 10^7, that is between 1 and 2 in the graft. It occurs in the majority of patients but still engrafting is very, very slow. However, overall survival at our institution is 44%; so despite the fact that there is very slow recovery they are getting able to get through it, but this is the reason for actually looking at the mixed cord blood. I have no idea whether it will work or not but clearly this is the strategy trying to make it possible for engraftment in adult recipients. So, we will know, I think, in the not too distant future whether we can make this work for adults on a more global scale, because surely the benefit of lower GvHD is a great benefit for adults. But right now it may be counteracted by the engraftment problems. So we will know soon but not yet.

W.G. Hughes: Well, what I was really asking was how you feel about the expansion of the primitive population and how that would go and how quickly. Even now you say the survival is reasonably good. It is that long period of engraftment that is so expensive and so terribly labour intensive for adult patients. That is what really makes the whole thing so difficult on the end of clinical setting.

J.E. Wagner: Well, yes the expansion is going to go quickly at our institution because of a commitment by the adult group, but you know in general the adult teams have been very reluctant to even in our own institution doing these cord blood transplants for such a fear of engrafting. What I do not understand, and you know I am talking as a paediatrician, that there is readiness to use umbilical cord blood for adult recipients even in patients for example with ALL. Adult patients receiving an unrelated marrow donor transplant for ALL do horribly, and yet why are people afraid to try something like this. We have engraftment as a problem, but on the other hand you cannot do much worse than you are already doing. So, therefore there are so many patient populations like relapsed high-risk ALLs and you can take some sort of sub-populations where you can actually then test this out; actually the same patients that are proposed to use mixed cord blood and we cannot do worse than we are already doing and hopefully we can have some positive impact.

R. Atamian (Beirut, Lebanon): Can we apply the cord cell transplantation to transfusion associated GvHD?

J.E. Wagner: Well, you have to remember malfailure was only observed as a positive aspect in terms of relapse; it had no impact on survival; it did not follow as being a significant variable. But you would have to give some type of preparative therapy or do something to eliminative that alloreactive clone, because even when it is causing a GvHD response in the recipient after a transfusion there surely is a capacity to reject the new cord blood. So, I am not sure that you want to follow through, you do not want to treat every patient with transfusion associated GvHD with a transplant, because many of those patients will eventually recover on their own. Maybe I am missing the question, but I mean I would not just use cord blood from the very outside. Maybe you were thinking also that there is something suppressive about cord blood; maybe you could suppress immune response, but still remember that is theoretical. Also when I showed you those studies that showed a variety of immunology characteristics, that is what I have glanced from the literature. That is not to say that I have proven that these things are all true; these are things that I have seen in people written in the literature that explain potentially the decreased immune reactivity of cord blood.

W.G. van Aken (Amsterdam, NL): Dr. Wagner, you mentioned the importance of the number of cells for successful transplantation. My question is in fact related to the yield of cells of cord blood. Is there further improvement in your view to increase the yield in order to make transplantation more successful or is the present technology already that advanced that you think there is only limited chance to improve.

J.E. Wagner: No, the technology of ex vivo expansion is still unknown; I mean it is all a goal, it may turn out that it is impossible to expand true stem cells, it remains to be proven. Certainly there are some ideas in ways of trying to improve the collections, but you know the different groups get different answers and so I do not know what the actual answer eventually is going to be. In our own experience we looked at and in utero collections compared to doing it the way blood centres have done it. Most blood banks have collected after the placenta is expelled. We found a significantly greater number of cells collected by doing it in utero. As you may know the St. Louis cord blood bank routinely collected in utero for purposes of unrelated transplant. They are the only group I know of that is doing it that way routinely and their collections in general seem to be better. But there is no sort of controlled trial on. Dr. Bertolini, a while ago, indicated that there was no difference[1]. But my inclination is that there is, but I cannot prove it to you.

D.J. Anstee (Bristol, UK): Are not you suggesting that blood banks involved in collecting cord blood should change their strategy dramatically? You are spending a lot of money on HLA typing and the focus should be on getting units of the right number of cells. But that has a big impact on resourcing. But first you want to optimise the chance to prove the success. Therefore what you want to do is you want to collect from the white Caucasian northern European descent, you know mother and child and use the one with the largest graft and only collect those. There you can have your highest probability of use to be able to answer the question is cord blood really going to work in adults, is it going to work for everyone. Instead, at least in the United States one has taken the focus of saying let us collect as many as we can particularly from ethnic minorities. That is only the future goal, but on the other hand it is not proof of principle of how good or bad this is going to be. So, I agree with you that we should just focus our efforts primarily on collecting the largest graft and not even do any HLA typing of the smaller grafts. Dr. Warwick in London, has made the comment that there is so many cord blood grafts and so many cord banks now around the world that perhaps it is an ethical issue because so many resources could be spent in other ways.

J.R. Aluoch (Nairobi, Kenia): In many parts of developing countries we have missed out on the bone marrow transplant for many reasons. Could you say that something like the UCBT, going by cost effectiveness projections, might offer some hope for some of these countries that are at economical disadvantage.

J.E. Wagner: It is an excellent question and I do not have a good economical analyst to tell you whether cord blood is economical or not. It is only a focus of one aspect of the trial at the NIH. You know I guess the first thing is to be able to demonstrate to everyone that there is a survival advantage. The second question is going to be is it economical. I mean can we do it the other way round. I have done the first case control comparison between cord blood and bone marrow, where patients always received exactly the same preparative therapy. They were all within one year of age and the case controlled comparison is a retrospective analysis. They all have the exact same disease. At the University of Minnesota we were able to find approximately 40 umbilical cord blood recipients that had exactly matched counterparts on the bone marrow side. We were able to show in our analysis that the survival is superior with cord blood. Now this is the first time this analysis has been done. I think it is quite encouraging that at least it is going to be potentially better compared to mismatched bone marrow. There was a significant survival advantage. To give you the numbers: for cord blood the overall one year survival was 76% compared to 42% for T-cell depleted bone marrow. Then in the analysis comparing cord blood to im-

1. Bertolini F, Battaglia M, De Iulio C, Sirchia G. Placental blood collection: Effects on newborns. Blood 1995;85:3361-62.

mune suppressive therapy and bone marrow transplant it was 66% for cord blood and 39% for immune suppressive therapy in terms of survival. Two years – this is now again looking at mismatched bone marrow as the comparison, at two years the cord blood patients remained still in the 76% when compared to T-cell depleted bone marrow, but the T-cell depleted bone marrow recipients had a survival of only 18%. Markedly different; remember this is mismatched bone marrow in occasions with high risk disease and so that 18% may seem low. It is low compared to our matched patients but in this case control comparison the p-value was 0.2. So, I think that is the first step, proven survival. If that is indeed the case then we can get at the economics of this.

H. Olthuis (Leiden, NL): Dr. Wagner, having heard that the quantity of cells transfused to the patients is the most important factor, would it be a strategy to make multiple dose transfusions and not only double but triple and more perhaps. The price could not be the problem I think.

J.E. Wagner: Well, you know I have been very negative about using multiple cord bloods; that is when you say multiple dose you mean multiple cord bloods together, but I think there is enough people interested in it. There is actually as far as I know 12 done around the world. Unfortunately there is no real data to be able to tell you as yet. I can give you some preliminary data, but nothing that is worthwhile really. But this strategy is going to be explored, it would be interesting to determine whether or not you go into engraftment from multiple sources, whether the engraftment is equal or does one win out in the end. It is going to be explored and well soon, but I do not know the answer yet. Because of the cell dose issue we are expanding and we are investigating multiple cord blood simultaneously.

C.Th. Smit Sibinga (Groningen, NL): Dr. Wagner, just to summarise a question which has been raised several times. We all now consistently see that the engraftment times are much different from when you use peripheral stem cells harvested from people that have already been in life for more than a couple of years; the mature stem cells so to speak, which have been stimulated. Is that delay because you deal in cord stem cells with more immature or earlier type of stem cells, which tend to first of all self renew before they start to differentiate or is there something different.

J.E. Wagner: Well, I think that might be the feeling and I do not have the answer. I think it is an interesting question. My feeling is that what you are implying is indeed correct, because when you take care of these patients and you look at their bone marrow's as a clinician what you find are very silent bone marrow's and this does not happen to everyone. But frequently you find very silent bone marrow biopsies on day 21 and yet they do not engraft until day 42. So, when you look at the bone marrow's in the very immature and fact in the original cohort of patients we looked at it. Actually we had phone calls from the pathologists telling me that my patients relapsed with this leukaemia when in fact they never had leukaemia. That is because you find sheets with very immature looking cells. In our institution they are called haematogens, these very primitive lymphoid looking cells. So indeed, there may be something different about the immaturity such that results in delay. Maybe we need to stimulate those with a variety of cytokines in vivo rather than expanding ex vivo.

C.Th. Smit Sibinga: Well thank you. As the session is called new horizons what is the last challenge. Also, because of the greater tolerance and because of the greater naiveness of these cells and other cells that grow along with the transplant like lymphocyte populations, it has been postulated that probably these cells could also be used for genetic manipulation and gene therapy. Could you say a few words about that?

J.E. Wagner: Well, it all depends I think first of the gene therapist who may come up with a strategy that avoids the need for dividing cells. Maybe there are ways of getting gene therapists to work for any stem cell source, but cord cells only has the advantage of being very proliferative. At least our own data suggest that we were able to get up to 50-60% of transduce LTCIC. Now it does that to a stem cell, but on the other hand it is a very primitive subset and so it is very positive you know for a step. So, I think if we are ever going to be able to get retroviral gene therapy it seems to be a reasonable choice using cord blood. Probably it is not going to be a good choice for peripheral blood stem cells, where it is going to be very difficult to put genes into those quiescent stem cells.

C.Th. Smit Sibinga: Although dr. Hospers has left she has taken care for an extremely competent discussant on her behalf: Mrs. Coba van Santen from the department of Clinical Immunology, who will be prepared to answer questions in relation to dr. Hospers presentation.

G.D.M. Beun (Groningen, NL): I have a question concerning the appropriate source of dendritic cells for clinical application. The group of Ronald Levy at Stanford* used dendritic cells derived from peripheral blood, where they are extremely scarce and this requires gradient processing. On the other hand there is the option of deriving monocytes or CD34 stem cells and drive them into differentiation using cytokines, but there is concern that they may lose their dendritic phenotype once they are no longer under the pressure of those cytokines and are infused. Could you comment on that.

C. van Santen (Groningen, NL): I think the most convenient way of obtaining dendritic cells is from the peripheral blood. Culturing and differentiating from monocytes to immature and going out to mature dendritic cells and in the way how stable they are in a mature state I think it depends also on the way they have been matured, depending on the kind of cytokines used. There are reports on that, but I think TNF-α and IL-1 will result in mature dendritic cells. Does that answer your question.

G.D.M. Beun: Yes, but there is one essential word lacking; are they stably differentiated cells?

C. van Santen: Yes, what I mentioned is that it is dependent on what you use but I think when you add IL-1 or TNF-α you get stable mature cells. Of course in vivo, how long they are stable is not known.

G.D.M. Beun: It is a scientific question that is difficult to investigate also; how do you follow those cells in vivo?

C. van Santen: Yes, first of all it is a very important issue to get mature cells, because it is not very easy to separate mature cells from immature dendritic cells. It depends on the donor, on the way you culture the cells and how you try to mature them. But to obtain a 100% population of mature dendritic cells is a very difficult topic up to date.

G.D.M. Beun: Is anyone of you aware of any animal experiments for instance where these dendritic cells have been followed in vivo?

M.S. Mincheff: I think Mark Jenkins' from Minnesota have presented good in vivo data for migration of antigen-pulsed dendritic cells to the regional lymph nodes. They did initially in a very elegant study conducted in mice[1]. Recently, they repeated this with dendritic cells pulsed with tumour associated anitgens in humans[2]. As far as dendritic cell generated ex vico with the help of growth factors are concerned, we do not know what the in vivo fate of these cells is. Furthermore, we have no knowledge on what the best route for administration of dendritic cells is – intradermal, subcutaneous or even intratumoural.

M.K. Elias: But do you really need all the population of mature dendritic cells, because when dendritic cells become mature and start to migrate they lose their ability to capture the antigen and they become accessory cells. So, maybe both populations, mature and immature dendritic cells, need to be used.

M.S. Mincheff: The way I understand your question is whether we know how stable the phenotype of the dendritic cells is after growing them in vitro and pulsing them with tumour antigens. We do not know. You are right that, for example, if we inject dendritic cells intra-tumoualy, we should not mature them in vitro. Immature dendritic cells capture and process antigen much more efficiently than mature ones. So, you are right – it depends on the final and the approach.

H.G. Klein: I would like to follow up on some of the questions about the tumour vaccines. This is obviously a very exciting area although it is very complex and it appears that at least in some instances the MHC molecules are extremely important in presenting antigen and that the tumour antigens almost have to be

1. Ingulli E, Mondino A, Khoruts A, Jenkins MK. In vivo detection of dendritic cell antigen presentation to CD4(+) T cells. J Exp Med 1997 185:2133-41.
2. Thomas R, Chambers M, Boytar R, Barker K et al. Immature human monocyte-derived dendritic cells migrate rapidly to draining lymph nodes after intradermal injection for melanoma immunotherapy. Melanoma Res 199;9:474-81.

designed for specific individuals. I wonder if you could comment on some of the silent mutations that perhaps are present in polymorphism's in the MHC complex; these polymorphisms may or may not allow someone to react to an antigen that you have used even though you have two people who appear phenotypically as if they have the same HLA type.

M.S. Mincheff: I probably do not have the knowledge to comment to this. Since the immune response is MHC restricted, probably there will be good and bad responders based on the ability of certain MHC to bind certain peptides and to present them. There is, however, something else that most immunologists do not seem to be concerned with, and this has to do with the lymphocyte traffic and recirculation. How do you direct a stimulated T cell towards the target organ? When a pathogen penetrates the skin, it creates locally an inflammatory response that later governs the extravasation of the stimulated T cells to that particular site. In cancer patients you immunise at a site distant to the tumour and T cells tend to accumulate at later vaccine applications at the immunisation site. In order to re-direct the stimulated T cells towards the target organ or tissue, you still have to create an inflammatory milieu in or around the target that will attract the immune cells. By inducing tumour cell death radiotherapy and chemotherapy may do exactly that. Another unresolved problem remains the phenomenon of down regulation of MHC molecules from tumour cells membranes. T cells recognise MHC-peptide complexes. Despite the success of eliciting a powerful immune response against a tumour antigen, tumour cells that have lost their MHC will not be destroyed because they will not be recognises. A cytotoxic antibody response against a membrane structure may solve the problem. For this particular reason we have selected the prostate specific membrane antigen as a target in our trial.

H.G. Klein: In some of the work for melanoma vaccines the metastases differ from the original tumour antigens that seem to be affected by the common vaccine. However, the metastases no longer seem to express the melanoma antigen. Have you looked at that in your prostate study? Do the metastases differ from the original tumour in some way?

M.S. Mincheff: It has been reported that tumour cells from metastases show greater loss of MHC molecules from their membrane, but we have not had the chance to look into that.

H.G. Klein: Finally, I think it is very interesting; again using melanoma as a model – when what looked like tumour specific antigens was used for vaccines, patients who responded also developed areas of vitiligo demonstrating that normal melanocytes were being killed by the immune response. Is this going to be a problem with your studies and others?

M.S. Mincheff: I think this is a potential problem. Strictly speaking only mutated antigens or viral antigens are strictly tumour-specific. All other antigens that are targeted in immunotherapy carry the potential risk of developing autoimmunity.

We have selected PSMA and PSMA-derived peptides as our targets. PSMA is expressed primarily in prostate tissue, but week expression has been found in tubular epithelial cells in the kidney and on duodenal epithelial cells. PSMA is expressed by endothelial cells in almost all tumour-associated neovasculature and this may open new avenues to therapy.

H.G. Klein: So, the short answer is yes there always exists the danger of developing an autoimmune response.

C. van Santen: About your question for the expression of the tumour associated antigens I know because I work with the major antigens. These antigens express higher in the original tumour compared to the metastases, much higher.

C.Th. Smit Sibinga: Looking into the future, to what extent do you think we might be able to recondition and use allogeneic dendritic cells, provided we stay within the limits of the standard rules of compatibility.

M.S. Mincheff: This is an exciting idea. You have to define ways, however, to control development of autoimmunity.

C.Th. Smit Sibinga: That is right, but if I listen well to you, you do not throw it away as an option.

M.S. Mincheff: Well, I think we should remain open to all options as we are still in the beginning. I think such ideas should be pursued, but you have to define what configuration the peptide will acquire after binding to the MHC groove and how much such configuration will resemble a self-MHC-tumour antigen complex.

C.Th. Smit Sibinga: Because here also dendrite cells, T-cell populations harvested from umbilical cord blood being naive, not being exposed yet might be a potential.

M.S. Mincheff: Certainly, however – I am not an expert on this. Maybe some of the colleagues?

M.K. Elias: Dr. Mincheff, already in 1996 dr. Murphy and his group[1] has used the prostatic antigen to pulse the dendritic cells, but got very limited response. Only some of the patients exhibited delay, type of hypersensitivity. You used a more sophisticated construction. Is that the reason why you have got a better response.

---

1. Murphy G, Tjoa B, Ragde H et al. Phase I clinical trial: T-cell therapy for prostate cancer using autologous dendritic cells pulsed with HLA-A0201- specific peptides from prostate-specific membrane antigen. Prostate 1996,29:371.

M.S. Mincheff: Well, Dr. Murphy is using a different approach. He is growing dendritic cells from the patient in vitro. He then loads them in vitro with PSMA-derived peptides know known to bind to HLA-A2 MHC. I do not know how well these peptides bind to other MHC molecules. The HLA-A2 haplotype is very common, but they have included also patients who are other HLA phenotypes. Also, Dr. Murphy's group is infusing het peptide-loaded dendritic cells intravenously and I do not see any reason why you should do this. My own experience from several years ago showed that if you present an alloantigen in the central lymphatics you generated aTh2 anti-allo response with the development of non-cytotoxic effectors and antibodies that did not fix compliment. In cancer immunotherapy we would like to do just the opposite – to stimulate cytotoxic responses and I will advice injection of antigen-loaded dendritic cells intradermally, subcutaneously, intratumouraly or even in the vicinity of a tumour-drainign lymph node. Back to the question – we do not use PSMA-derived peptides, we use DNA, encoding for the PSMA. When a dendritic cell is transfected with our construct numerous PSMA peptides suitable for MHC presentation are formed in teh proteasome. This may explain the higher incidence of DTH (delayed type hypersensitivity) responses.

H. Olthuis: The outcome of the cord blood transplantation with respect to HLA disparity presented in this and last years symposium have shown that there was a difference in outcome between the 0 and 1 or 2 disparities and the 3 and 4 HLA groups.

H.G. Klein: I am increasingly convinced that that is less important than we had thought previously. But there are possibly other factors as well. We spent a lot of time and effort again at the molecular level determining what the HLA types are and how closely they are matched. I think one issue that is extremely important is whether both class I and class II of these matches were done by high resolution molecular techniques and what that means. Certainly bone marrow transplantation and the results of bone marrow transplantation have been redefined when we go back and look with better techniques of determining what HLA is. But I am becoming increasingly convinced by the data that Dr. Wagner and others were presenting on umbilical cord blood. I do not think there is as much a difference as we have felt there was in the past.

INDEX

DEVELOPMENTS IN HEMATOLOGY AND IMMUNOLOGY

1. H.R. Lijnen, D. Collen and M. Verstraete (eds.): *Synthetic Substrates in Clinical Blood Coagulation Assays*. 1980 ISBN 90-247-2409-0
2. C.Th. Smit Sibinga, P.C. Das and J.O. Forfar (eds.): *Paediatrics and Blood Transfusion*. Proceedings of the 5th Annual Symposium on Blood Transfusion, organized by the Red Cross Bloodbank Groningen-Drenthe, The Netherlands (1980). 1982 ISBN 90-247-2619-0
3. N. Fabris (ed.): *Immunology and Ageing*. 1982 ISBN 90-247-2640-9
4. G. Hornstra: *Dietary Fats, Prostanoids and Arterial Thrombosis*. With an Introductory Chapter by A. Bleakley Chandler. 1982 ISBN 90-247-2667-0
5. C.Th. Smit Sibinga, P.C. Das and J.J. van Loghem (eds.): *Blood Transfusion and Problems of Bleeding*. Proceedings of the 6th Annual Symposium on Blood Transfusion, organized by the Red Cross Bloodbank Groningen-Drenthe (1981). 1982 ISBN 90-247-3058-9
6. J. Dormandy (ed.): *Red Cell Deformability and Filterability*. 1983 ISBN 0-89838-578-4
7. C.Th. Smit Sibinga, P.C. Das and H.F. Taswell (eds.): *Quality Assurance in Blood Banking and Its Clinical Impact*. Proceedings of the 7th Annual Symposium on Blood Transfusion, organized by the Red Cross Bloodbank Groningen-Drenthe (1982). 1984 ISBN 0-89838-618-7
8. A.M.H.P. van den Besselaar, H.R. Gralnick and S.M. Lewis (eds.): *Thromboplastin Calibration and Oral Anticoagulant Control*. 1984 ISBN 0-89838-637-3
9. P. Fondu and O. Thijs (eds.): *Haemostatic Failure in Liver Disease*. 1984 ISBN 0-89838-640-3
10. C.Th. Smit Sibinga, P.C. Das and G. Opelz (eds.): *Transplantation and Blood Transfusion*. Proceedings of the 8th Annual Symposium on Blood Transfusion, organized by the Red Cross Bloodbank Groningen-Drenthe (1983). 1984 ISBN 0-89838-686-1
11. H. Schmid-Schnbein, L.J. Wurzinger and R.E. Zimmerman (eds.): *Enzyme Activation in Blood-perfused Artificial Organs*. 1985 ISBN 0-89838-704-3
12. J. Dormandy (ed.): *Blood Filtration and Blood Cell Deformability*. 1985 ISBN 0-89838-714-0
13. C.Th. Smit Sibinga, P.C. Das and E. Seidl (eds.): *Plasma Fractionation and Blood Transfusion*. Proceedings of the 9th Annual Symposium on Blood Transfusion, organized by the Red Cross Bloodbank Groningen-Drenthe (1984). 1985 ISBN 0-89838-761-2
14. S. Dawids and A. Bantjes (eds.): *Blood Compatible Materials and their Testing*. 1986 ISBN 0-89838-813-9
15. C.Th. Smit Sibinga, P.C. Das and T.J. Greenwalt (eds.): *Future Developments in Blood Banking*. Proceedings of the 10th Annual Symposium on Blood Transfusion, organized by the Red Cross Bloodbank Groningen-Drenthe (1985). 1986 ISBN 0-89838-824-4
16. A. Berlin, J. Dean, M.H. Draper, E.M.B. Smith and F. Spreafico (eds.): *Immunotoxicology*. Proceedings of the International Seminar on the Immunological System as a Target for Toxic Damage – Present Status, Open Problems and Future Perspectives, in collaboration with the UNEP-ILO-WHO IPCS and the Commission of the European Communities. 1987 ISBN 0-89838-843-0
17. T. Ottenhoff and R. de Vries: *Recognition of M. leprae Antigens*. 1987 ISBN 0-89838-887-2
18. J.-L. Touraine, R.P. Gale and V. Kochupillai (eds.): *Fetal Liver Transplantation*. 1987 ISBN 0-89838-975-5
19. C.Th. Smit Sibinga P.C. Das and C.P. Engelfriet (eds.): *White Cells and Platelets in Blood Transfusion*. Proceedings of the 11th Annual Symposium on Blood Transfusion, organized by the Red Cross Bloodbank Groningen-Drenthe (1986). 1987 ISBN 0-89838-976-3
20. C.F.M. Hendriksen: *Laboratory Animals in Vaccine Production and Control*. 1988 ISBN 0-89838-398-6

21. C.Th. Smit Sibinga, P.C. Das and L.R. Overby (eds.): *Biotechnology in Blood Transfusion.* Proceedings of the 12th Annual Symposium on Blood Transfusion, organized by the Red Cross Bloodbank Groningen-Drenthe (1987). 1988 ISBN 0-89838-404-4
22. C.Th. Smit Sibinga, P.C. Das and C.F. Hgman (eds.): *Automation in Blood Transfusion.* Proceedings of the 13th Annual Symposium on Blood Transfusion, organized by the Red Cross Bloodbank Groningen-Drenthe (1988). 1989 ISBN 0-7923-0488-8
23. S. Dawids (ed.): *Polymers: Their Properties and Blood Compatibility.* 1989 ISBN 0-7923-0491-8
24. C.Th. Smit Sibinga, P.C. Das and H.T. Meryman (eds.): *Cryopreservation and Low Temperature Biology in Blood Transfusion.* Proceedings of the 14th Annual Symposium on Blood Transfusion, organized by the Red Cross Bloodbank Groningen-Drenthe (1989). 1990 ISBN 0-7923-0908-1
25. C.Th. Smit Sibinga and L. Kater (eds.): *Advances in Haemapheresis.* Proceedings of the 3rd International Congress of the World Apheresis Association (Amsterdam, The Netherlands, 1990). 1991 ISBN 0-7923-1312-7
26. C.Th. Smit Sibinga, P.C. Das and P.M. Mannucci (eds.): *Coagulation and Blood Transfusion.* Proceedings of the 15th Annual Symposium on Blood Transfusion, organized by the Red Cross Bloodbank Groningen-Drenthe (1990). 1991 ISBN 0-7923-1331-3
27. C.Th. Smit Sibinga, P.C. Das and J.D. Cash (eds.): *Transfusion Medicine: Fact and Fiction.* Proceedings of the 16th International Symposium on Blood Transfusion, organized by the Red Cross Bloodbank Groningen-Drenthe (1991). 1992 ISBN 0-7923-1732-7
28. C.Th. Smit Sibinga, P.C. Das and T.H. The (eds.): *Immunology and Blood Transfusion.* Proceedings of the 17th International Symposium on Blood Transfusion, organized by the Red Cross Bloodbank Groningen-Drenthe (1992). 1993 ISBN 0-7923-2380-7
29. C.Th. Smit Sibinga, P.C. Das and H.J. Heiniger (eds.): *Good Manufacturing Practice in Transfusion Medicine.* Proceedings of the 18th International Symposium on Blood Transfusion, organized by the Red Cross Blood Bank Groningen-Drenthe (1993). 1994 ISBN 0-7923-3009-9
30. C.Th. Smit Sibinga, P.C. Das and E. Brižt (eds.): *Hereditary Diseases and Blood Transfusion.* Proceedings of the 19th International Symposium on Blood Transfusion, organized by the Red Cross Blood Bank Groningen-Drenthe (1994). 1995 ISBN 0-7923-3694-1
31. C.Th. Smit Sibinga, P.C. Das and E.L. Snyder (eds.): *Trigger Factors in Transfusion Medicine.* Proceedings of the 20th International Symposium on Blood Transfusion, organized by the Red Cross Blood Bank Noord-Nederland (1995). 1996 ISBN 0-7923-4225-0
32. C.Th. Smit Sibinga, P.C. Das and B. Lwenberg (eds.): *Cytokines and Growth Factors in Blood Transfusion.* Proceedings of the 21th International Symposium on Blood Transfusion, organized by the Red Cross Blood Bank Noord-Nederland (1996). 1997 ISBN 0-7923-4787-0
33. C.Th. Smit Sibinga, P.C. Das and J.C. Fratantoni (eds.): *Alternative Approaches to Human Blood Resources in Clinical Practice.* 1998 ISBN 0-7923-8305-2
34. C.Th. Smit Sibinga and H.J. Alter (eds.): *Risk Management in Blood Transfusion: The Virtue of Reality.* 1999 ISBN 0-7923-6054-0

KLUWER ACADEMIC PUBLISHERS – DORDRECHT / BOSTON / LONDON

www.ingramcontent.com/pod-product-compliance
Ingram Content Group UK Ltd.
Pitfield, Milton Keynes, MK11 3LW, UK
UKHW020244290926
15470UKWH00002B/739

9780792365341